Systems Modeling and Architecting

-- Structure-Behavior Coalescence for Systems Architecture --

William S. Chao

Structure-Behavior Coalescence

Systems Architecture = Systems Structure + Systems Behavior

SBC Architecture

(1) SBC Architecture Description Language

(2) SBC Architecture Development Method

(3) SBC View Model

CONTENTS

CONTENTS 7
PREFACE 13
ABOUT THE AUTHOR 15
PART I: BASIC IDEAS 17
    Chapter 1: Introduction to Systems Architecture..19
        1-1 Multiple Views of a System..20
        1-2 Systems Model...22
        1-3 Non-Architectural Approaches vs. Architectural Approaches 25
        1-4 Definition of Systems Architecture 26
        1-5 Architecture Description Language 28
        1-6 Systems Architecture as a Knowledge Repository 29
        1-7 Systems Architects, Architecting and Systems Architectures 31
        1-8 Architecting the Systems Architecture Iteratively and Evolutionally ...31
        1-9 Architecture Development Method 32
        1-10 View Model 34
        1-11 Is Architecting a Procedure or an Architecture?................................37
        1-12 Maintainability of Systems Architecture 38
        1-13 Applications of Systems Architectures...39
    Chapter 2: Structure-Behavior Coalescence for Systems Architecture........41
       2-1 Multiple Views Coalescence to Achieve the Systems Architecture......41
       2-2 Integrating the Systems Structures and Systems Behaviors 42
       2-3 Structure-Behavior Coalescence to Facilitate Multiple Views
       Coalescence 45
       2-4 Structure-Behavior Coalescence to Achieve the Systems Architecture 45
    Chapter 3: Introduction to SBC Architecture..49
       3-1 Definition of SBC Architecture 49
       3-2 SBC Architecture Description Language 50
           3-2-1 Systems Structure Is Composed of Components........................53
           3-2-2 Interactions among Components and Actors to Draw forth the
           Systems Behavior...57
       3-3 SBC Architecture Development Method...60
       3-4 SBC View Model 62
PART II: SBC ARCHITECTURE DESCRIPTION LANGUAGE ..69
    Chapter 4: Systems Structure 71

4-1 Architecture Hierarchy Diagram 71
    4-1-1 Decomposition and Composition 71
    4-1-2 Multi-Level Decomposition and Composition 74
    4-1-3 Aggregated and Non-Aggregated Systems 76
4-2 Framework Diagram 76
    4-2-1 Multi-Layer Decomposition and Composition 77
    4-2-2 Only Non-Aggregated Systems Appearing in Framework Diagrams 78
4-3 Component Operation Diagram 79
    4-3-1 Operations of Components 79
    4-3-2 Drawing the Component Operation Diagram 82
4-4 Component Connection Diagram 85
    4-4-1 Essence of a Connection 85
    4-4-2 Drawing the Component Connection Diagram 87

Chapter 5: Systems Behavior 89
5-1 Structure-Behavior Coalescence Diagram 89
    5-1-1 Purpose of Structure-Behavior Coalescence Diagram 89
    5-1-2 Drawing the Structure-Behavior Coalescence Diagram 90
5-2 Interaction Flow Diagram 92
    5-2-1 Individual Systems Behavior Represented by Interaction Flow Diagram 92
    5-2-2 Drawing the Interaction Flow Diagram 95

PART III: CONCEPT VIEW 101
Chapter 6: Principle of the Concept View 103
6-1 Architecting the Concept View 103
6-2 Concept's Systems Structure 104
6-3 Concept's Systems Behavior 105

Chapter 7: Concept's Systems Structure 109
7-1 Concept's AHD 109
7-2 Concept's FD 110
7-3 Concept's COD 111
7-4 Concept's CCD 112

Chapter 8: Concept's Systems Behavior 115
8-1 Concept's SBCD 115
8-2 Concept's IFD 117

PART IV: ANALYSIS VIEW 119
Chapter 9: Principle of the Analysis View 121
9-1 Architecting the Analysis View 121

  9-2 Analysis' Systems Structure 123

  9-3 Analysis' Systems Behavior 124

 Chapter 10: Analysis' Systems Structure 127

  10-1 Analysis' AHD 127

  10-2 Analysis' FD 128

  10-3 Analysis' COD 129

  10-4 Analysis' CCD 130

 Chapter 11: Analysis' Systems Behavior 133

  11-1 Analysis' SBCD 133

  11-2 Analysis' IFD 135

PART V: DESIGN VIEW 137

 Chapter 12: Principle of the Design View 139

  12-1 Architecting the Design View 139

  12-2 Design's Systems Structure 140

  12-3 Design's Systems Behavior 141

 Chapter 13: Design's Systems Structure 145

  13-1 Design's AHD 145

  13-2 Design's FD 146

  13-3 Design's COD 147

  13-4 Design's CCD 148

 Chapter 14: Design's Systems Behavior 151

  14-1 Design's SBCD 151

  14-2 Design's IFD 153

PART VI: IMPLEMENTATION VIEW 155

 Chapter 15: Principle of the Implementation View 157

  15-1 Architecting the Implementation View 157

  15-2 Implementation's Systems Structure 158

  15-3 Implementation's Systems Behavior 160

 Chapter 16: Implementation's Systems Structure 163

  16-1 Implementation's AHD 163

  16-2 Implementation's FD 164

  16-3 Implementation's COD 165

  16-4 Implementation's CCD 166

 Chapter 17: Implementation's Systems Behavior 169

  17-1 Implementation's SBCD 169

  17-2 Implementation's IFD 171

PART VII: CASE STUDIES 173

 Chapter 18: Automobile --Hardware Architecture 175

18-1 Analysis View of the Automobile 176
    18-1-1 Analysis' Systems Structure 177
    18-1-2 Analysis' Systems Behavior 179
18-2 Design View of the Automobile 181
    18-2-1 Design's Systems Structure 182
    18-2-2 Design's Systems Behavior 185

Chapter 19: Multi-Tier Personal Data System -- Software Architecture ... 189
19-1 Analysis View of the Multi-Tier Personal Data System 193
    19-1-1 Analysis' Systems Structure 194
    19-1-2 Analysis' Systems Behavior 197
19-2 Design View of the Multi-Tier Personal Data System 200
    19-2-1 Design's Systems Structure 201
    19-2-2 Design's Systems Behavior 207

Chapter 20: Stanford University -- Enterprise Architecture 211
20-1 Concept View of the Stanford University 212
    20-1-1 Concept's Systems Structure 213
    20-1-2 Concept's Systems Behavior 217
20-2 Analysis View of the Stanford University 224
    20-2-1 Analysis' Systems Structure 225
    20-2-2 Analysis' Systems Behavior 231
20-3 Design View of the Stanford University 241
    20-3-1 Design's Systems Structure 242
    20-3-2 Design's Systems Behavior 249
20-4 Implementation View of the Stanford University 259
    20-4-1 Implementation's Systems Structure 260
    20-4-2 Implementation's Systems Behavior 268

Chapter 21: Robot -- Knowledge Architecture 279
21-1 Analysis View of the Robot 280
    21-1-1 Analysis' Systems Structure 281
    21-1-2 Analysis' Systems Behavior 283
21-2 Design View of the Robot 286
    21-2-1 Design's Systems Structure 286
    21-2-2 Design's Systems Behavior 289

Chapter 22: Strategic Thinking of an Airline -- Thinking Architecture 291
22-1 Concept View of the Strategic Thinking of an Airline 292
    22-1-1 Concept's Systems Structure 293
    22-1-2 Concept's Systems Behavior 295
22-2 Analysis View of the Strategic Thinking of an Airline 298

  22-2-1 Analysis' Systems Structure 299
  22-2-2 Analysis' Systems Behavior 302
 22-3 Design View of the Strategic Thinking of an Airline 305
  22-3-1 Design's Systems Structure 306
  22-3-2 Design's Systems Behavior 310
 22-4 Implementation View of the Strategic Thinking of an Airline 313
  22-4-1 Implementation's Systems Structure 314
  22-4-2 Implementation's Systems Behavior 318

APPENDIX A: SBC ARCHITECTURE DESCRIPTION LANGUAGE (SBC-ADL) 323

APPENDIX B: SBC ARCHITECTURE DEVELOPMENT METHOD (SBC-ADM) 329

APPENDIX C: SBC VIEW MODEL (SBC-VM) 331

BIBLIOGRAPHY 341

INDEX 347

PREFACE

A system comprises multiple views such as strategy/version n, strategy/version n+1, concept, analysis, design, implementation, structure, behavior and input/output data views. A systems model is required to describe and represent all these multiple views. The systems model describes and represents the system multiple views possibly using two different approaches. The first one is the non-architectural approach and the second one is the architectural approach. The non-architectural (model multiplicity) approach respectively picks a model for each view. The architectural (model singularity) approach, instead of picking many heterogeneous and unrelated models, will use only one single multiple views coalescence (MVC) architecture model.

In general, MVC architecture is synonymous with the systems architecture. Since structure and behavior views are the two most prominent ones among multiple views, integrating the structure and behavior views becomes a superb approach for integrating multiple views of a system. In other words, structure-behavior coalescence (SBC) leads to the coalescence of multiple views. Therefore, we conclude that SBC architecture is also synonymous with the systems architecture.

Systems architecture is emerging as an important discipline for hardware, software, enterprise, knowledge, or thinking systems modeling and architecting. This book focuses on the SBC architecture which consists of a) SBC architecture description language (SBC-ADL), b) SBC architecture development method (SBC-ADM) and c) SBC view model (SBC-VM).

An architecture description language is a special kind of language used to describe the architecture of a system. SBC-ADL uses six fundamental diagrams to formally grasp the essence of a system and its details at the same time. These diagrams are: a) architecture hierarchy diagram, b) framework diagram, c) component operation diagram, d) component connection diagram, e) structure-behavior coalescence diagram and f) interaction flow diagram.

The iterative and cyclic ADM, being used by a systems architect to accomplish each version management of the systems architecture, shall do the strategic management first and then go through the concept, analysis, design and implementation phases of systems architecture construction. Every phase checks with the requirements to make sure that each version of the constructed systems architecture is what the users want.

A view model is a three-dimensional matrix representation of a system's multiple views. In the SBC view model, dimension 1 stands for the evolution&motivation view which contains the strategy/version 1, strategy/version 2, strategy/version 3, strategy/version 4 and strategy/version ∞ views; dimension 2 stands for the multi-level (hierarchical) view which contains the concept, analysis, design and implementation views; dimension 3 stands for the systemic view which contains the structure, behavior, input/output data views.

Systems could be hardware, software, enterprise, knowledge, or thinking systems. Accordingly, systems architectures could also be hardware, software, enterprise, knowledge, or thinking systems architectures depending on what systems are applied to.

Systems architecture used for hardware, software, enterprise, knowledge, or thinking systems modeling and architecting is on the rise. By this book's penetrating introduction and elaboration, all readers shall clearly understand how the SBC architecture helps systems architects effectively perform architecting, in order to productively construct fruitful hardware, software, enterprise, knowledge, or thinking systems architectures.

ABOUT THE AUTHOR

Dr. William S. Chao is the CEO & founder of SBC Architecture International®. SBC (Structure-Behavior Coalescence) architecture is a systems architecture which demands the integration of systems structure and systems behavior of a system. SBC architecture applies to hardware architecture, software architecture, enterprise architecture, knowledge architecture and thinking architecture. The core theme of SBC architecture is: "Architecture = Structure + Behavior."

William S. Chao received his bachelor degree (1976) in telecommunication engineering and master degree (1981) in information engineering, both from the National Chiao-Tung University, Taiwan. From 1976 till 1983, he worked as an engineer at Chung-Hwa Telecommunication Company, Taiwan.

William S. Chao received his master degree (1985) in information science and Ph.D. degree (1988) in information science, both from the University of Alabama at Birmingham, USA. From 1988 till 1991, he worked as a computer scientist at GE Research and Development Center, Schenectady, New York, USA.

Dr. William S. Chao has been teaching at National Sun Yat-Sen University, Taiwan since 1992 and now serves as the president of Association of Enterprise Architects, Taiwan Chapter. His research covers: systems architecture, hardware architecture, software architecture, enterprise architecture, knowledge architecture and thinking architecture.

PART I: BASIC IDEAS

Chapter 1: Introduction to Systems Architecture

A system comprises multiple views such as strategy/version n, strategy/version n+1, concept, analysis, design, implementation, structure, behavior and input/output data views. A systems model is required to describe and represent all these multiple views.

The systems model describes and represents the system multiple views possibly using two different approaches. The first one is the non-architectural approach and the second one is the architectural approach. The non-architectural approach respectively picks a model for each view. The architectural approach, instead of picking many heterogeneous and unrelated models, will use only one single coalescence model.

When used as a knowledge repository of a system, systems architecture becomes a communicating tool for comprehension enhancement, internal collaboration and interworking with partners. Systems architecture also supplies documented systems structures and systems behaviors.

Systems architects construct systems architectures in a sense just like that systems architects perform architecting and architecting results in systems architectures. Systems architecture should not be constructed in one step. On the contrary, systems architects will iteratively and evolutionally construct each version of the systems architecture. Iterations and evolutions allow systems architects to demonstrate incremental value of their works and obtain early feedback of the systems architecture.

Architecting can be either a procedure or an architecture. If the work of architecting is not complicated, then describing and representing the architecting by a procedure is a preferred choice. If the work of architecting is complicated, then describing and representing the architecting by an architecture is a desired choice.

Higher maintainability means a better construction result of systems architecture. Lower maintainability indicates a worse construction result of systems architecture. Therefore, to achieve high maintainability of systems architecture is always the first goal to accomplish when a systems architect is constructing the systems architecture.

There are five different applications of systems architectures. Hardware, software, enterprise, knowledge and thinking systems architectures are the application of systems architecture to the hardware, software, enterprise, knowledge and thinking systems respectively.

1-1 Multiple Views of a System

In general, a system is extremely complex that it consists of several evolution&motivation views such as strategy/version n and strategy/version n+1 views; it also consists of various multi-level (hierarchical) views such as concept, analysis, design and implementation views; it also consists of many systemic views such as structure, behavior and input/output data views [Kend10, Pres09, Somm06].

Figure 1-1 shows that in a system all these strategy/version n, strategy/version n+1, concept, analysis, design, implementation, structure, behavior and input/output data views represent the multiple views of a system.

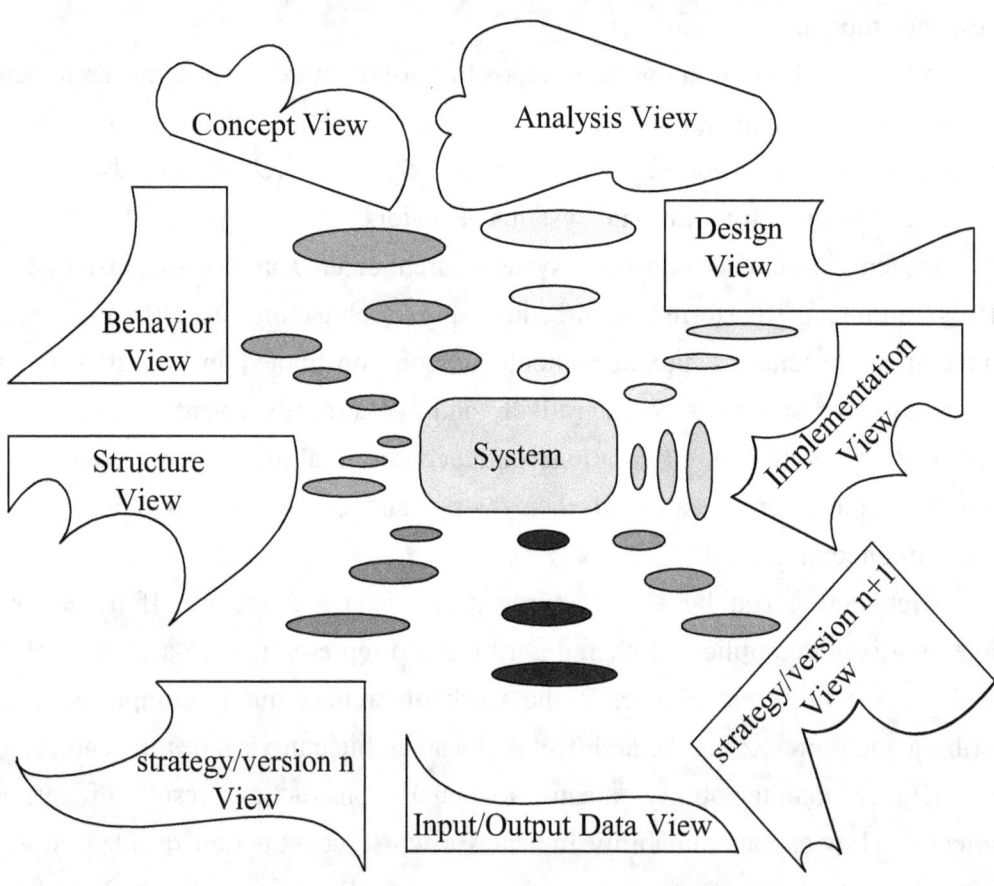

Figure 1-1   Multiple Views of a System

Among the above multiple views, the structure and behavior views are perceived as the two prominent ones. The structure view focuses on the systems structure which is described by components and their composition while the behavior view concentrates on the systems behavior which involves interrelationships among the external environment's actors and components. Strategy/version n, strategy/version n+1, concept, analysis, design, implementation and input/output data views are considered to be other views as shown in Figure 1-2.

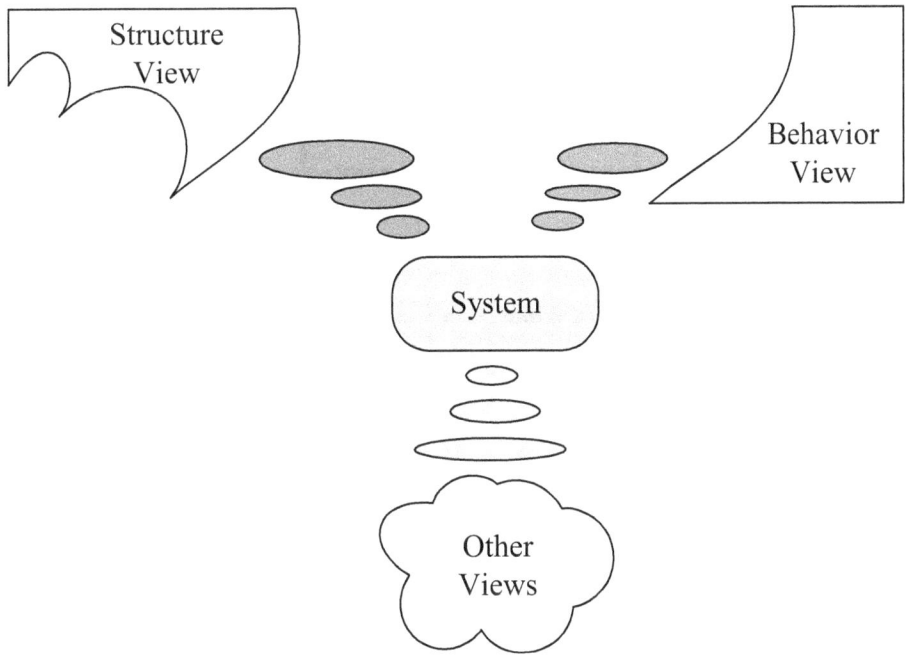

Figure 1-2   Structure, Behavior and Other Views

Either Figure 1-1 or Figure 1-2 represents the multiple views of a system. In some situations Figure 1-1 is used and in other situations Figure 1-2 is used.

Accordingly, a system is defined, in Figure 1-3, hopefully to be an integrated whole of that system's multiple views, i.e., structure, behavior and other views, embodied in its assembled components, their interrelationships with each other and the environment. Components are sometimes labeled as non-aggregated systems, parts, entities, objects and building blocks [Chao14a, Chao14b].

> A system, hopefully is an integrated whole of that system's multiple views, i.e., structure, behavior, and other views, embodied in its assembled components, their interrelationships with each other and the environment.

Figure 1-3   Definition of a System

Since multiple views are embodied in a system's assembled components which belong to the systems structure, they shall not exist alone. Multiple views must be loaded on the systems structure just like a cargo is loaded on a ship as shown in Figure 1-4. There will be no multiple views if there is no systems structure. Stand-alone multiple views are not meaningful.

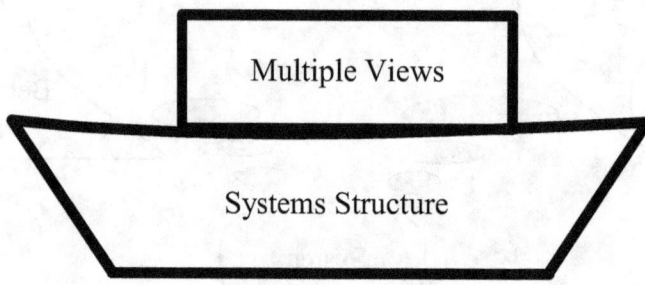

Figure 1-4   Multiple Views Must be Loaded on the Systems Structure

1-2 Systems Model

A systems model (SM) is a virtual system, distinguished from a physical system, used to describe and represent either the physical or virtual systems. A physical system, e.g., house, tree, river, airplane, etc., exists in the physical world. A virtual system, e.g., symbol, language, diagram, software, virtual reality, thought, etc., exists in the virtual world.

Figure 1-5 shows a physical system in which there are two buildings located in the upper left side and right underneath. The upper left building is Tuscaloosa Hotel and the right underneath building is Auburn Theater.

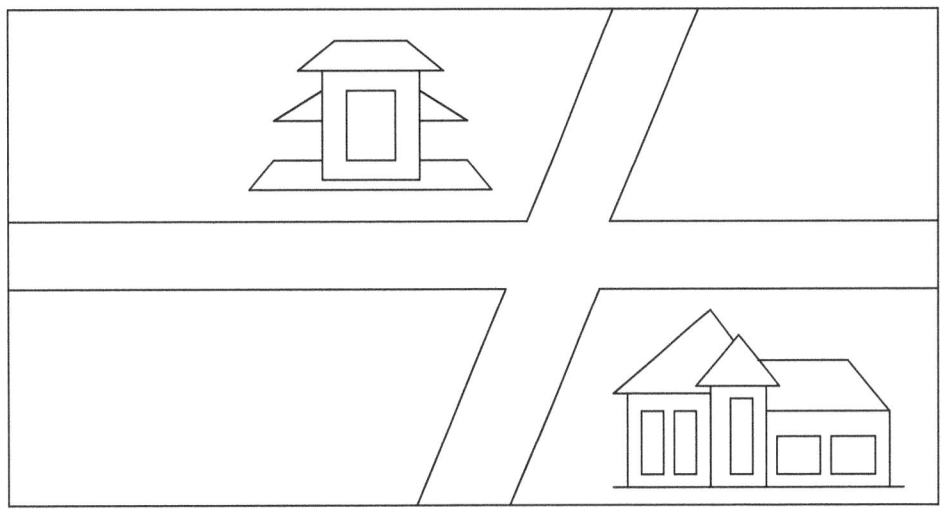

Figure 1-5 A Physical System

To model the physical system in Figure 1-5 we may then obtain a map as shown in Figure 1-6. The map is a kind of systems model used to describe and represent the physical system.

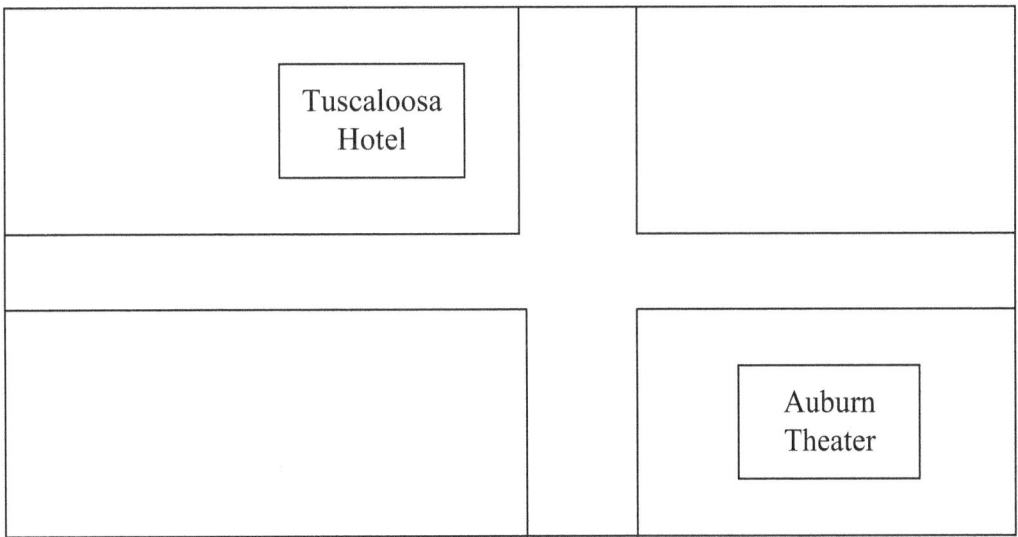

Figure 1-6 Map as a Systems Model

Besides describing and representing systems in the physical world, a systems model can also describe and represent systems in the virtual world. Virtual world includes a software system, virtual reality, or the thought inside a person's mind, and so on. Figure 1-7 shows that a fashion designer is designing a new suit of clothes. Designing a suit of clothes, being the thought inside a person's mind, belongs to the virtual world.

Figure 1-7　Thought inside a Person's Mind

To model the thought inside a person's mind in Figure 1-7, we may then use a clothes design diagram as shown in Figure 1-8. The clothes design diagram is a kind of systems model used to describe and represent a person's thought.

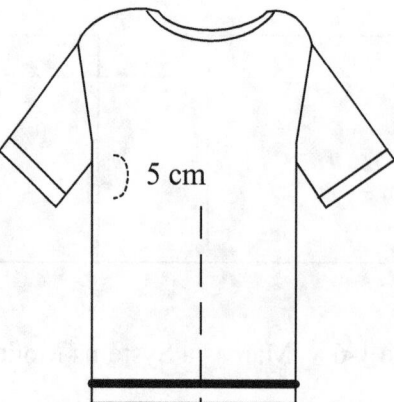

Figure 1-8　Clothes Design Diagram as a System Model

1-3 Non-Architectural Approaches vs. Architectural Approaches

A system is exceptionally complex that it includes multiple views such as strategy/version n, strategy/version n+1, concept, analysis, design, implementation, structure, behavior and input/output data views.

The systems model describes and represents the system multiple views possibly using two different approaches. The first one is the non-architectural approach and the second one is the architectural approach.

The non-architectural approach, also known as the model multiplicity approach [Dori95, Dori02, Dori16], respectively picks a model for each view as shown in Figure 1-9, the strategy/version n view has the strategy/version n model, the strategy/version n+1 view has the strategy/version n+1 model, the concept view has the concept model, the analysis view has the analysis model, the design view has the design model, the implementation view has the implementation model, the structure view has the structure model, the behavior view has the behavior model, and the input/output data view has the input/output data model. These multiple models, are heterogeneous and not related to each other, and thus become the primary cause of model multiplicity problems [Dori95, Dori02, Dori16, Pele02, Sode03].

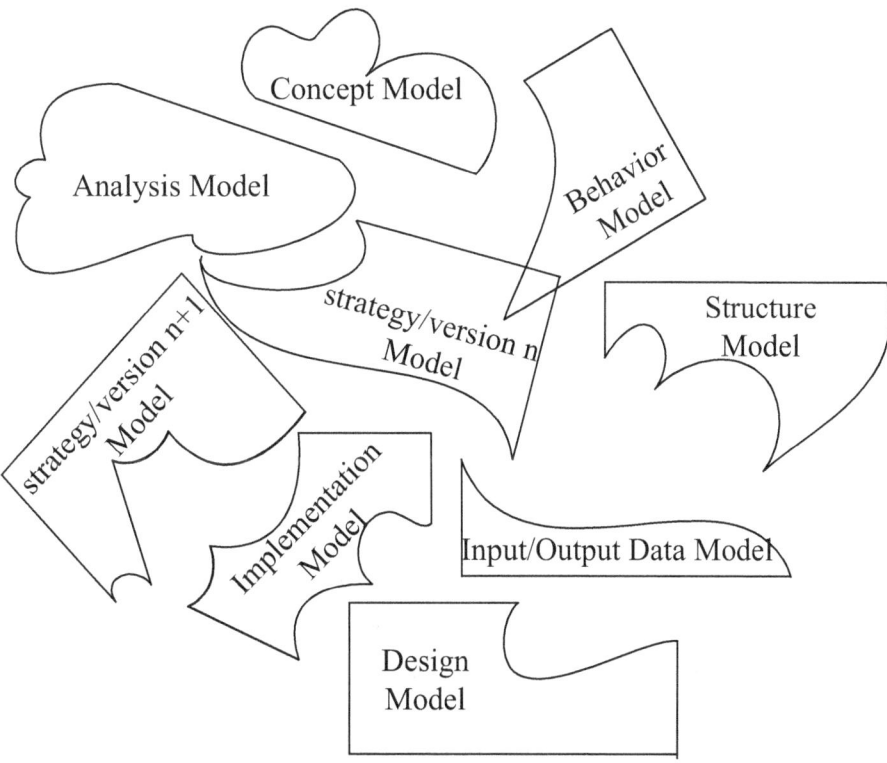

Figure 1-9   The Non-architectural Approach Picks a Model for Each View

The architectural approach, also known as the model singularity approach [Dori95, Dori02, Dori16, Pele02, Sode03], instead of picking many heterogeneous and unrelated models, will use only one single model as shown in Figure 1-10. The strategy/version n, strategy/version n+1, concept, analysis, design, implementation, structure, behavior and input/output data views are all integrated in this multiple views coalescence (MVC) model of systems architecture (SA) [Bass03, Chao14a, Chao14b, Chao15a, Chao15b, Chao16, Chao17a, Chao17b, Chao17c, Chao17d, Chao17e, Chao17f, Clem02, Clem10, Dike01, Dori95, Dori02, Putm00, Roza05, Shaw96, Tayl09, Wang99].

Figure 1-10   Systems Architecture Uses a Single Model

Figure 1-9 has many models. Figure 1-10 has only one model. Comparing Figure 1-9 with Figure 1-10, we unquestionably conclude that an integrated, holistic, united, coordinated, coherent and coalescence model is more favorable than a collection of many heterogeneous and unrelated models.

1-4 Definition of Systems Architecture

Involved systems are extremely complex in every aspect so that each stakeholder needs a blueprint or model to capture their essential structures and behaviors. Systems architecture is such a blueprint or model.

There are several well-know definitions of systems architecture [Bass03, Burd10, Dam06, Mino08, O'Rou03, Paul01, Roza05]. ANSI/IEEE 1471-2000 defines systems architecture as: "the fundamental organization of a system, embodied in its components, their relationships to each other and the environment, and the principles governing its design and evolution." The Open Group defines systems architecture as either "a formal description of a system, or a detailed plan of the system at component level to guide its implementation," or as "the structure of components, their interrelationships, and the principles and guidelines governing their design and evolution over time" [Rayn09, Toga08].

Concluding the above definitions, we now give systems architecture a definition of our own as shown in Figure 1-11.

> Systems architecture is an integrated whole of a system's multiple views, i.e., structure, behavior and other views, embodied in its assembled components, their interactions with each other and the environment, and the principles and guidelines governing its design and evolution.

Figure 1-11   Definition of Systems Architecture

From the above definition, we find out that systems architecture is an integrated whole of a system's multiple views, i.e., structure, behavior and other views, embodied in its assembled components, their interactions (or handshakes) with each other and the environment, and the principles and guidelines governing its design and evolution. That is, systems architecture is an integrated and coalescence model of multiple views. In this coalescence model, structure, behavior and other views are all included in it as shown in Figure 1-12. We do not supply each view a respective model in this systems architecture coalescence model.

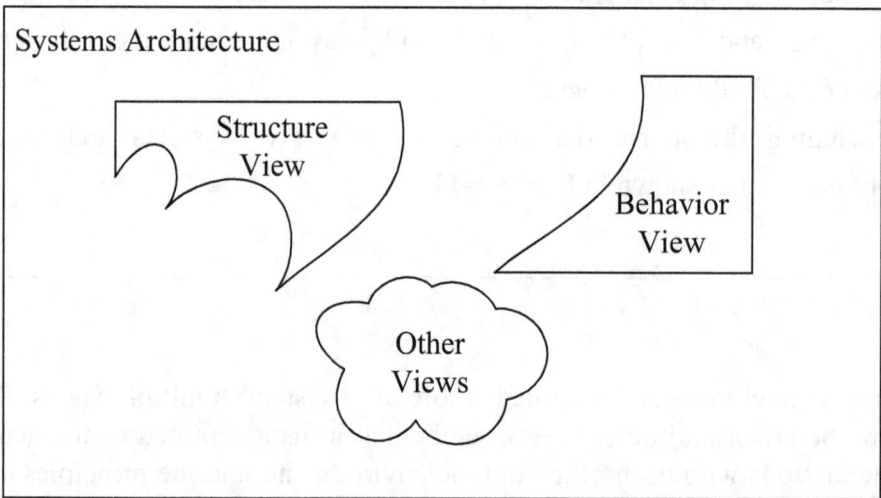

Figure 1-12 All Multiple Views are Included in This Systems Architecture

Since multiple views are embodied in a system's assembled components which belong to the structure view, they shall not exist alone. Multiple views must be loaded on the structure view just like a cargo is loaded on a ship as shown in Figure 1-13. There will be no multiple views if there is no structure view. Stand-alone multiple views are not meaningful.

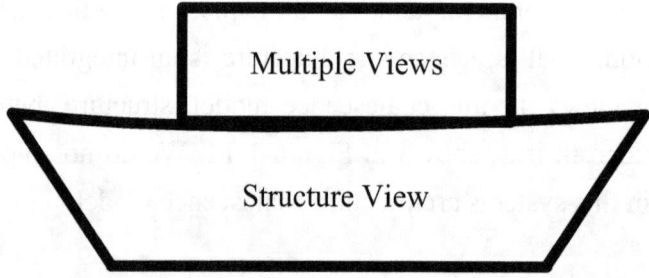

Figure 1-13 Multiple Views Must be Loaded on the Structure View

1-5 Architecture Description Language

An architecture description is a formal description and representation of a system. A description of the systems architecture has to grasp the essence of the system and its details at the same time. In other words, an architecture description not only provides an overall picture that summarizes the whole system, but also contains enough detail that the system can be constructed and validated.

The language for architecture description is called the architecture description language (ADL) [Bass03, Clem02, Clem10, Dike01, Roza05, Shaw96, Tayl09]. An ADL is a special kind of language used in describing the architecture of a system.

Since the architectural approach uses a coalescence model for all multiple views of a system, the foremost duty of ADL is to make the strategy/version n, strategy/version n+1, concept, analysis, design, implementation, structure, behavior and input/output data views all integrated and coalesced within this architecture description.

1-6 Systems Architecture as a Knowledge Repository

Based on its definition, systems architecture can be regarded as a knowledge repository of a system. Each stakeholder, through structure, behavior and other views, submits his own knowledge and expertise to this repository when the systems architecture is built up, as shown in Figure 1-14.

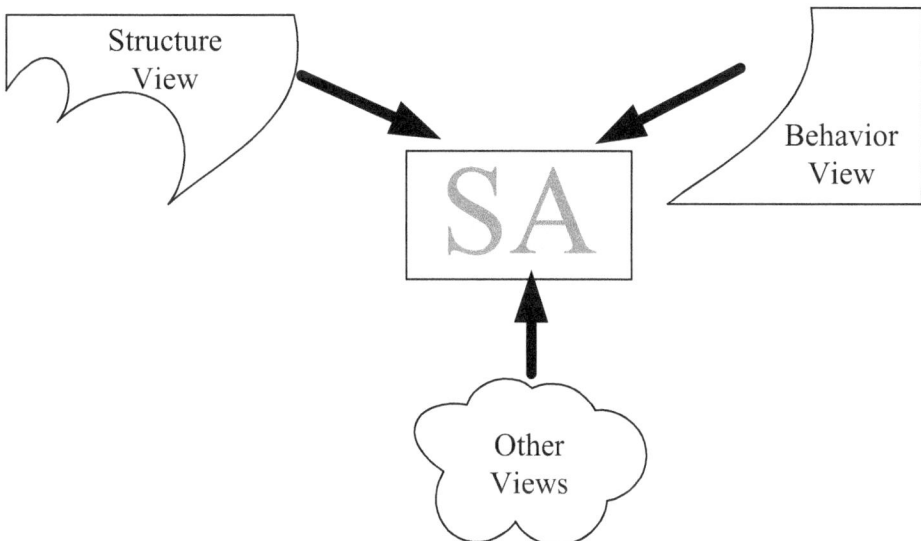

Figure 1-14 Each Stakeholder Submits His Own Knowledge and Expertise

On the other hand, any stakeholder, if there is any request then he would query the systems architecture. The result of the query is gathered into a view for stakeholders to see or read, as shown in Figure 1-15.

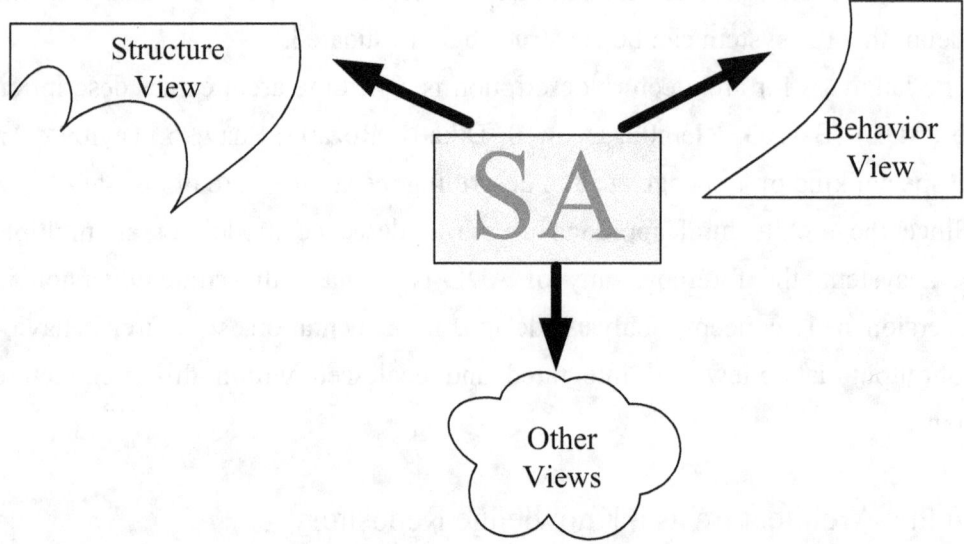

Figure 1-15   Views for Stakeholders to See or Read

Combining the above two figures, Figure 1-16 tells us that systems architecture is exactly a knowledge repository of a system. Stakeholders can submit and acquire knowledge to and from the systems architecture.

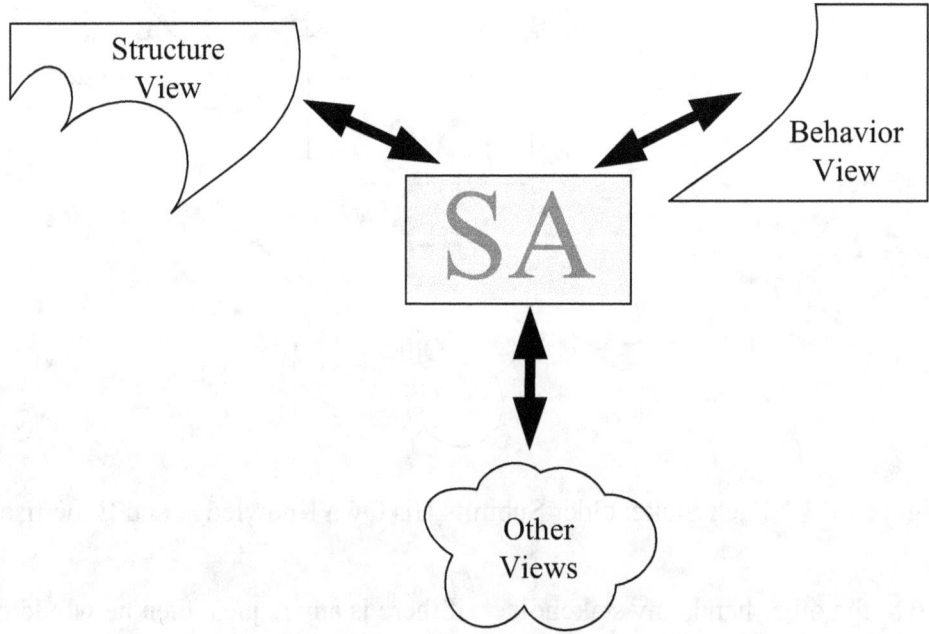

Figure 1-16   SA as a Knowledge Repository of a System

When used as a knowledge repository of a system, systems architecture becomes a communicating tool for comprehension enhancement, internal collaboration and interworking with partners. Systems architecture also supplies documented systems structures and systems behaviors.

1-7 Systems Architects, Architecting and Systems Architectures

A systems architect is the person, team, or organization responsible for systems architectures. In a sense, we say that systems architects construct systems architectures.

Architecting represents the activities of defining, documenting, maintaining, improving and certifying proper implementation of systems architecture. In a sense, we declare that systems architects perform architecting and architecting results in systems architectures [Eele09, Maie09].

The relationship among systems architects, architecting and systems architectures is shown in Figure 1-17.

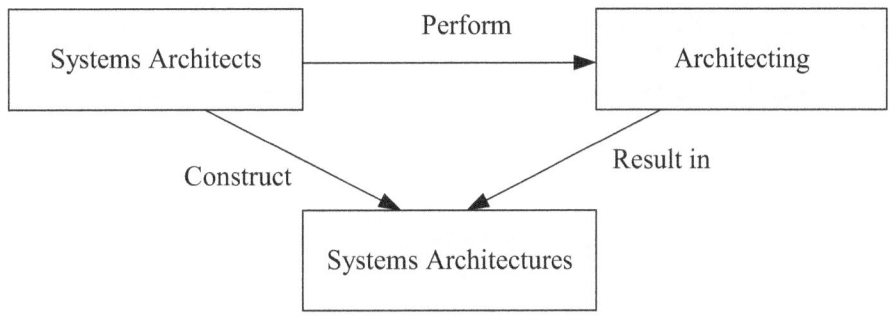

Figure 1-17 Relationship among Systems Architects, Architecting and Systems Architectures

1-8 Architecting the Systems Architecture Iteratively and Evolutionally

Systems architecture shall not be architected in one step. On the contrary, systems architects must architect the systems architecture iteratively and evolutionally. Iterations and evolutions allow systems architects to demonstrate incremental values of their works and obtain early feedback of the systems architecture.

Figure 1-18 shows that the systems architecture *version 1*, *version 2*, *version 3*, *version 4*,…, and *version ∞* are architected iteratively and evolutionally by a systems architect.

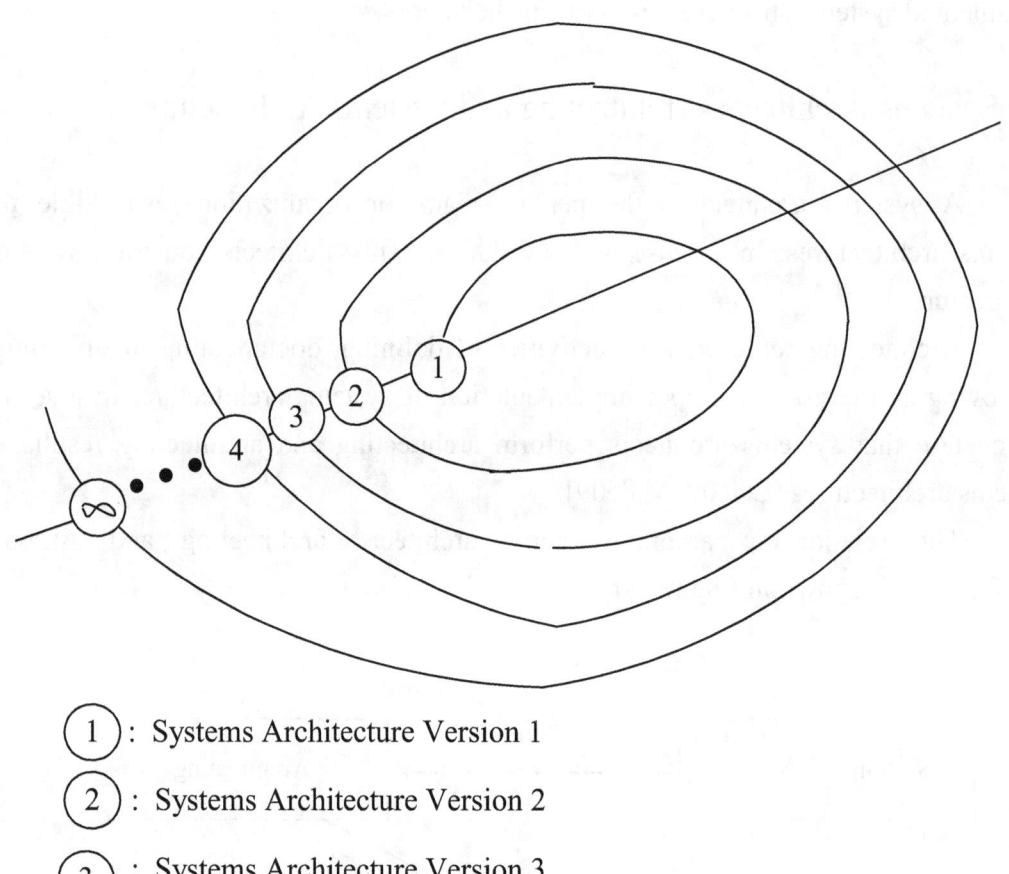

① : Systems Architecture Version 1

② : Systems Architecture Version 2

③ : Systems Architecture Version 3

④ : Systems Architecture Version 4

• • •

∞ : Systems Architecture Version ∞

Figure 1-18　Systems Architecture is Architected Iteratively and Evolutionally

Systems architecture *version n* is sometimes referred to as the baseline (As-Is) architecture which represents the current system that has been formally reviewed and agreed upon. On the other hand, systems architecture *version n+1* is sometimes referred to as the target (To-Be) architecture which represents the goal system that will be formally architected.

1-9 Architecture Development Method

If we adopt the iterative and evolutional construction of systems architecture approach, then we would obtain the architecture development method (ADM) [Toga08] as shown in Figure 1-19.

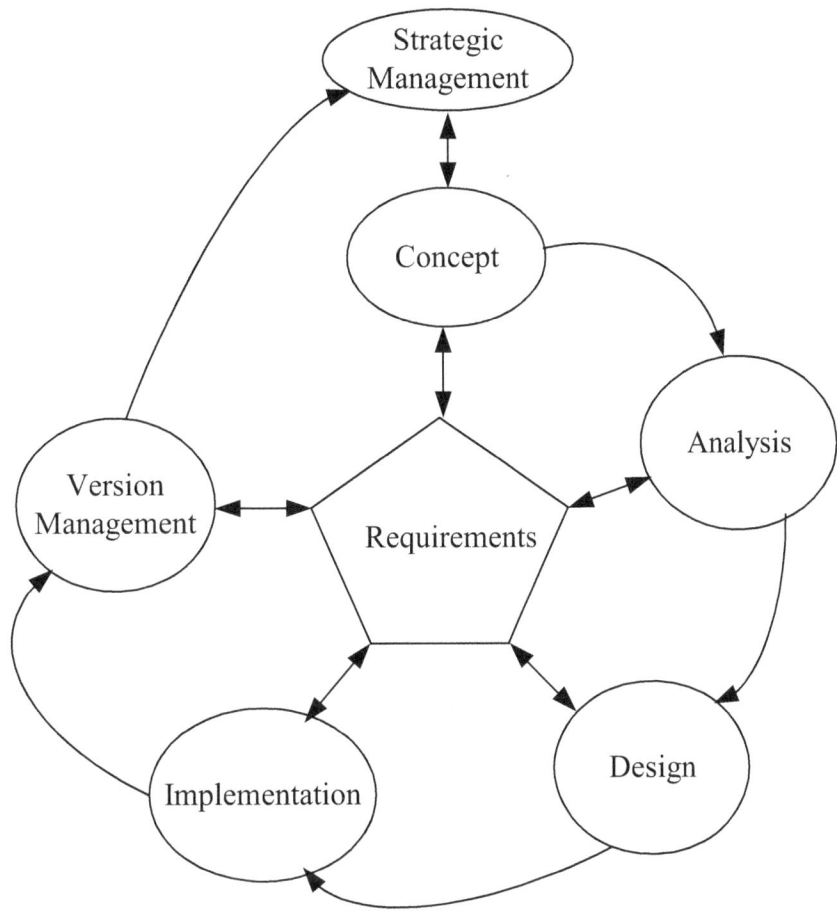

Figure 1-19　Architecture Development Method

　　The iterative and cyclic ADM, being used by a systems architect to accomplish each version management of systems architecture, shall do the strategic management first and then go through the concept, analysis, design and implementation phases of systems architecture construction. Every phase checks with the requirements to make sure that each version of the constructed systems architecture is what the users want.

　　Each strategic management generates a strategy and each version management generates a version of systems architecture. Therefore, each strategy is mapped to a version of systems architecture as shown in Figure 1-20.

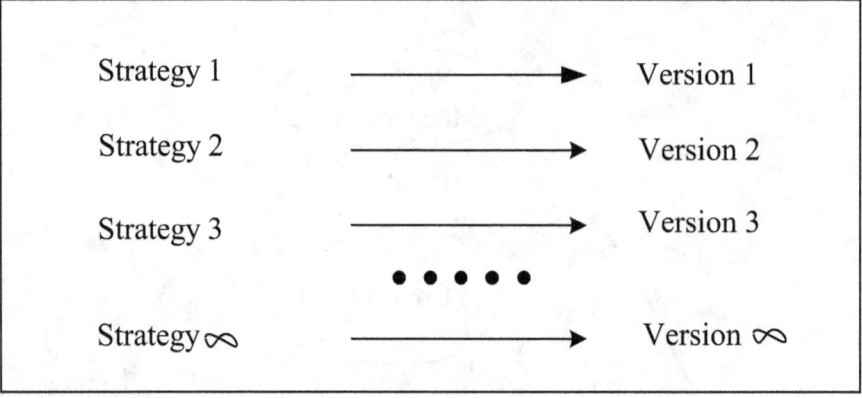

Figure 1-20   Each Strategy is Mapped to a Version of Systems Architecture

1-10 View Model

A system comprises multiple views such as strategy/version n, strategy/version n+1, concept, analysis, design, implementation, structure, behavior and input/output data views. We can represent all these multiple views in a one-dimensional array as shown in Figure 1-21.

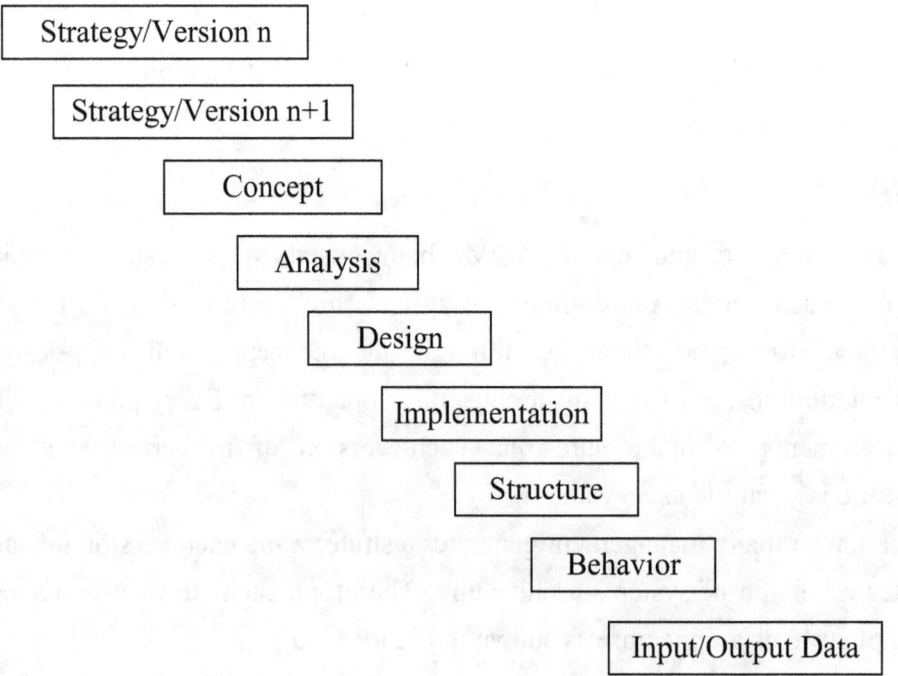

Figure 1-21   Array Representation of Multiple Views

We can also describe and represent all these multiple views in a three-dimensional matrix as shown in Figure 1-22.

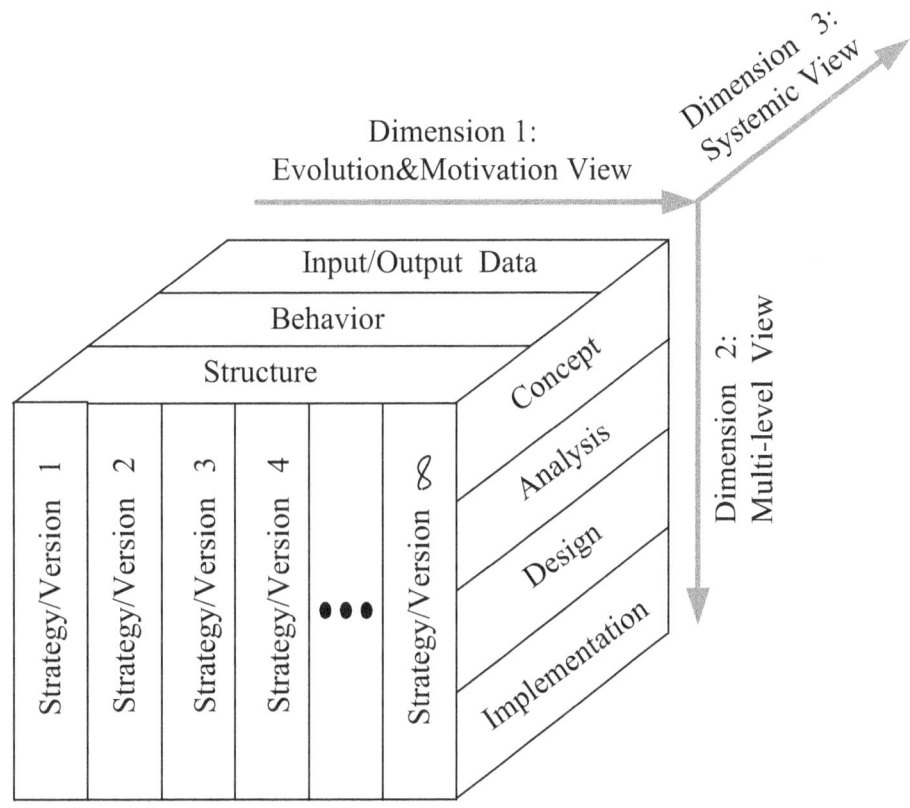

Figure 1-22   Three-dimensional Matrix Representation of Multiple Views

In the matrix representation of multiple views, dimension 1 stands for the evolution&motivation view which contains the strategy/version 1, strategy/version 2, strategy/version 3, strategy/version 4 and strategy/version ∞ views; dimension 2 stands for the multi-level (hierarchical) view which contains the concept, analysis, design and implementation views; dimension 3 stands for the systemic view which contains the structure, behavior, input/output data views. The matrix representation of multiple views is also called a view model (VM) or architecture framework (AF) [Chao14a, Chao14b, Dam06, Mino08, O'Rou03].

There are various view models. The simplest one is a three-dimensional matrix of infinite evolution&motivation views and two multi-level (hierarchical) views and two systemic views, as shown in Figure 1-23.

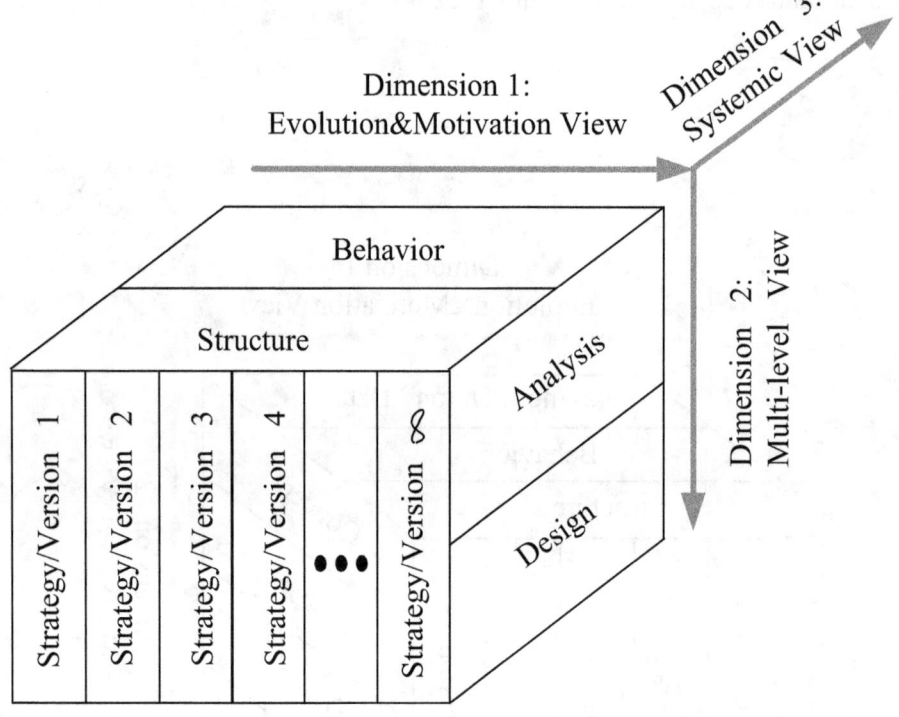

Figure 1-23   The Simplest View Model

A second example of view model is a three-dimensional matrix of infinite evolution&motivation views and four multi-level (hierarchical) views and two systemic views, as shown in Figure 1-24.

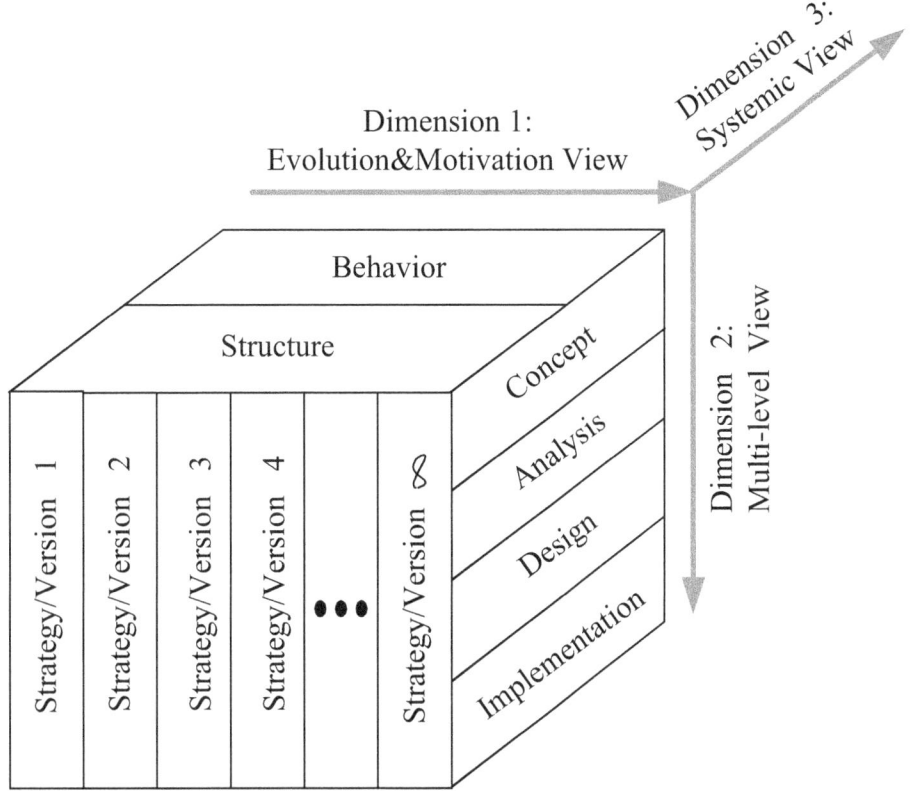

Figure 1-24 A Second Example of View Model

1-11 Is Architecting a Procedure or an Architecture?

Systems architects perform architecting and architecting results in systems architectures. Is architecting a procedure or an architecture? Both are the answer. Architecting can be either a procedure or an architecture.

In most cases, architecting is described and represented by a process (behavior view) as shown Figure 1-25. If the work of architecting is not complicated, then describing and representing the architecting by a procedure is a preferred choice.

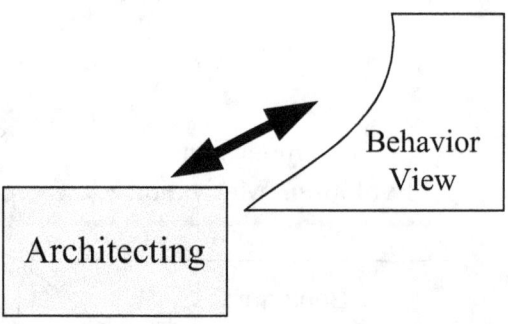

Figure 1-25   Architecting is Described and Represented by a Procedure

In some advanced cases, architecting is described and represented by an architecture as shown Figure 1-26. If the work of architecting is complicated, then describing and representing the architecting by an architecture is a desired choice.

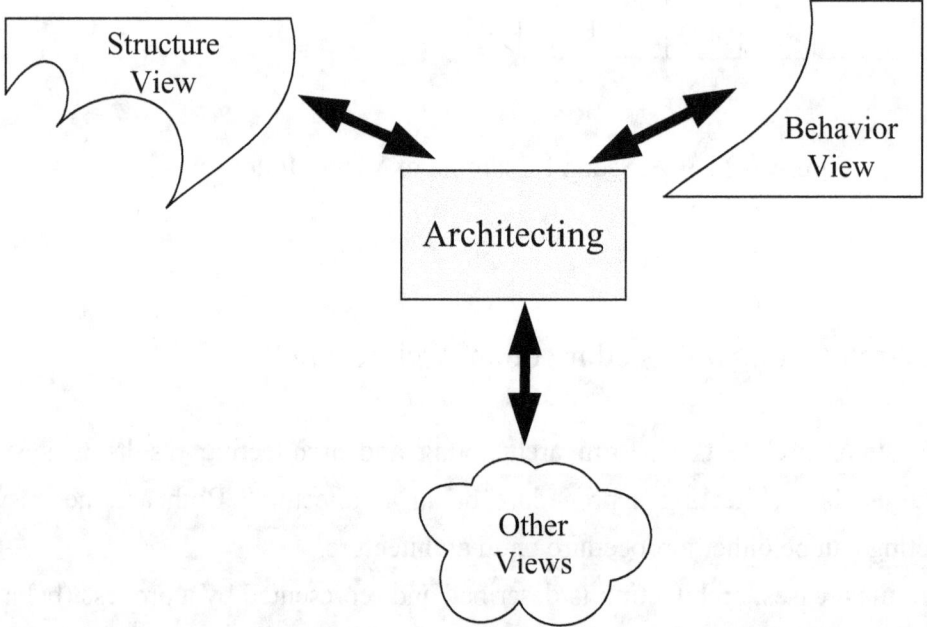

Figure 1-26   Architecting is Described and Represented by an Architecture

1-12 Maintainability of Systems Architecture

Since systems architecture is architected iteratively and evolutionally, its maintainability grows into a key issue. Systems architects certainly incline to spend less effort when they architect each version of systems architecture. Maintainability of systems architecture is measured by the (inverse proportional) mean effort to change and update, as shown in Figure 1-27.

$$\text{Maintainability} = 1 / \text{Mean Effort to Change}$$

Figure 1-27   Measurement of the Maintainability

From the above figure, we know that high maintainability of systems architecture implies less effort to change, update, amend, correct, improve, revise and modify each version of systems architecture.

Besides, we also have the following conclusions. Higher maintainability means a better construction result of systems architecture. Instead, lower maintainability indicates a worse construction result of systems architecture. Therefore, to achieve high maintainability of systems architecture should always stay inside the mind of a systems architect whenever he or she is constructing the systems architecture.

1-13 Applications of Systems Architectures

There are five different applications of systems architectures, as shown in Figure 1-28. Hardware architecture, abbreviation of hardware systems architecture, is an application of systems architecture to the hardware system. Software architecture, abbreviation of software systems architecture, is an application of systems architecture to the software system. Enterprise architecture, abbreviation of enterprise systems architecture, is an application of systems architecture to the enterprise system. Knowledge architecture, abbreviation of knowledge systems architecture, is an application of systems architecture to the knowledge system. Thinking architecture, abbreviation of thinking systems architecture, is an application of systems architecture to the thinking system.

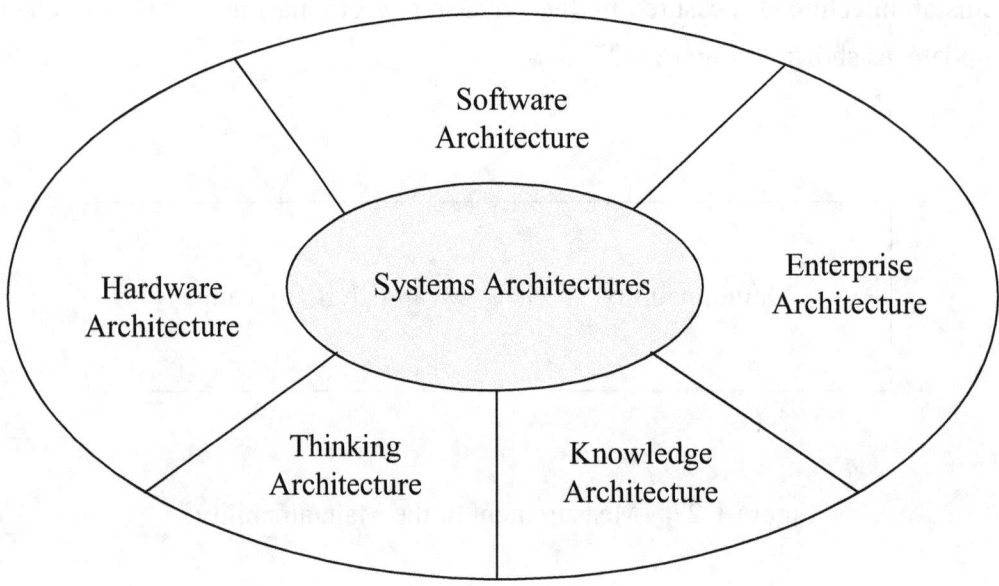

Figure 1-28   Five Kinds of Systems Architectures

Chapter 2: Structure-Behavior Coalescence for Systems Architecture

In general, multiple views coalescence (MVC) architecture is synonymous with the systems architecture. Since structure and behavior views are the two most prominent ones among multiple views, integrating the structure and behavior views becomes a superb approach for integrating multiple views of a system. In other words, structure-behavior coalescence (SBC) results in the coalescence of multiple views. Therefore, we conclude that SBC architecture is also synonymous with the systems architecture.

2-1 Multiple Views Coalescence to Achieve the Systems Architecture

Systems architecture has been defined as a coalescence model of multiple views. Multiple views coalescence uses only a single coalescence model as shown in Figure 2-1. Strategy/version n, strategy/version n+1, concept, analysis, design, implementation, structure, behavior and input/output data views are all integrated in this MVC architecture.

Figure 2-1   MVC Architecture

Generally, MVC architecture is synonymous with the systems architecture. In other words, multiple views coalescence sets a path to achieve the systems architecture as shown in Figure 2-2.

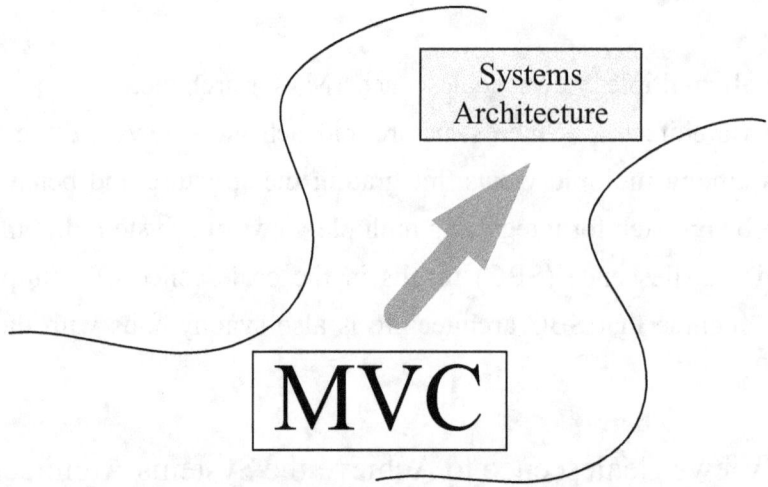

Figure 2-2   MVC to Achieve the Systems Architecture

In the MVC architecture, multiple views must be attached to or built on the systems structure. In other words, multiple views shall not exist alone; they must be loaded on the systems structure just like a cargo is loaded on a ship as shown in Figure 2-3. There will be no multiple views if there is no systems structure. Stand-alone multiple views are not meaningful.

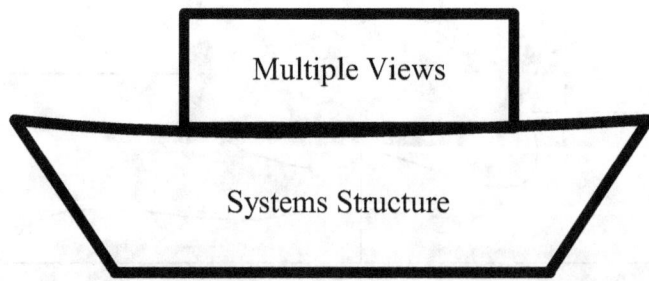

Figure 2-3   Multiple Views Must be Loaded on the Systems Structure

2-2 Integrating the Systems Structures and Systems Behaviors

By integrating the systems structure and systems behavior, we obtain structure-behavior coalescence (SBC) within the system as shown in Figure 2-4.

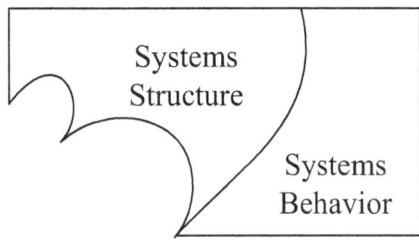

Figure 2-4 Structure-Behavior Coalescence

Structure-behavior coalescence has never been used in any systems model (SM) for systems development except the SBC architecture. There are many advantages to use the structure-behavior coalescence approach to integrate the systems structure and systems behavior.

SBC architecture uses a single model as shown in Figure 2-5. Systems structures and systems behaviors are integrated in this SBC architecture.

Figure 2-5 SBC Architecture

Since systems structures and systems behaviors are so tightly integrated, we sometimes claim that the core theme of SBC architecture is: "Systems Architecture = Systems Structure + Systems Behavior," as shown in Figure 2-6.

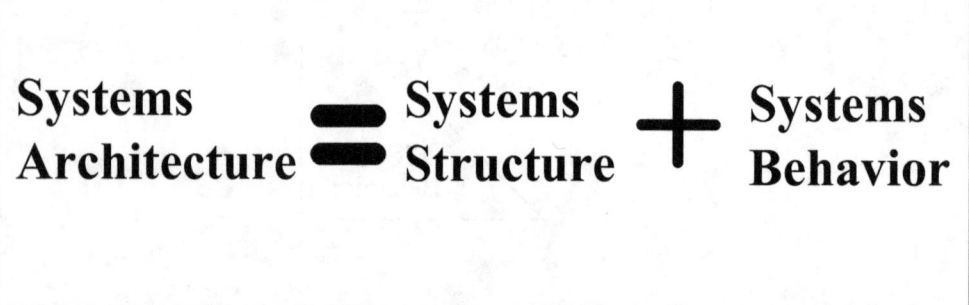

Figure 2-6 Core Theme of SBC Architecture

So far, systems behaviors are separated from systems structures in most cases [Pres09, Roza05, Somm06]. For example, the well-known structured systems analysis and design (SSA&D) approach uses structure charts (SC) to represent the systems structure and data flow diagrams (DFD) to represent the systems behavior [Denn08, Kend10, Your99]. SC and DFD are two heterogeneous and separated models. They are so separated like that there is "Atlantic Ocean" between them, as shown in Figure 2-7.

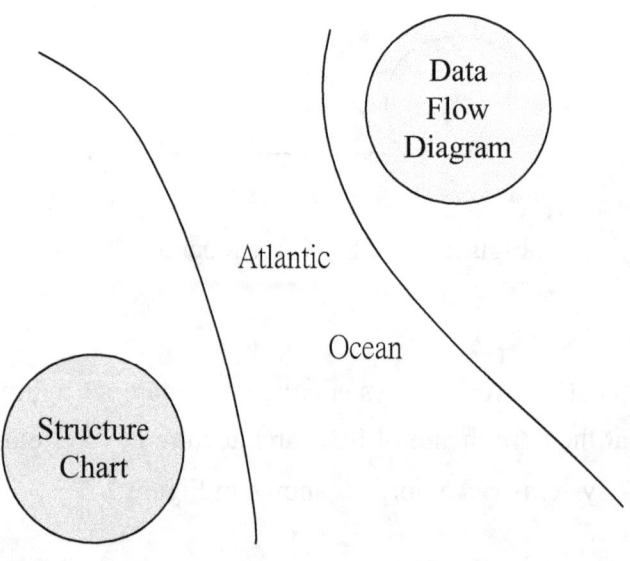

Figure 2-7 Two Heterogeneous and Separated Models

2-3 Structure-Behavior Coalescence to Facilitate Multiple Views Coalescence

Since structure and behavior views are the two most prominent ones among multiple views, integrating the structure and behavior views is clearly the best way to integrate multiple views of a system. In other words, structure-behavior coalescence facilitates multiple views coalescence as shown in Figure 2-8.

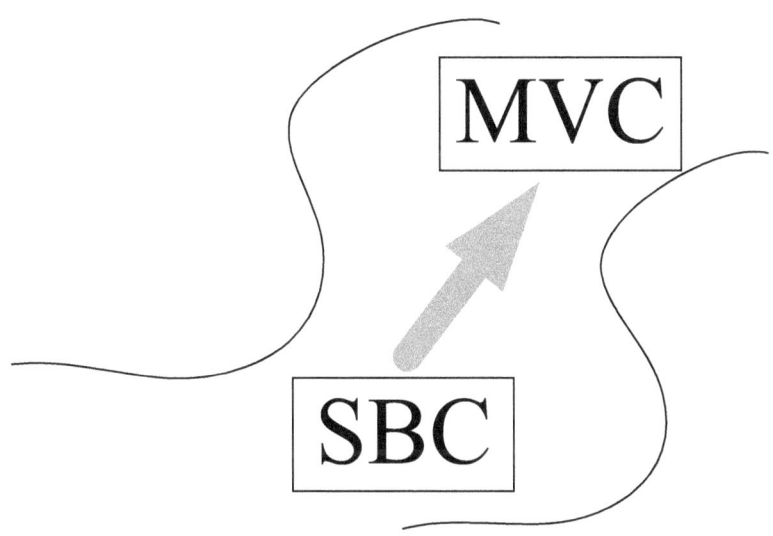

Figure 2-8　SBC Facilitates MVC

2-4 Structure-Behavior Coalescence to Achieve the Systems Architecture

Figure 2-2 declares that multiple views coalescence sets a path to achieve the desired systems architecture with the most efficient approach. Figure 2-8 declares that structure-behavior coalescence facilitates multiple views coalescence.

Combining the above two declarations, we conclude that structure-behavior coalescence sets a path to achieve the systems architecture as shown in Figure 2-9. In this case, SBC architecture is also synonymous with the systems architecture.

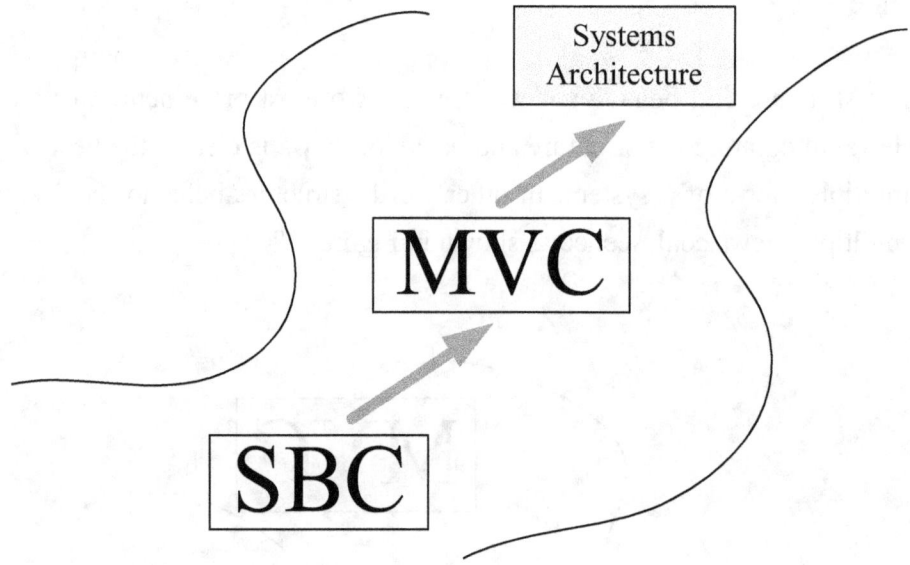

Figure 2-9   SBC to Achieve the Systems Architecture

SBC architecture strongly demands that the structure and behavior views must be coalesced and integrated. This never happens in other architectural approaches such as Zachman Framework [O'Rou03], The Open Group Architecture Framework (TOGAF) [Rayn09, Toga08], Department of Defense Architecture Framework (DoDAF) [Dam06] and Unified Modeling Language (UML) [Rumb91]. Zachman Framework does not offer any mechanism to integrate the structure and behavior views. TOGAF, DoDAF and UML do not, either.

In the SBC architecture, the systems behavior must be attached to or built on the systems structure. In other words, the systems behavior can not exist alone; it must be loaded on the systems structure just like a cargo is loaded on a ship as shown in Figure 2-10. There will be no systems behavior if there is no systems structure. A stand-alone systems behavior is not meaningful.

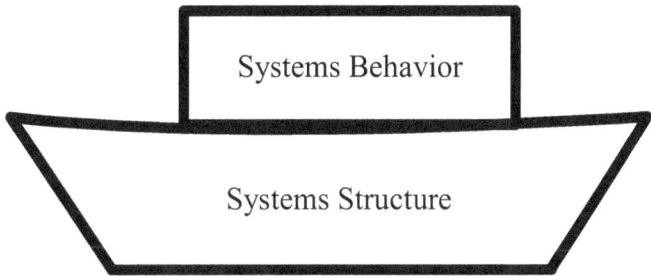

Figure 2-10   Systems Behavior Must be Loaded on the Systems Structure

Chapter 3: Introduction to SBC Architecture

As discussed in the previous chapter, SBC (i.e. structure-behavior coalescence) architecture is the recommended systems architecture approach. SBC architecture includes: a) SBC process algebra (SBC-PA) or SBC architecture description language (SBC-ADL), b) SBC architecture development method (SBC-ADM) and c) SBC view model (SBC-VM).

An architecture description language (ADL) is a special kind of language used in describing the architecture of a system. SBC-ADL uses six fundamental diagrams to formally describe the integration of systems structure and systems behavior of a system.

Through the iterative and cyclic architecture development method (ADM), a systems architect is able to construct the systems architecture smoothly. SBC-ADM, being used by a systems architect to accomplish each version management of the systems architecture, shall do the strategic management first and then go through the concept, analysis, design and implementation phases of systems architecture construction. Every phase checks with the requirements to make sure that each version of the constructed systems architecture is what the users want.

View model (VM) is a three-dimensional matrix representation of multiple views of a system. In the SBC view model, dimension 1 stands for the evolution&motivation view which contains the strategy/version 1, strategy/version 2, strategy/version 3, strategy/version 4, strategy/version ∞ views; dimension 2 stands for the multi-level (hierarchical) view which contains the concept, analysis, design and implementation views; dimension 3A stands for the SBC-ADL view which contains the architecture hierarchy diagram (AHD), framework diagram (FD), component operation diagram (COD), component connection diagram (CCD), structure-behavior coalescence diagram (SBCD) and interaction flow diagram (IFD) views; dimension 3B stands for the multi-layer view which contains the business layer, application layer, data layer and technology layer views.

3-1 Definition of SBC Architecture

Here, let us first give the SBC architecture a definition of its own as shown in Figure 3-1.

> SBC architecture,
> through structure-behavior coalescence,
> truly is an integrated whole of a system's multiple views, i.e., structure, behavior, and other views, embodied in its components, their interactions with each other and the environment, and the principles and guidelines governing its design and evolution.

Figure 3-1　Definition of SBC Architecture

From the above definition, we find out that SBC architecture, through structure-behavior coalescence, is a truly integrated whole of a system's multiple views, i.e., structure, behavior and other views, embodied in its assembled components, their interactions (or handshakes) with each other and the environment, and the principles and guidelines governing its design and evolution.

SBC architecture includes: a) SBC process algebra (SBC-PA) [Chao17a, Chao17b, Chao17c, Chao17d, Chao17e, Chao17f] or SBC architecture description language (SBC-ADL), b) SBC architecture development method (SBC-ADM) and c) SBC view model (SBC-VM).

3-2 SBC Architecture Description Language

An architecture description language is a special kind of language used in describing the architecture of a system [Shaw96, Tayl09].

A description of the systems architecture has to grasp the essence of a system and its details at the same time. In other words, a systems architecture description not only provides an overall picture that summarizes the system, but also contains enough detail that the system can be constructed and validated.

SBC-ADL uses six fundamental diagrams to describe the integration of systems structure and systems behavior of a system. These diagrams, as shown in Figure 3-2, are: a) architecture hierarchy diagram (AHD), b) framework diagram (FD), c) component operation diagram (COD), d) component connection diagram (CCD), e) structure-behavior coalescence diagram (SBCD) and f) interaction flow diagram (IFD).

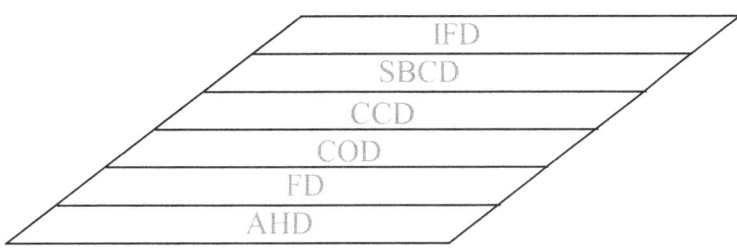

Figure 3-2   Six Fundamental Diagrams of SBC-ADL

SBC-ADL uses AHD, FD, COD, CCD, SBCD and IFD to depict the systems structure and systems behavior of a system as shown in Figure 3-3.

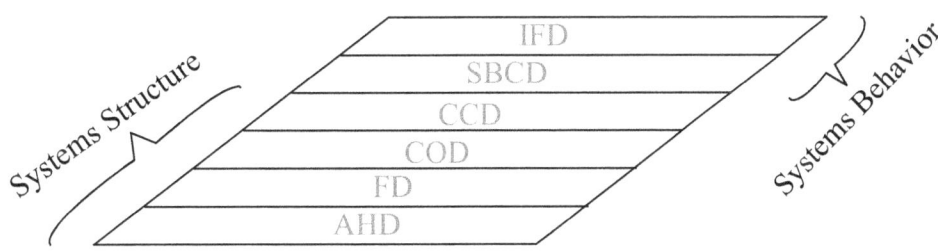

Figure 3-3   Systems Structure and Systems Behavior of a System

Examining the SBC-ADL approach, we find out that it depicts the systems structure first and then depicts the systems behavior later, not the other way around. The reason SBC-ADL does so lies in that the systems behavior must be attached to or built on the systems structure. With the systems structure and attached systems behavior, then, we can smoothly get the systems architecture as shown in Figure 3-4.

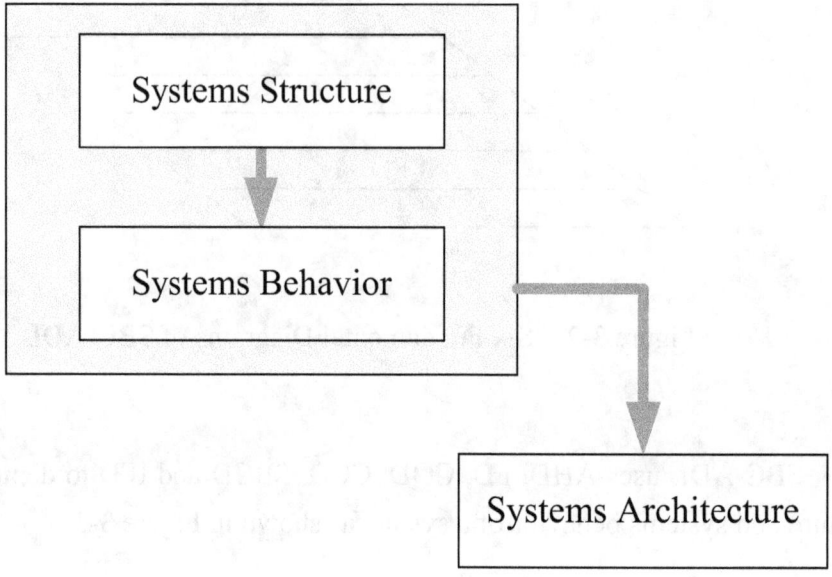

Figure 3-4   Systems Behavior Must be Attached to the Systems Structure

Let us ask the opposite question. Can the systems structure be attached to or built on the systems behavior? The answer is "No" as shown in Figure 3-5.

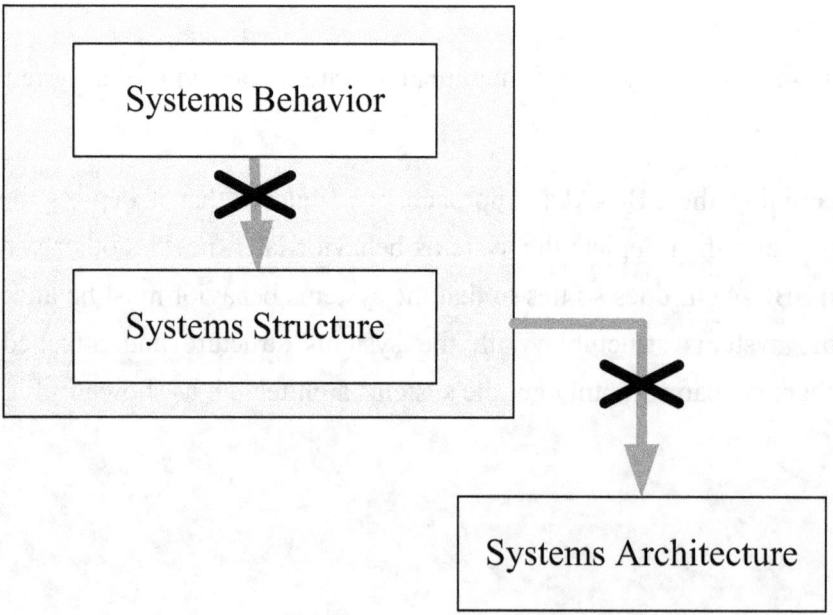

Figure 3-5   Systems Structure is not Attached to the Systems Behavior

In the SBC-ADL, the systems behavior must be attached to or built on the systems structure. In other words, the systems behavior shall not exist alone; it must be loaded on the systems structure just like a cargo is loaded on a ship as shown in Figure 3-6. There will be no systems behavior if there is no systems structure. A stand-alone systems behavior is not meaningful.

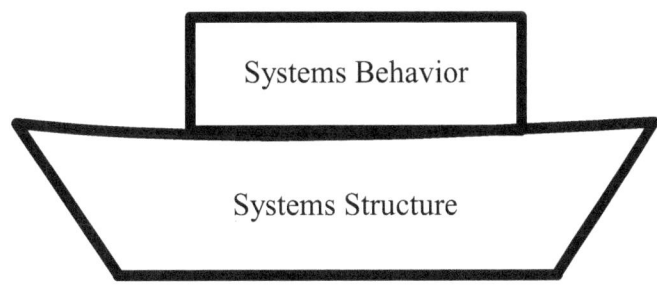

Figure 3-6　Systems Behavior Must be Loaded on the Systems Structure

AHD, FD, COD and CCD belong to systems structure. SBCD and IFD belong to systems behavior. Concluding the above discussion, we perceive that SBC-ADL will describe AHD, FD, COD and CCD first then describe SBCD and IFD later when it constructs a systems architecture.

3-2-1 Systems Structure Is Composed of Components

Through architecture hierarchy diagram (AHD), systems architects shall clearly observe the multi-level (hierarchical) decomposition and composition of a system. AHD is the first fundamental diagram to achieve structure-behavior coalescence.

As an example, Figure 3-7 shows that the *QQQ* system is composed of *B1*, *B2*, *A1*, *D1*, *D2* and *T1*; *A1* is composed of *A11*; *T1* is composed of *T11* and *T12*; *A11* is composed of *A111*; *A111* is composed of *A1111* and *A1112*. Among them, *QQQ*, *A1*, *T1*, *A11* and *A111* are aggregated systems while *B1*, *B2*, *A1111*, *A1112*, *D1*, *D2*, *T11* and *T12* are non-aggregated systems.

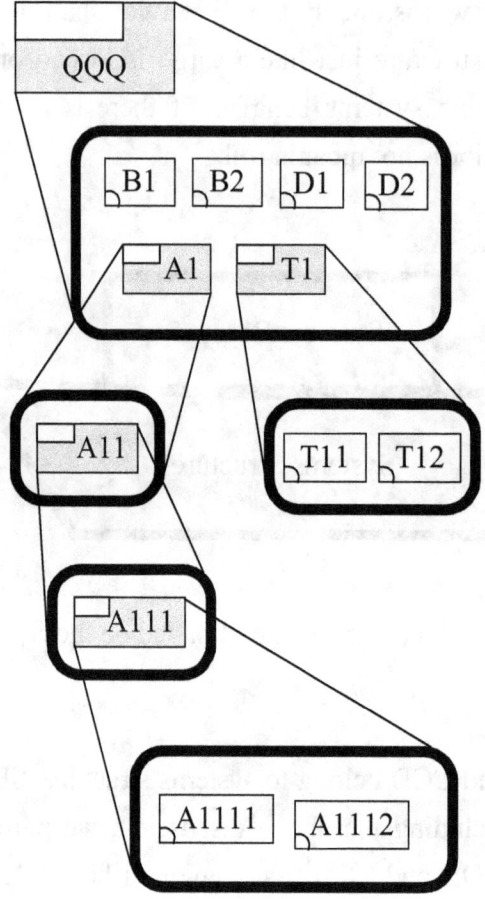

Figure 3-7   AHD of the *QQQ* System

There are aggregated and non-aggregated systems in an architecture hierarchy diagram. Non-aggregated systems are sometimes labeled as components, parts, entities, objects and building blocks [Chao14a, Chao14b].

Framework diagram (FD) represents the decomposition and composition of a system in a multi-layer (also referred to as multi-tier) manner. Only components will appear in a FD. FD is the second fundamental diagram to achieve structure-behavior coalescence.

As an example, Figure 3-8 shows a FD of the *QQQ* system. In the figure, *Business_Layer* contains the *B1* and *B2* components; *Application_Layer* contains the *A1111* and *A1112* components; *Data_Layer* contains the *D1* and *D2* components; *Technology_Layer* contains the *T11* and *T12* components.

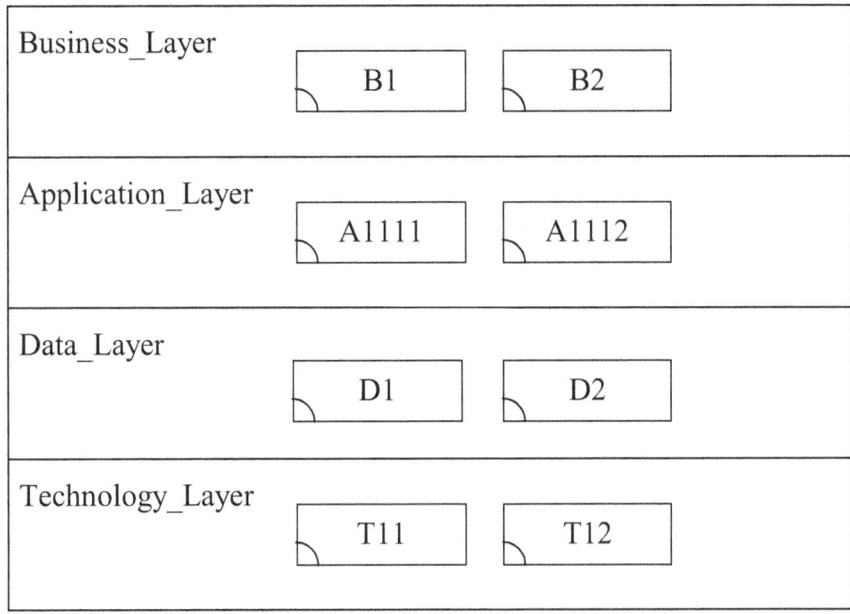

Figure 3-8　FD of the *QQQ* System

Systems structure, apart from outlining the components, shall also describe all operations for each component and all connections among components and actors in the external environment.

For a system, we use a component operation diagram (COD) to demonstrate all components' operations. COD is the third fundamental diagram to achieve structure-behavior coalescence. Figure 3-9 shows a COD of the *QQQ* system. In the figure, component *B1* has two operations: *op_01* and *op_02*; component *B2* has one operation: *op_03*; component *A1111* has one operation: *op_04*; component *A1112* has two operations: *op_05* and *op_06*; component *D1* has two operations: *op_07* and *op_08*; component *D2* has one operation: *op_9*; component *T11* has one operation: *op_10*; component *T12* has one operation: *op_11*.

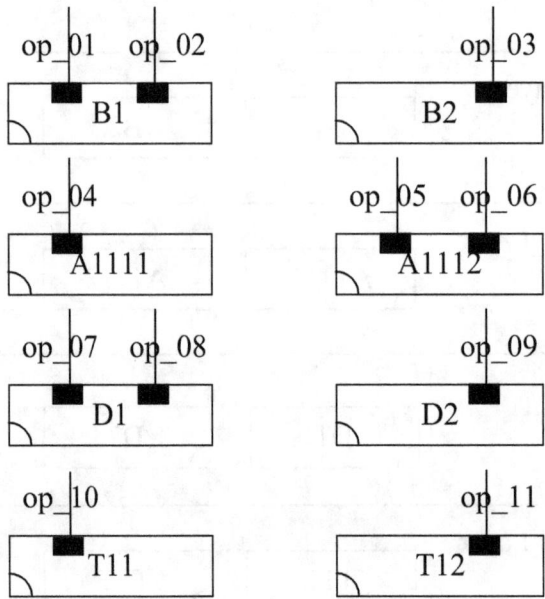

Figure 3-9   COD of the *QQQ* System

We use a component connection diagram (CCD) to describe how the components and actors (in the external environment) are connected within a system. CCD is the fourth fundamental diagram to achieve structure-behavior coalescence. Figure 3-10 exhibits a CCD of the *QQQ* system.

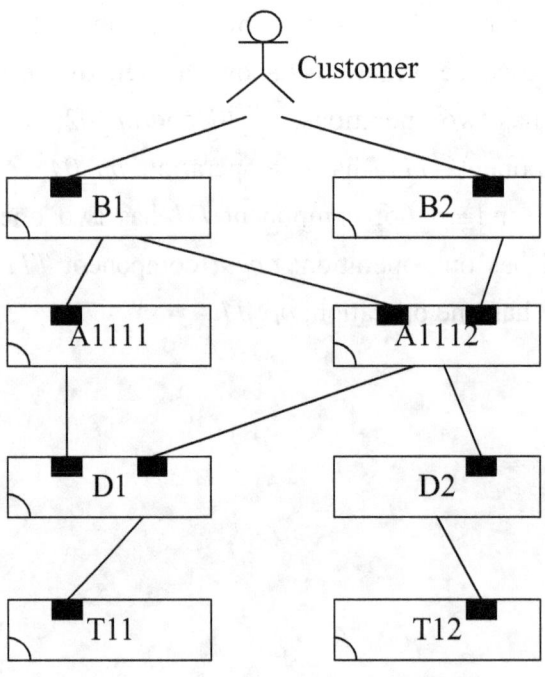

Figure 3-10   CCD of the *QQQ* System

In Figure 3-10, we see that actor *Customer* has a connection with each of the *B1* and *B2* components; component *B1* has a connection with each of the *A1111* and *A1112* components; component *B2* has a connection with the *A1112* component; component *A1111* has a connection with the *D1* component; component *A1112* has a connection with each of the *D1* and *D2* components; component *D1* has a connection with the *T11* component; component *D2* has a connection with the *T12* component.

3-2-2 Interactions among Components and Actors to Draw forth the Systems Behavior

In a system, if the components, and among them and the external environment's actors to interact (or handshake), these interactions will draw forth the systems behavior. That is, "interaction" plays an important factor in integrating the systems structure and systems behavior for a system.

We use a structure-behavior coalescence diagram (SBCD) to describe how the systems structure and systems behavior are integrated within a system. SBCD is the fifth fundamental diagram to achieve structure-behavior coalescence. Figure 3-11 exhibits a SBCD of the *QQQ* system. In this example, an actor interacting with eight components shall describe the overall systems behavior. Interactions among the *Customer* actor and the *B1*, *A1111*, *D1* components draw forth the *qqq_1* behavior. Interactions among the *Customer* actor and the *B1*, *A1112*, *D1*, *T11* components draw forth the *qqq_2* behavior. Interactions among the *Customer* actor and the *B2*, *A1112*, *D2*, *T12* components draw forth the *qqq_3* behavior.

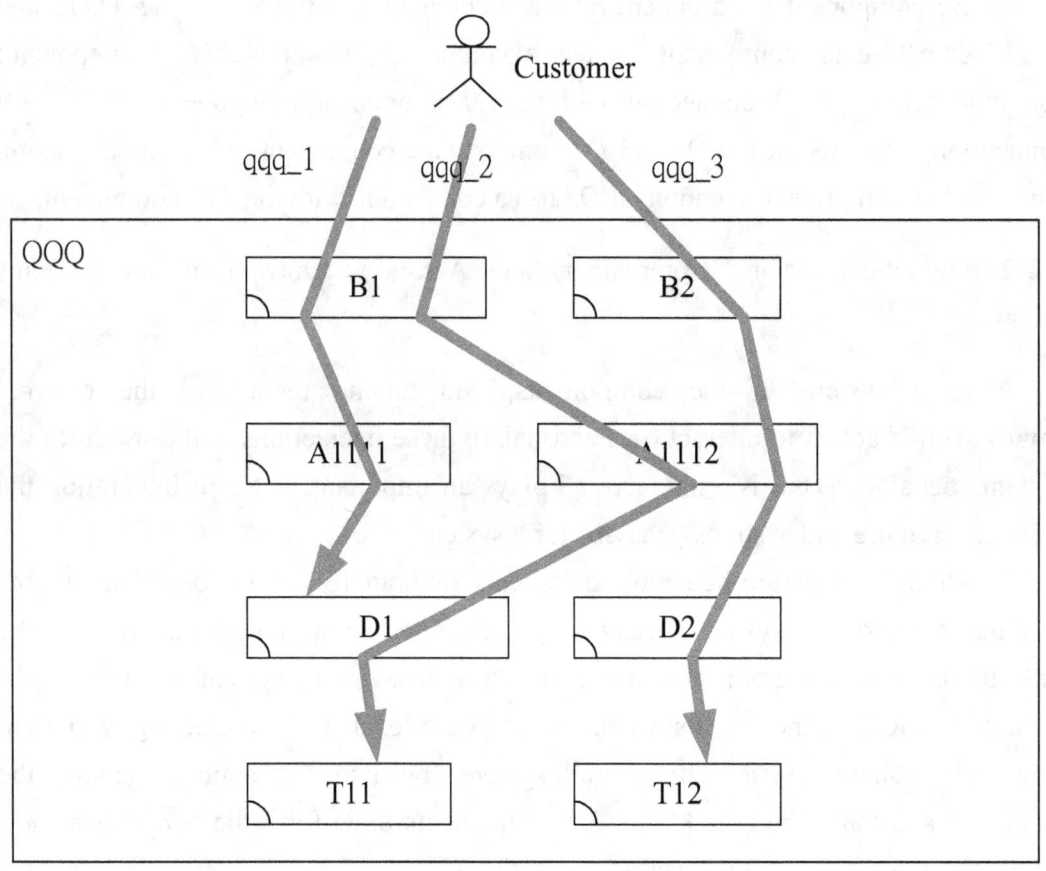

Figure 3-11   SBCD of the *QQQ* System

The overall behavior of a system is a collection of all its individual behaviors. All individual behaviors are mutually independent of each other. They tend to be executed concurrently [Hoar85, Miln89, Miln99]. For example, the overall the *QQQ* system's behavior includes the *qqq_1*, *qqq_2* and *qqq_3* behaviors. In other words, the *qqq_1*, *qqq_2* and *qqq_3* behaviors are combined to produce the overall behavior of the the *QQQ* system.

The major purpose of using the architectural approach, instead of separating the structure model from the behavior model, is to achieve one single coalesced model. In Figure 3-11, systems architects are able to see that the systems structure and systems behavior coexist in a SBCD. That is, in the SBCD of the *QQQ* system, architects not only see its systems structure but also see (at the same time) its systems behavior.

The overall behavior of a system consists of many individual behaviors. Each individual behavior represents an execution path. We use an interaction flow diagram (IFD) to demonstrate this individual behavior. IFD is the sixth fundamental diagram utilized to achieve structure-behavior coalescence. The *QQQ* system's overall behavior includes three behaviors: *qqq_1*, *qqq_2* and *qqq_3*. Each of them is described by an individual IFD.

Figure 3-12 shows an IFD of the *qqq_1* behavior. First, actor *Customer* interacts with the *B1* component through the *op_01* operation call interaction. Next, component *B1* interacts with the *A1111* component through the *op_04* operation call interaction. Finally, component *A1111* interacts with the *D1* component through the *op_07* operation call interaction.

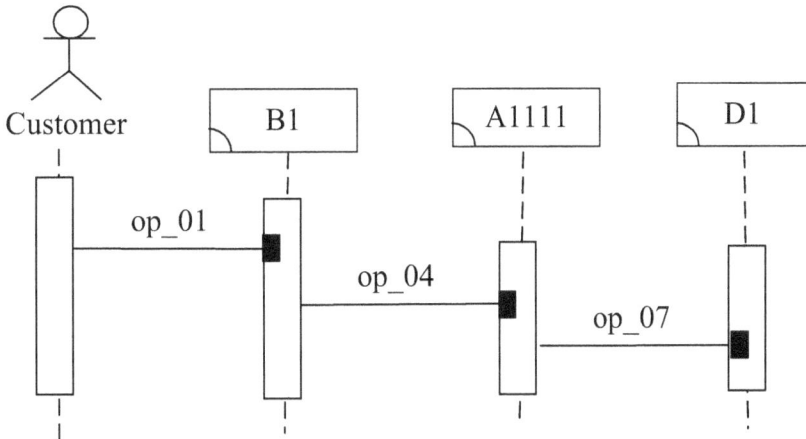

Figure 3-12   IFD of the *qqq_1* Behavior

Figure 3-13 shows an IFD of the *qqq_2* behavior. First, actor *Customer* interacts with the *B1* component through the *op_02* operation call interaction. Next, component *B1* interacts with the *A1112* component through the *op_05* operation call interaction. Continuingly, component *A1112* interacts with the *D1* component through the *op_08* operation call interaction. Finally, component *D1* interacts with the *T11* component through the *op_10* operation call interaction.

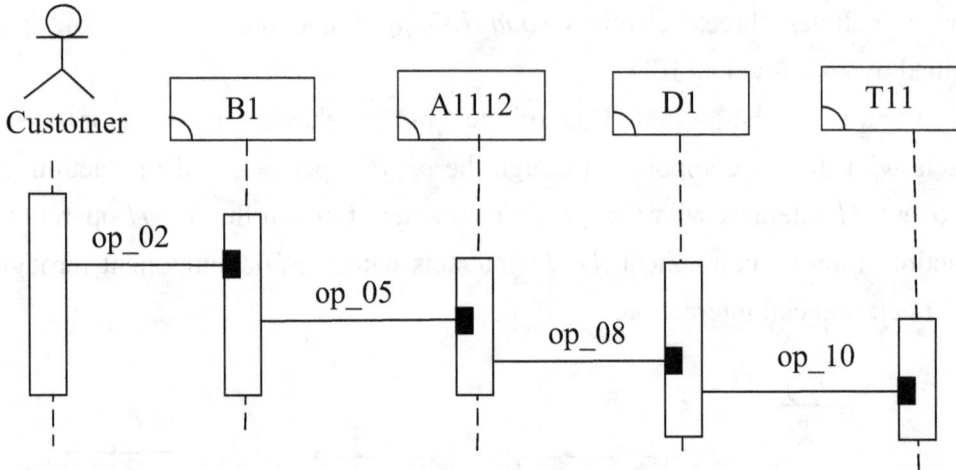

Figure 3-13   IFD of the *qqq_2* Behavior

Figure 3-14 shows an IFD of the *qqq_3* behavior. First, actor *Customer* interacts with the *B2* component through the *op_03* operation call interaction. Next, component *B2* interacts with the *A1112* component through the *op_06* operation call interaction. Continuingly, component *A1112* interacts with the *D2* component through the *op_09* operation call interaction. Finally, component *D2* interacts with the *T12* component through the *op_11* operation call interaction.

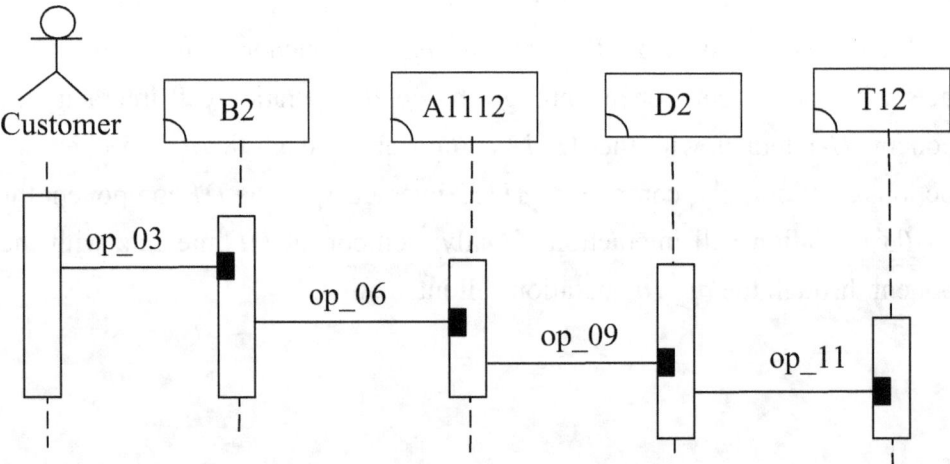

Figure 3-14   IFD of the *qqq_3* Behavior

3-3 SBC Architecture Development Method

The term of architecture development method (ADM) is first used in the open group architecture framework [Toga08]. Through the iterative and cyclic ADM, a systems architect is able to architect the systems architecture smoothly.

SBC architecture development method (SBC-ADM), being used by a systems architect to accomplish each version management of the systems architecture, shall do the strategic management first and then go through the concept, analysis, design and implementation phases of systems architecture construction. Every phase shall check with the requirements to make sure that each version of the constructed systems architecture is what the users want as shown in Figure 3-15.

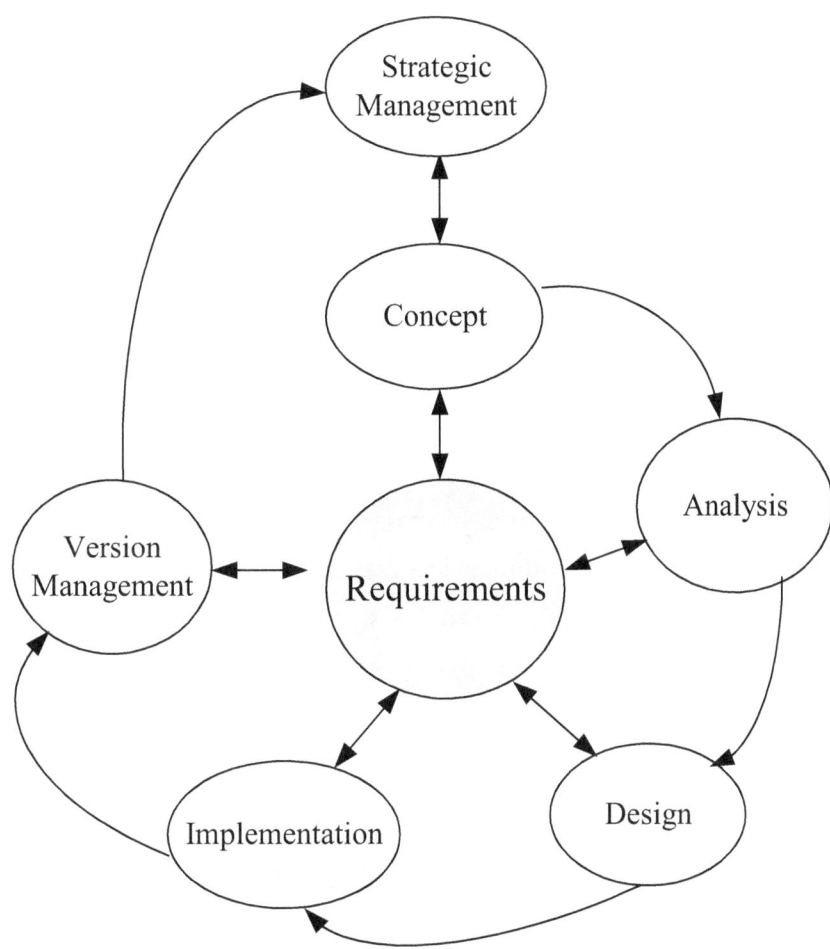

Figure 3-15 SBC Architecture Development Method

Each strategic management generates a strategy and each version management generates a version of systems architecture. Therefore, each strategy is mapped to a version of systems architecture as shown in Figure 3-16.

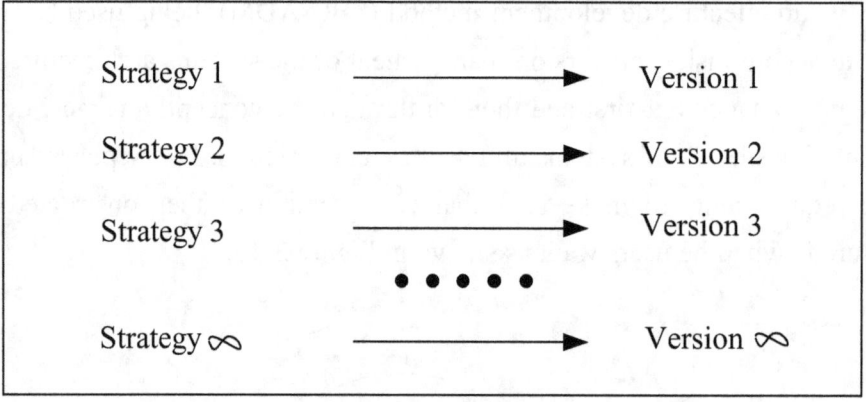

Figure 3-16   Each Strategy is Mapped to a Version of Systems Architecture

3-4 SBC View Model

View model [Chao14a, Chao14b, Dam06, Mino08, O'Rou03] is a three-dimensional matrix representation of a system's multiple views as shown in Figure 3-17. In the figure, dimension 1 stands for the evolution&motivation view which contains the strategy/version 1, strategy/version 2, strategy/version 3, strategy/version 4 and strategy/version ∞ views; dimension 2 stands for the multi-level (hierarchical) view which contains the concept, analysis, design and implementation views; dimension 3 stands for the systemic view which contains the structure, behavior, input/output data views.

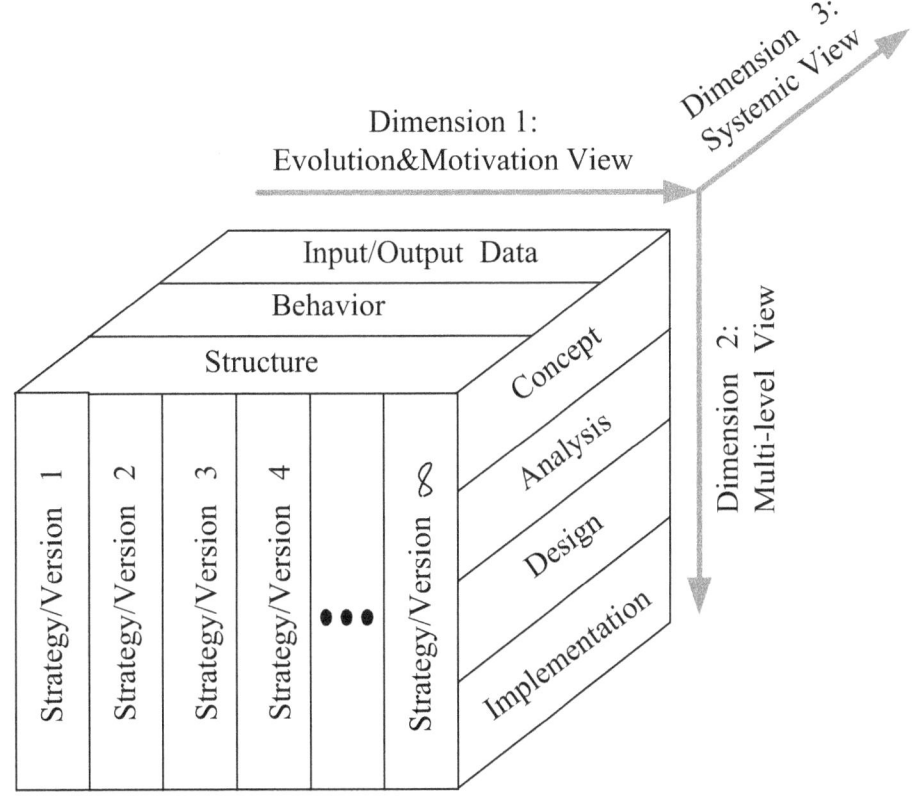

Figure 3-17   View Model

SBC view model (SBC-VM) [Chao14a, Chao14b, Dam06, Mino08, O'Rou03] adopts a simplifier representation of multiple views as shown in Figure 3-18.

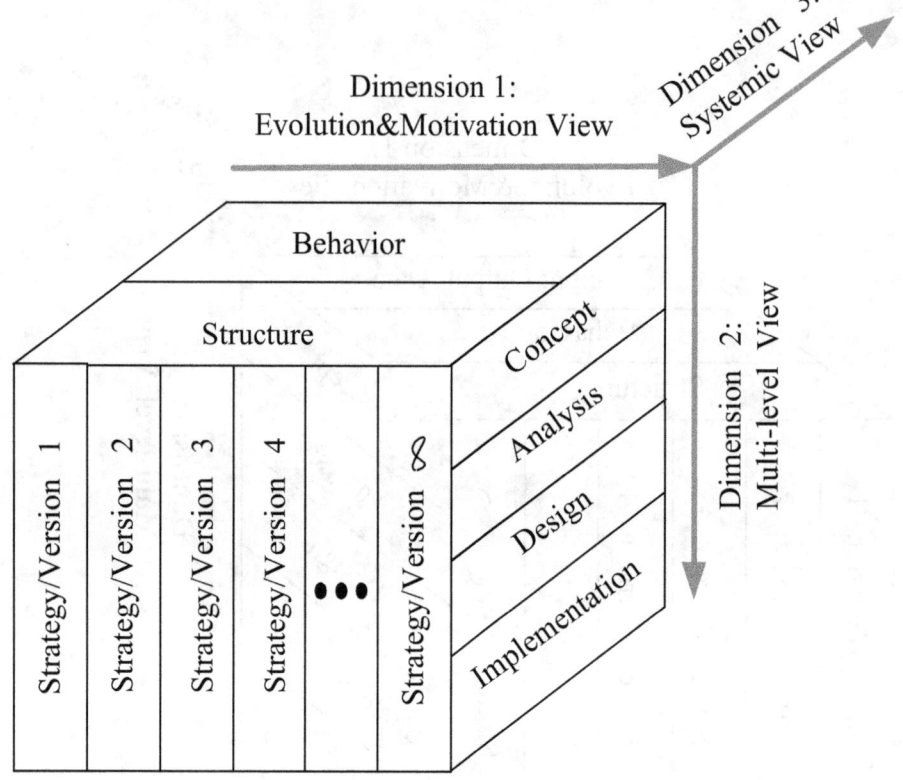

Figure 3-18 SBC View Model

In the SBC view model, dimension 1 stands for the evolution&motivation view which contains the strategy/version 1, strategy/version 2, strategy/version 3, strategy/version 4, strategy/version ∞ views; dimension 2 stands for the multi-level (hierarchical) view which contains the concept, analysis, design and implementation views; dimension 3 stands for the systemic view which contains the structure and behavior views. According to the approach of SBC-ADL, the structure view consists of AHD, FD, COD and CCD; the behavior view consists of SBCD and IFD. Also, FD consists of business layer, application layer, data layer and technology layer. Adding these ideas to Figure 3-18, we then get the complete SBC view model as shown in Figure 3-19.

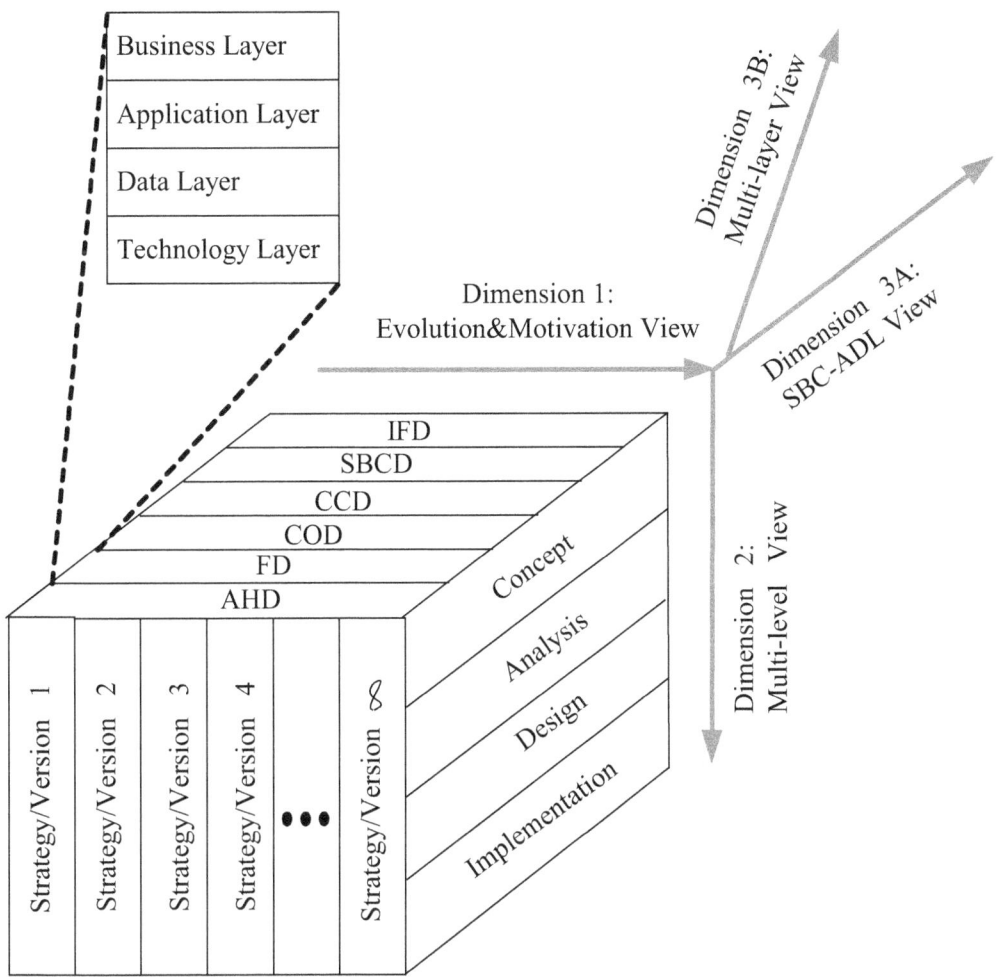

Figure 3-19　Complete SBC View Model

We have already discussed SBC-ADL in the previous sections. Let us now explain how to represent those multi-level (hierarchical) views. Architectural construction involves multi-level decisions (i.e. views) made regarding how the systems architecture should be constructed. These multi-level decisions may lead us to the following multi-level architecture diagram, as shown in Figure 3-20. In the figure, *QQQ* is composed of *B1*, *B2*, *A1*, *D1*, *D2* and *T1*; *A1* is composed of *A11*; *T1* is composed of *T11* and *T12*; *A11* is composed of *A111*; *A111* is composed of *A1111* and *A1112*.

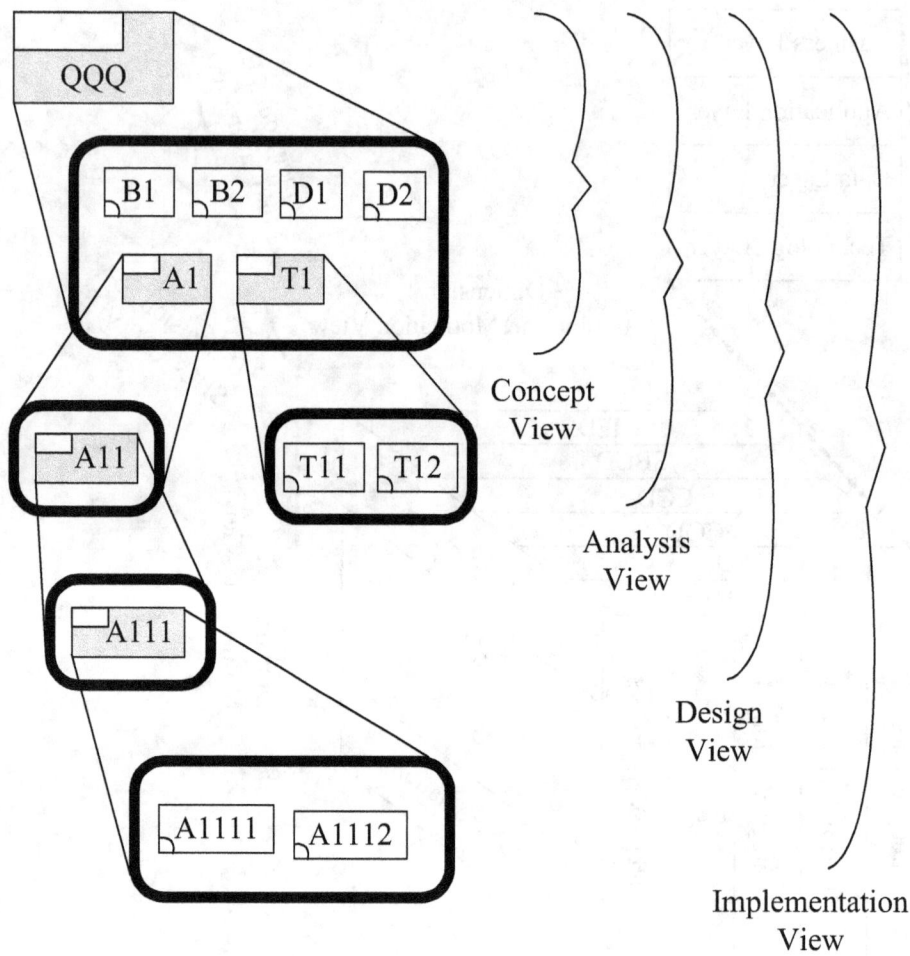

Figure 3-20 Multi-Level Architecture Diagram

There are aggregated and non-aggregated systems in Figure 3-20. Systems *QQQ*, *A1*, *T1*, *A11* and *A111* are categorized into aggregated systems while systems *B1*, *B2*, *A1111*, *A1112*, *D1*, *D2*, *T11* and *T12* are categorized into non-aggregated systems.

Concept view corresponds to an executive summary for an administrator who wants an estimate of the scope of the system, what it would cost, and how it would relate to the general environment in which it will operate. Analysis view corresponds to a summary for an analyzer who works on the analysis of a system. Analysis view is one level down structural decomposition (with observation congruence verification) of the concept view [Chao15a, Chao15b, Chao15c, Chao15d, Chao15e]. Design view describes what a designer has accomplished for his task. Design view is one level down structural decomposition (with observation congruence verification) of the analysis view. Implementation view shows what an implementer has done for his work. Implementation view is one level down structural decomposition (with observation congruence verification) of the design view.

Figures 3-20 gives us some ideas of architecture construction in brief, but is short of systemic views. Figure 3-21 demonstrates a multi-level architecture relating to the systemic view.

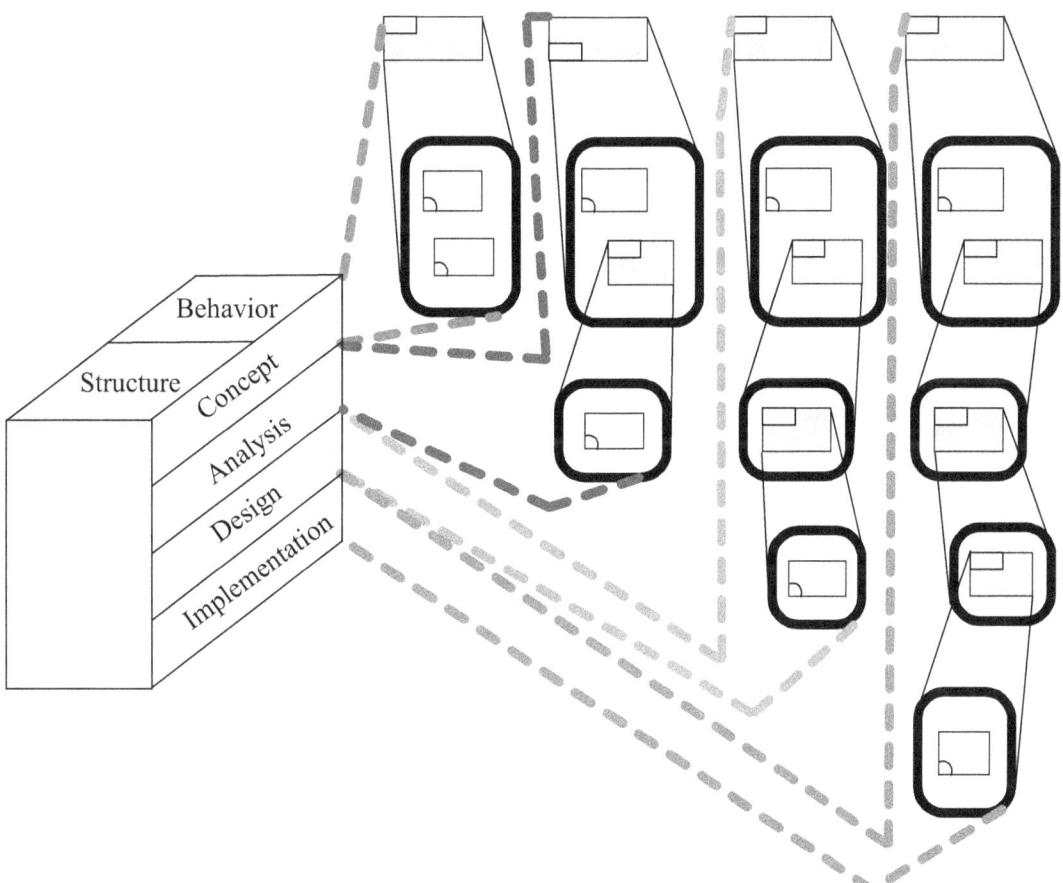

Figure 3-21 Multi-Level Architecture Relating to the Systemic View

In Figure 3-21, we observe that *Concept view* contains the *Concept's Systems Structure* and *Concept's Systems Behavior*; *Analysis View* contains the *Analysis' Systems Structure* and *Analysis' Systems Behavior*; *Design View* contains the *Design's Systems Structure* and *Design's Systems Behavior*; *Implementation View* contains the *Implementation's Systems Structure* and *Implementation's Systems Behavior*.

According to the approach of SBC-ADL, systems structure consists of AHD, FD, COD and CCD; systems behavior consists of SBCD and IFD. Adding these SBC-ADL systemic views to Figure 3-21, we then get the SBC multi-level (hierarchical) view as shown in Figure 3-22.

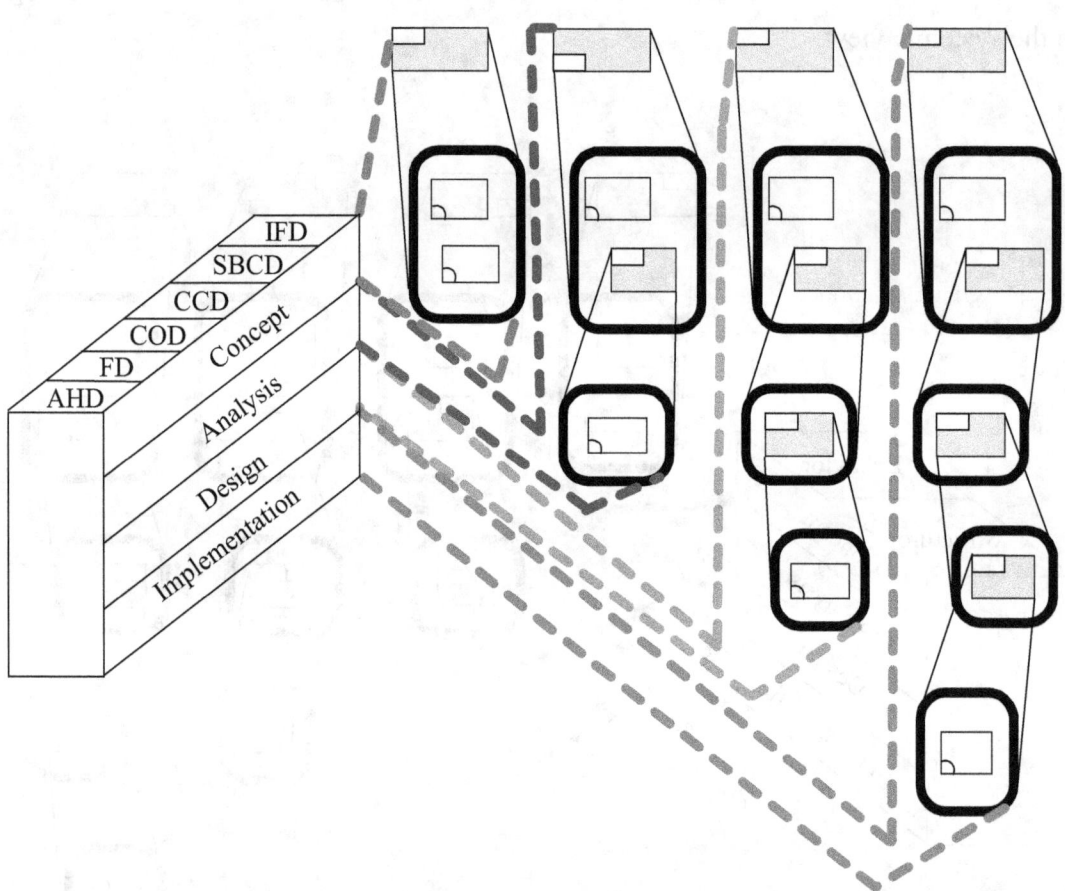

Figure 3-22　　SBC Multi-Level (Hierarchical) View

PART II: SBC ARCHITECTURE DESCRIPTION LANGUAGE

Chapter 4: Systems Structure

SBC-ADL uses the architecture hierarchy diagram, framework diagram, component operation diagram and component connection diagram to depict the systems structure of a system.

4-1 Architecture Hierarchy Diagram

Systems architects use an architecture hierarchy diagram (AHD) to define the multi-level (hierarchical) decomposition and composition of a system. AHD is the first fundamental diagram to achieve structure-behavior coalescence.

4-1-1 Decomposition and Composition

The following is an example of systems decomposition and composition. The *Computer* system consists of *Monitor*, *Keyboard*, *Mouse* and *Case*, as shown in Figure 4-1. The *Monitor*, *Keyboard*, *Mouse* and *Case* are subsystems comprising the *Computer* system.

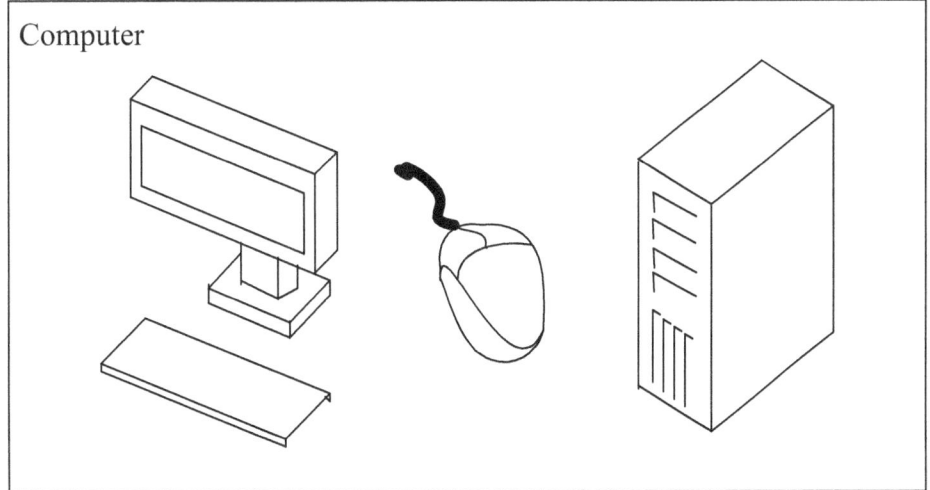

Figure 4-1 Decomposition and Composition of the *Computer* System

Another example indicates that the *Tree* system is composed of *Root* and *Stem*, as shown in Figure 4-2. In this example, we would say that the *Root* and *Stem* are subsystems, respectively, while the *Tree* system consists of its subsystems.

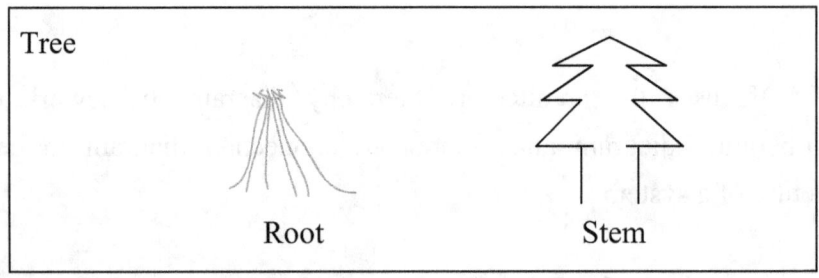

Figure 4-2   Decomposition and Composition of the *Tree* System

The last example demonstrates that the *Architecture_Book* system is composed of *Chapter_1*, *Part_1* and *Part_2*, as shown in Figure 4-3. In this example, we would say that the *Chapter_1*, *Part_1* and *Part_2* are subsystems, respectively while the *Architecture_Book* system consists of its subsystems.

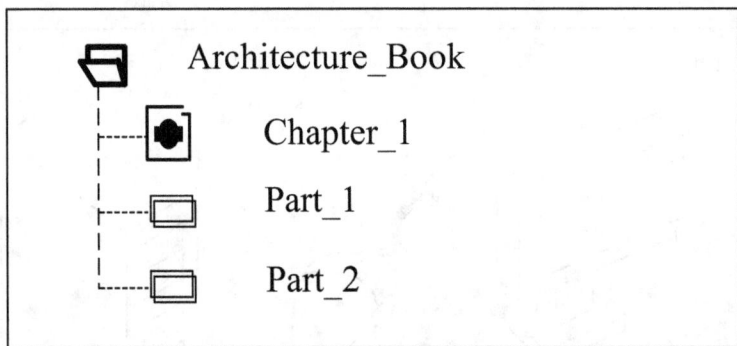

Figure 4-3   Decomposition and Composition of the *Architecture_Book* System

Architecture hierarchy diagram (AHD) is used to represent the decomposition and composition of a system. As an example, an AHD of the *Computer* system is shown in Figure 4-4. Systems architects clearly observe that the *Computer* system is composed of *Monitor*, *Keyboard*, *Mouse* and *Case*.

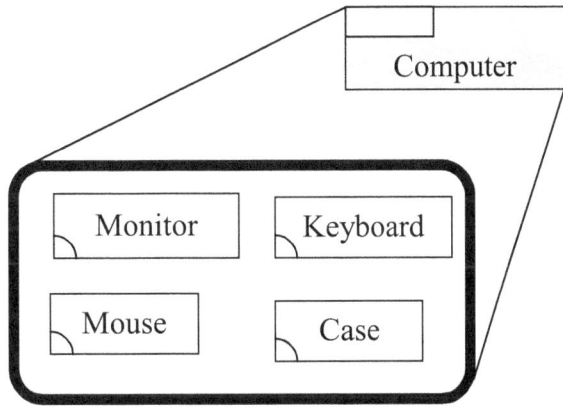

Figure 4-4   AHD of the *Computer* System

As a second example, Figure 4-5 shows an AHD of the *Tree* system. Systems architects clearly observe that the *Tree* system is composed of *Root* and *Stem*.

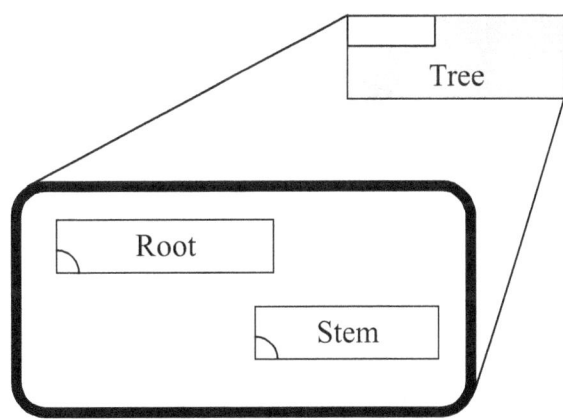

Figure 4-5   AHD of the *Tree* System

As a third example, Figure 4-6 shows an AHD of the Architecture_Book system. Systems architects clearly observe that the *Architecture_Book* is composed of *Chapter_1*, *Part_1* and *Part_2*.

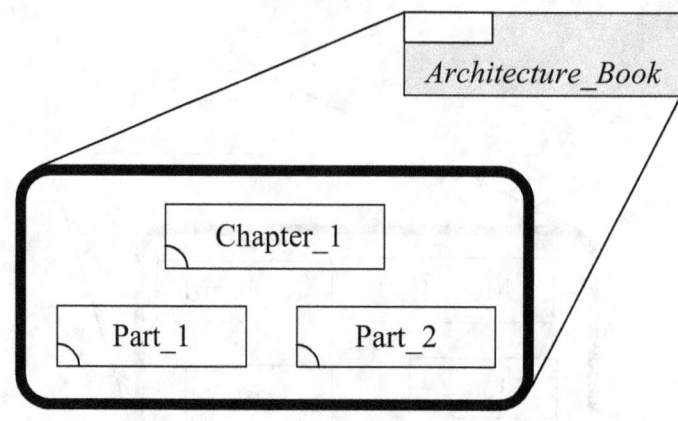

Figure 4-6   AHD of the *Architecture_Book* System

4-1-2 Multi-Level Decomposition and Composition

The subsystem may also contain subsystems as we further decompose it. For example, *Case* is a subsystem of the *Computer*, and we further decompose it into *Motherboard*, *Hard_Disk*, *Power_Supply* and *DVD_Disk*, as shown in Figure 4-7.

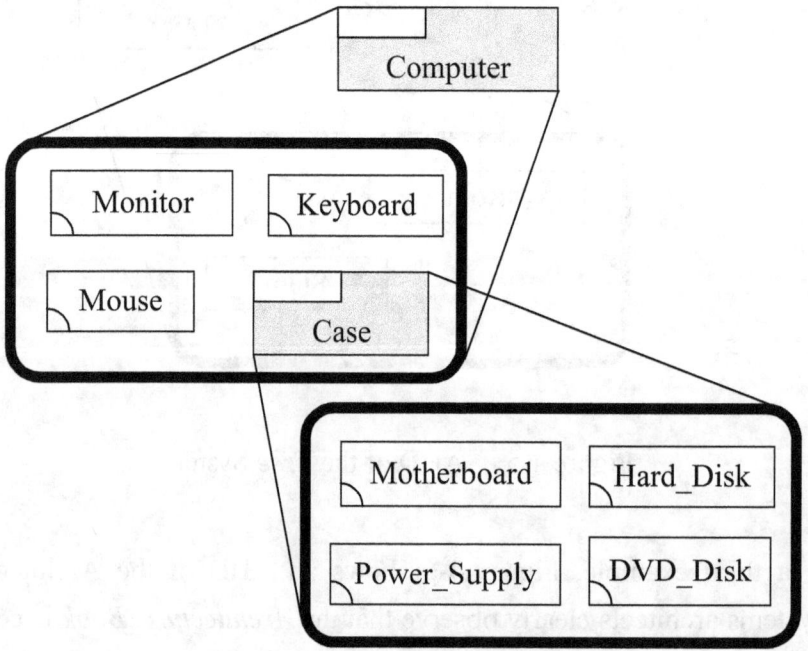

Figure 4-7 Multi-Level Decomposition/Composition of the *Computer* System

As a second example, *Stem* is a subsystem of the *Tree*, and we further decompose it into *Trunk* and *Leaf*, as shown in Figure 4-8.

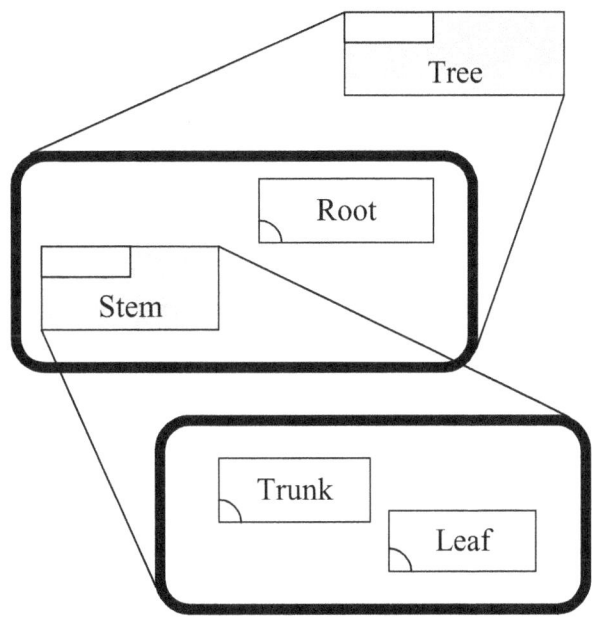

Figure 4-8 Multi-Level Decomposition/Composition of the *Tree* System

As a third example, *Part_1* is a subsystem of the *Architecture_Book*, and we further decompose it into *Chapter_2* and *Chapter_3*; *Part_2* is also a subsystem of the *Architecture_Book*, and we further decompose it into *Chapter_4* and *Chapter_5*, as shown in Figure 4-9.

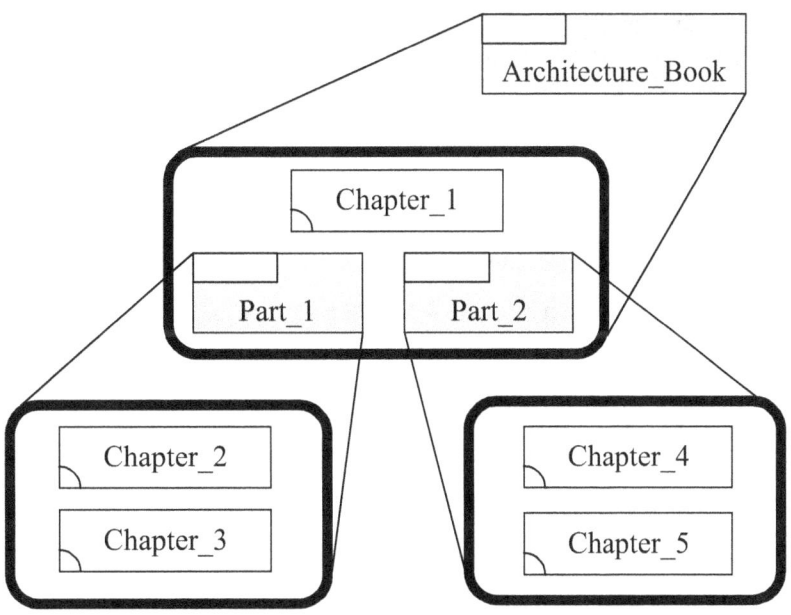

Figure 4-9 Multi-Level Decomposition/Composition of the *Architecture_Book* System

Generally speaking, multi-level (hierarchical) decomposition and composition of a system is applied often in architecting its architecture. To make a complex system look simple, the mechanism of multi-level composition and decomposition should be utilized.

4-1-3 Aggregated and Non-Aggregated Systems

Any system (at any level) involved with multi-level decomposition and composition of a system is either aggregated or non-aggregated. The definition of aggregated and non-aggregated systems is shown in Figure 4-10.

Definition of Aggregated Systems:
A system (within an AHD) is aggregated if it is composed of any sub-system.
Definition of Non-aggregated Systems
A system (within an AHD) is non-aggregated if it is NOT composed of any sub-system.

Figure 4-10   Definition of Aggregated and Non-aggregated Systems

Non-aggregated systems are sometimes referred to as components, parts, entities, objects and building blocks [Chao14a, Chao14b].

In the multi-level (hierarchical) decomposition and composition, any system is either aggregated or non-aggregated, but not both. For example, in Figure 4-4, *Case* is a non-aggregated system, not an aggregated system. As an interesting contrast, in Figure 4-7, *Case* is an aggregated system, not a non-aggregated system.

As a second example, in Figure 4-5, *Stem* is a non-aggregated system, not an aggregated system. As an interesting contrast, in Figure 4-8, *Stem* is an aggregated system, not a non-aggregated system.

As a third example, in Figure 4-6, *Part_1* and *Part_2* are non-aggregated systems, not aggregated systems. As an interesting contrast, in Figure 4-9, *Part_1* and *Part_2* are aggregated systems, not non-aggregated systems.

4-2 Framework Diagram

Framework diagram (FD) enables systems architects to examine the multi-layer (also referred to as multi-tier) decomposition and composition of a system. FD is the second fundamental diagram to achieve structure-behavior coalescence.

4-2-1 Multi-Layer Decomposition and Composition

Decomposition and composition of a system can also be represented in a multi-layer (or multi-tier) manner. We draw a framework diagram (FD) for the multi-layer decomposition and composition of a system.

As an example, Figure 4-11 shows a FD of the *Computer* system. In the figure, *Technology_SubLayer_2* contains *Monitor*, *Keyboard* and *Mouse*; *Technology_SubLayer_1* contains *Motherboard*, *Hard_Disk*, *Power_Supply* and *DVD_Disk*.

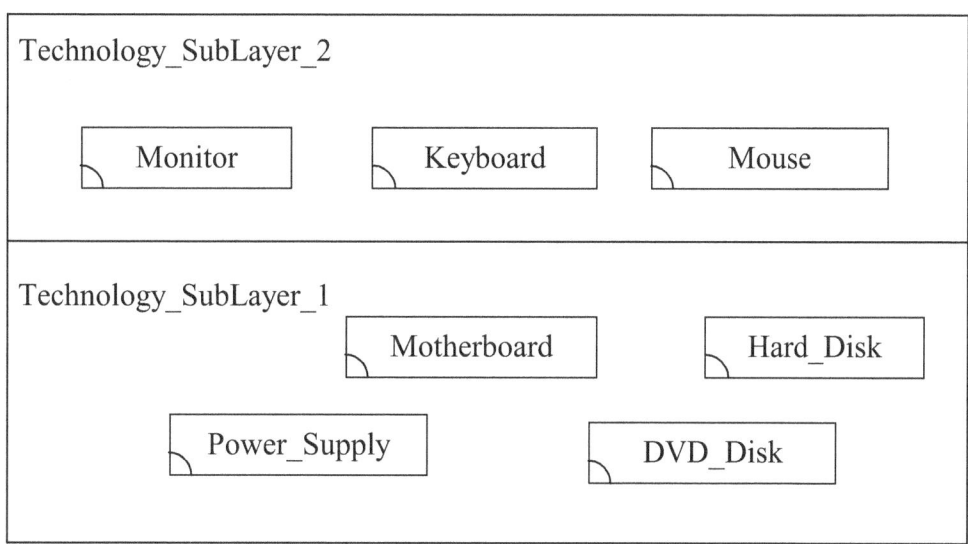

Figure 4-11   FD of the *Computer* System

As a second example, Figure 4-12 shows a FD of the *Tree* system. In the figure, *Technology_SubLayer_2* contains *Root*; *Technology_SubLayer_1* contains *Trunk* and *Leaf*.

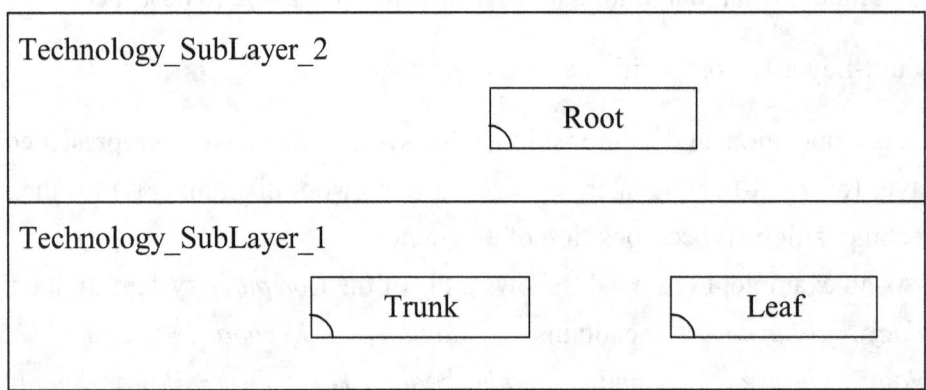

Figure 4-12   FD of the *Tree* System

As a third example, Figure 4-13 shows a FD of the *Architecture_Book* system. In the figure, *Technology_SubLayer_2* contains *Chapter_1*; *Technology_SubLayer_1* contains *Chapter_2*, *Chapter_3*, *Chapter_4* and *Chapter_5*.

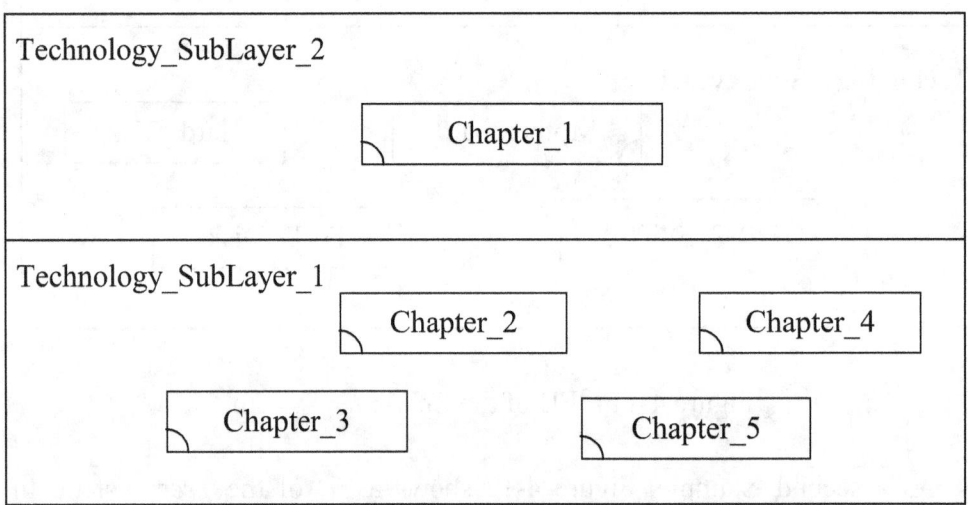

Figure 4-13   FD of the *Architecture_Book* System

4-2-2 Only Non-Aggregated Systems Appearing in Framework Diagrams

Both aggregated and non-aggregated systems are displayed in the multi-level (hierarchical) AHD decomposition and composition of a system. As an interesting contrast, only non-aggregated systems shall appear in the multi-layer FD decomposition and composition of a system.

For example, Figure 4-7 in the previous section shows an AHD of the *Computer* system in which both aggregated systems such as *Computer*, *Case* and non-aggregated systems such as *Monitor*, *Keyboard*, *Mouse*, *Motherboard*, *Hard_Disk*,

Power_Supply, *DVD_Disk* are displayed. As an interesting contrast, Figure 4-11 in the previous section shows a FD of the *Computer* system in which only non-aggregated systems such as *Monitor*, *Keyboard*, *Mouse*, *Motherboard*, *Hard_Disk*, *Power_Supply* and *DVD_Disk* are displayed.

For a second example, Figure 4-8 in the previous section shows an AHD of the *Tree* system in which both aggregated systems such as *Tree*, *Stem* and non-aggregated systems such as *Root*, *Trunk*, *Leaf* are displayed. As an interesting contrast, Figure 4-12 in the previous section shows a FD of the *Tree* system in which only non-aggregated systems such as *Root*, *Trunk* and *Leaf* are displayed.

For a third example, Figure 4-9 in the previous section shows an AHD of the *Architecture_Book* system in which both aggregated systems such as *Architecture_Book*, *Part_1*, *Part_2* and non-aggregated systems such as *Chapter_1*, *Chapter_2*, *Chapter_3*, *Chapter_4*, *Chapter_5* are displayed. As an interesting contrast, Figure 4-13 in the previous section shows a FD of the *Architecture_Book* system in which only non-aggregated systems such as *Chapter_1*, *Chapter_2*, *Chapter_3*, *Chapter_4* and *Chapter_5* are displayed.

4-3 Component Operation Diagram

Systems architects use a component operation diagram (COD) to display all components' operations of a system. COD is the third fundamental diagram to achieve structure-behavior coalescence.

4-3-1 Operations of Components

An operation provided by each component represents a procedure, or method, or function of the component. If other systems request this component to perform an operation, then shall use it to accomplish the operation request.

Each component in a system must possess at least one operation. A component should not exist in a system if it does not possess any operation. Figure 4-14 shows that component *SalePurchase_GUI* has four operations: *SaleInputClick*, *SalePrintClick*, *PurchaseInputClick* and *PurchasePrintClick*.

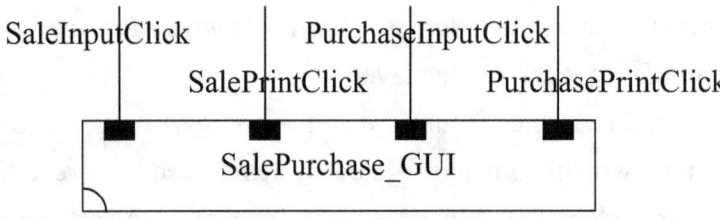

Figure 4-14   Four Operations of the *SalePurchase_GUI* Component

An operation formula is utilized to fully represent an operation. An operation formula includes a) operation name, b) input parameters and c) output parameters as shown in Figure 4-15.

$$\text{Operation\_Name (In } a_1, a_2, ..., a_M \text{ ; Out } a_{M+1}, a_{M+2}, ..., a_{M+N})$$

Figure 4-15   Operation Formula

Operation name is the name of this operation. In a system, every operation name should be unique. Duplicate operation names shall not be allowed in any system.

An operation may have several input and output parameters. The input and output parameters, gathered from all operations, represent the input data and output data views of a system [Date03, Elma10]. As shown in Figure 4-16, component *SalePrint_GUI* possesses the *ShowModal* operation which has no input/output parameter; component *SalePrint_GUI* also possesses the *SalePrintButtonClick* operation which has the *sDate* and *sNo* input parameters (with the arrow direction pointing to the component) and the *s_report* output parameter (with the arrow direction opposite to the component).

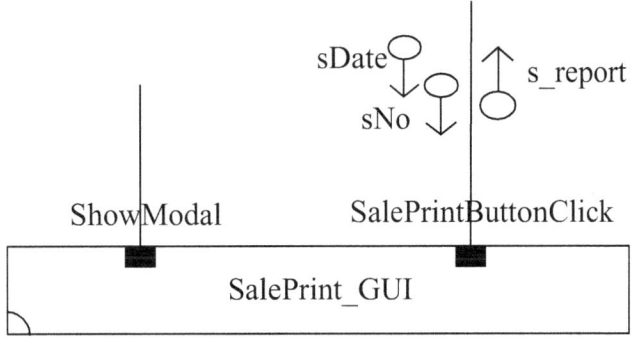

Figure 4-16   Input/Output Parameters of *SalePrintButtonClick*

Data formats of input and output parameters can be described by data type specifications. There are two sets of data types: primitive and composite [Date03, Elma10]. Figure 4-17 shows the primitive data type specification of the *sDate* and *sNo* input parameters occurring in the *SalePrintButtonClick(In sDate, sNo; Out s_report)* operation formula.

Parameter	Data Type	Instances
sDate	Text	20100517, 20100612
sNo	Text	001, 002

Figure 4-17   Primitive Data Type Specification

Figure 4-18 shows the composite data type specification of the *s_report* output parameter occurring in the *SalePrintButtonClick(In sDate, sNo; Out s_report)* operation formula.

Parameter	*s_report*

Data Type	TABLE of
	  Sale Date : Text
	  Sale No : Text
	  Customer : Text
	  ProductNo : Text
	  Quantity : Integer
	  UnitPrice : Real
	  Total : Real
	End TABLE;

Instances	Sale Date : 20100517 Sale No : 001 Customer : Larry Fink | ProductNo | Quantity | UnitPrice | | A12345 | 400 | 100.00 | | A00001 | 300 | 200.00 | Total : 100,000.00

Figure 4-18   Composite Data Type Specification

4-3-2 Drawing the Component Operation Diagram

For a system, COD is used to display all components' operations. Figure 4-19 shows the *Multi-Tier Personal Data System's COD*. In the figure, component *MTPDS_GUI* has two operations: *Calculate_AgeClick* and *Calculate_OverweightClick*; component *Age_Logic* has one operation: *Calculate_Age*; component *Overweight_Logic* has one operation: *Calculate_Overweight*; component *Personal_Database* has two operations: *Sql_DateOfBirth_Select* and *Sql_SexHeightWeight_Select*.

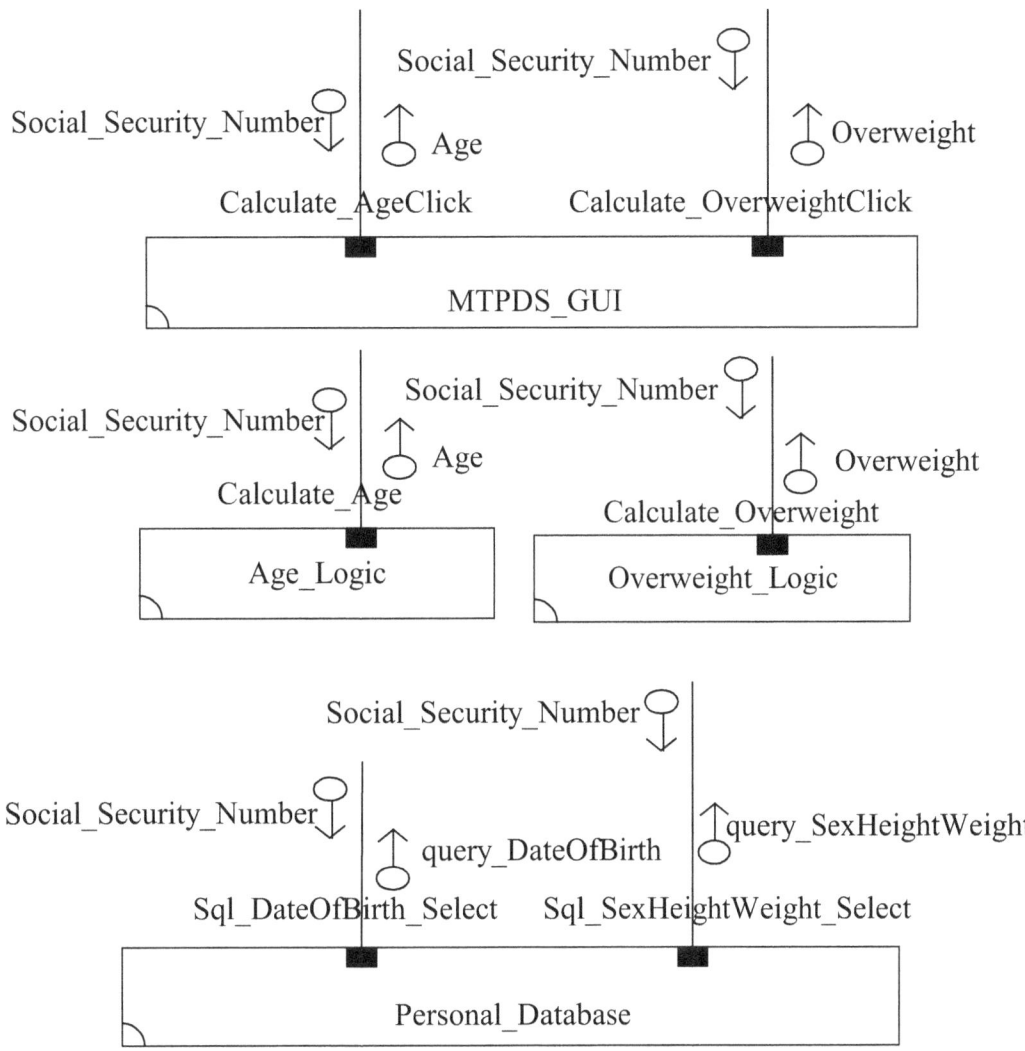

Figure 4-19   COD of the *Multi-Tier Personal Data System*

The operation formula of *Calculate_AgeClick* is *Calculate_AgeClick(In Social_Security_Number; Out Age)*. The operation formula of *Calculate_OverweightClick* is *Calculate_OverweightClick(In Social_Security_Number; Out Overweight)*. The operation formula of *Calculate_Age* is *Calculate_Age(In Social_Security_Number; Out Age)*. The operation formula of *Calculate_Overweight* is *Calculate_Overweight(In Social_Security_Number; Out Overweight)*. The operation formula of *Sql_DateOfBirth_Select* is *Sql_DateOfBirth_Select(In Social_Security_Number; Out query_DateOfBirth)*. The operation formula of *Sql_SexHeightWeight_Select* is *Sql_SexHeightWeight_Select(In Social_Security_Number; Out query_SexHeightWeight)*.

Figure 4-20 shows the primitive data type specification of the *Social_Security_Number* input parameter and the *Age, Overweight* output parameters.

Parameter	Data Type	Instances
Social_Security_Number	Text	424-87-3651, 512-24-3722
Age	Integer	28, 56
Overweight	Boolean	Yes, No

Figure 4-20   Primitive Data Type Specification

Figure 4-21 shows the composite data type specification of the *query_DateOfBirth* output parameter occurring in the *Sql_DateOfBirth_Select(In Social_Security_Number; Out query_DateOfBirth)* operation formula.

Parameter	*query_DateOfBirth*

Data Type	TABLE of
	   Social_Security_Number : Text
	   Age : Integer
	End TABLE ;

Instances	424-87-3651 28 512-24-3722 56

Figure 4-21   Composite Data Type Specification

Figure 4-22 shows the composite data type specification of the *query_SexHeightWeight* output parameter occurring in the *Sql_SexHeightWeight_Select(In Social_Security_Number; Out query_SexHeightWeight)* operation formula.

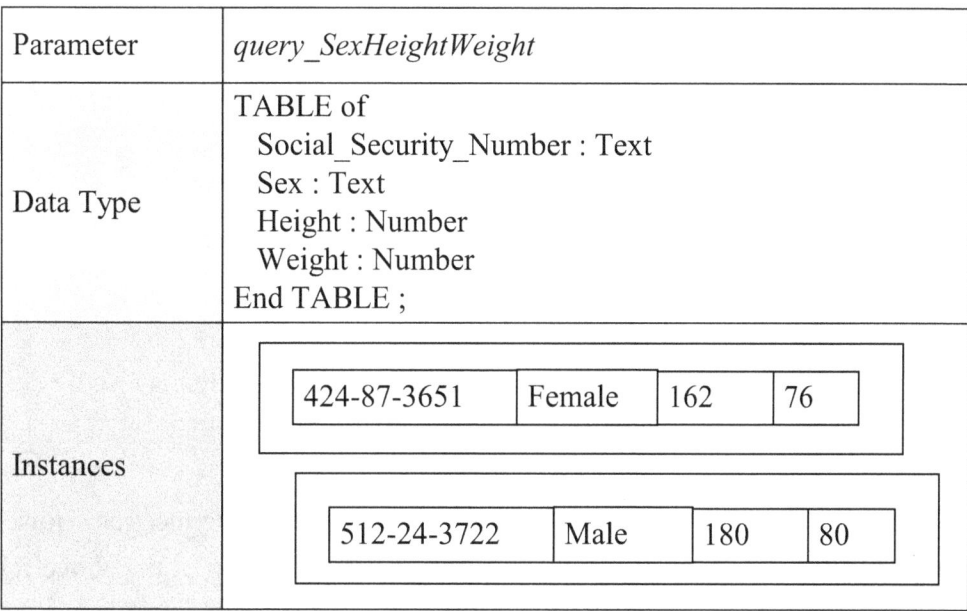

Figure 4-22   Composite Data Type Specification

4-4 Component Connection Diagram

A component connection diagram (CCD) is utilized to describe how all components and actors are connected within a system. CCD is the fourth fundamental diagram to achieve structure-behavior coalescence.

4-4-1 Essence of a Connection

A connection implies an operation request. When an operation is used by another subsystem then a connection appears. Accordingly, a connection is defined as the linkage that is constructed when an operation is used by another subsystem. Figure 4-23 shows that Subs*ystem_A* uses the *Salary_Calculation* operation provided by the *Component_B* component.

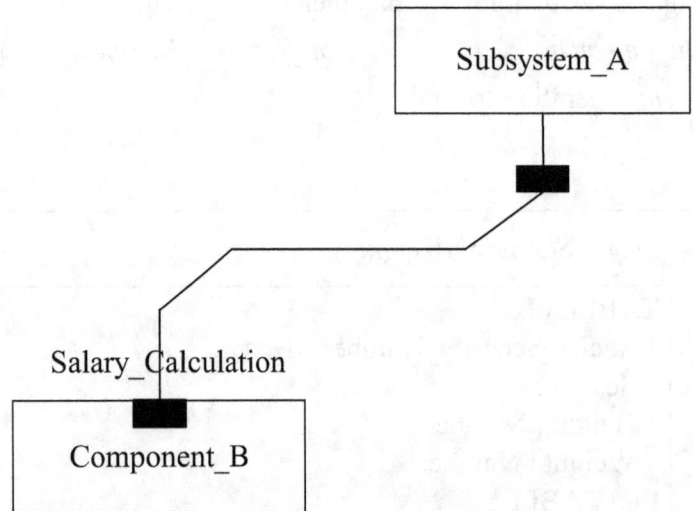

Figure 4-23   A Connection Appears When an Operation is Used

The above figure describes, sufficiently, the essence of a connection. However, we seldom use this kind of drawing. Instead, a simplified drawing of the above figure is often used as shown in Figure 4-24.

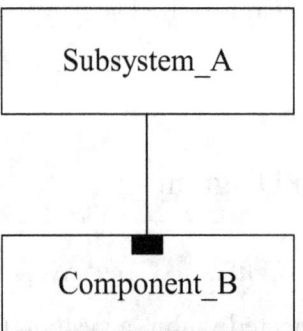

Figure 4-24   Simplified Drawing of a Connection

Since an operation is always provided by a component, there is no doubt that the *Component_B* operation provider is a component. On the contrary, the *Subsystem_A* operation user can be either a component (e.g., *Component_A*) or an actor (e.g., *Actor_A*) as shown in Figure 4-25. An actor belongs to the external environment of a system.

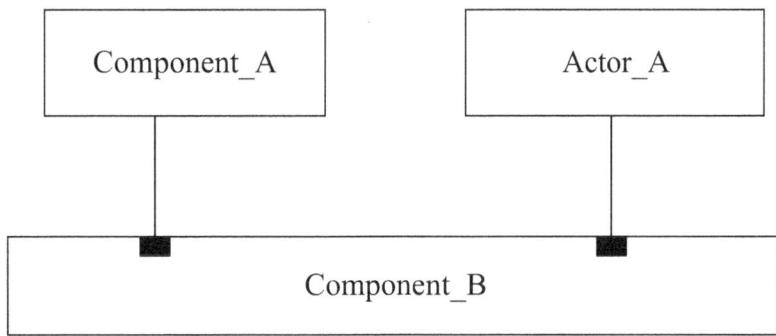

Figure 4-25   Operation User is Either a Component Or an Actor

Within a connection the subsystem (either a component or an actor) using the operation is always entitled the *Client* and the component which provides the operation is always entitled the *Server* as Figure 4-26 shows.

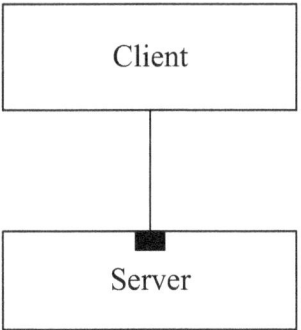

Figure 4-26   Roles of Client and Server Within a Connection

4-4-2 Drawing the Component Connection Diagram

A component connection diagram (CCD) is utilized to describe how all components and actors (in the external environment) are connected within a system. Figure 4-27 exhibits the *Multi-Tier Personal Data System's* COD.

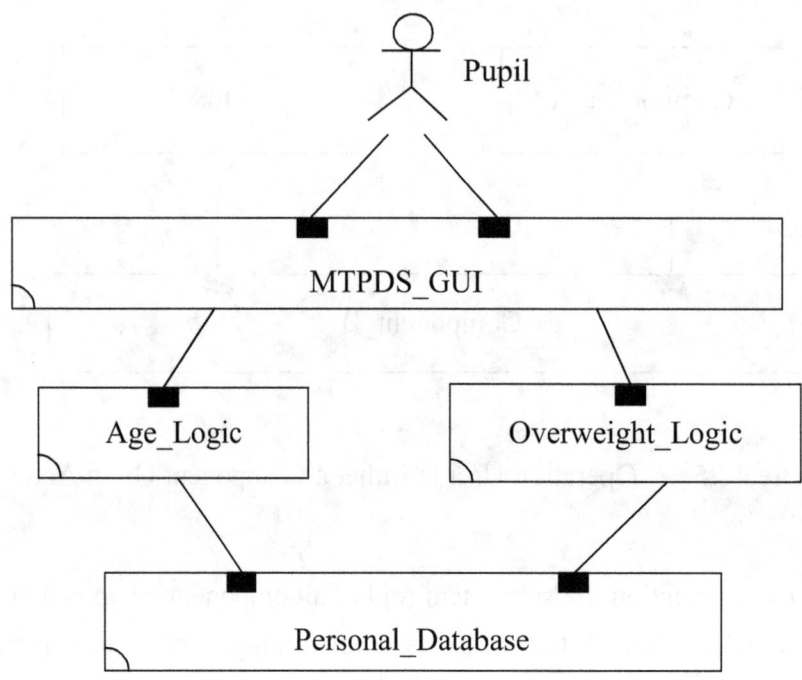

Figure 4-27   CCD of the *Multi-Tier Personal Data System*

In Figure 4-27, actor *Pupil* has two connections with the *MTPDS_GUI* component; component *MTPDS_GUI* has one connection with each of the *Age_Logic* and *Overweight_Logic* components; component *Age_Logic* has a connection with the *Personal_Database* component; component *Overweight_Logic* has a connection with the *Personal_Database* component.

After finishing the CCD, the formation pattern of the *Multi-Tier Personal Data System* will be constructed; thus the systems structure of the *Multi-Tier Personal Data System* becomes more transparent.

Chapter 5: Systems Behavior

SBC-ADL uses the structure-behavior coalescence diagram and interaction flow diagram to delineate the systems behavior of a system.

5-1 Structure-Behavior Coalescence Diagram

Structure-behavior coalescence diagram (SBCD) enables a systems architect to observe the structure and behavior coexisting in a system. SBCD is the fifth fundamental diagram to achieve structure-behavior coalescence.

5-1-1 Purpose of Structure-Behavior Coalescence Diagram

The major aim of the SBC-ADL approach is to achieve the integration of systems structure and systems behavior within a system. SBCD enables a systems architect to observe the systems structure and systems behavior coexisting in a system. This is the purpose of utilizing SBCD when architecting the systems architecture.

Figure 5-1 exhibits the *Multi-Tier Personal Data System*'s SBCD In this example, interactions among the *Pupil* actor and the *MTPDS_GUI*, *Age_Logic*, *Overweight_Logic* and *Personal_Database* components shall draw forth the *AgeCalculation* and *OverweightCalculation* behaviors.

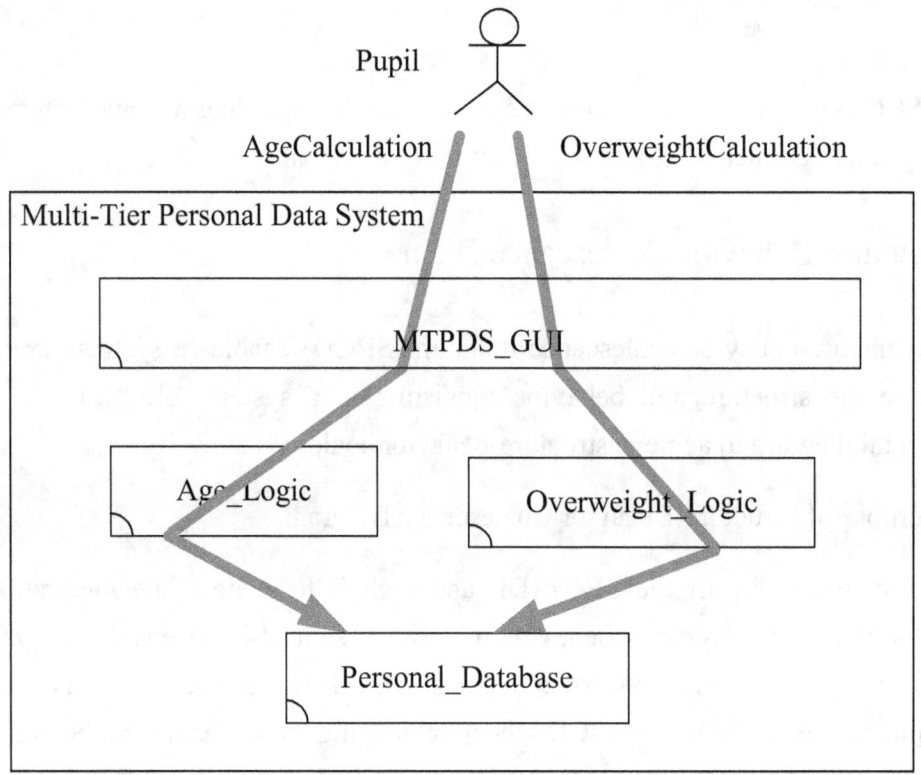

Figure 5-1   SBCD of the *Multi-Tier Personal Data System*

The overall behavior of a system is the aggregation of all its individual behaviors. All individual behaviors are mutually independent of each other. They tend to be executed concurrently [Hoar85, Miln89, Miln99]. For example, the overall behavior of the *Multi-Tier Personal Data System* includes the *AgeCalculation* and *OverweightCalculation* behaviors. In other words, the *AgeCalculation* and *OverweightCalculation* behaviors are combined to produce the overall behavior of the *Multi-Tier Personal Data System*.

The major purpose of using the architectural approach, instead of separating the structure model from the behavior model, is to achieve a coalesced model. In Figure 5-1, systems architects are able to see the systems structure and systems behavior coexisting in a SBCD. That is, in the *Multi-Tier Personal Data System's* SBCD, we not only see its systems structure but also see (at the same time) its systems behavior.

5-1-2 Drawing the Structure-Behavior Coalescence Diagram

Let us now explain the usage of SBCD by constructing a SBCD step by step. The goal of having a SBCD is enabling systems architects to see both the structure and behavior, simultaneously. In order to achieve this goal, a SBCD is drawn by first constructing all of the components, then describing the external environment's actors, and finally describing the interactions among these components and the external environment's actors.

For example, the *Multi-Tier Personal Data System* has two behaviors: *AgeCalculation* and *OverweightCalculation*. After constructing the *Multi-Tier Personal Data System* with all its components, the external environment's actors and the *AgeCalculation* behavior, we obtain the graphical representation as shown in Figure 5-2. In this Figure, the *AgeCalculation* behavior indicates that actor *Pupil* interacts with the *MTPDS_GUI* component first, then component *MTPDS_GUI* interacts with the *Age_Logic* component later, then component *Age_Logic* interacts with the *Personal_Database* component finally.

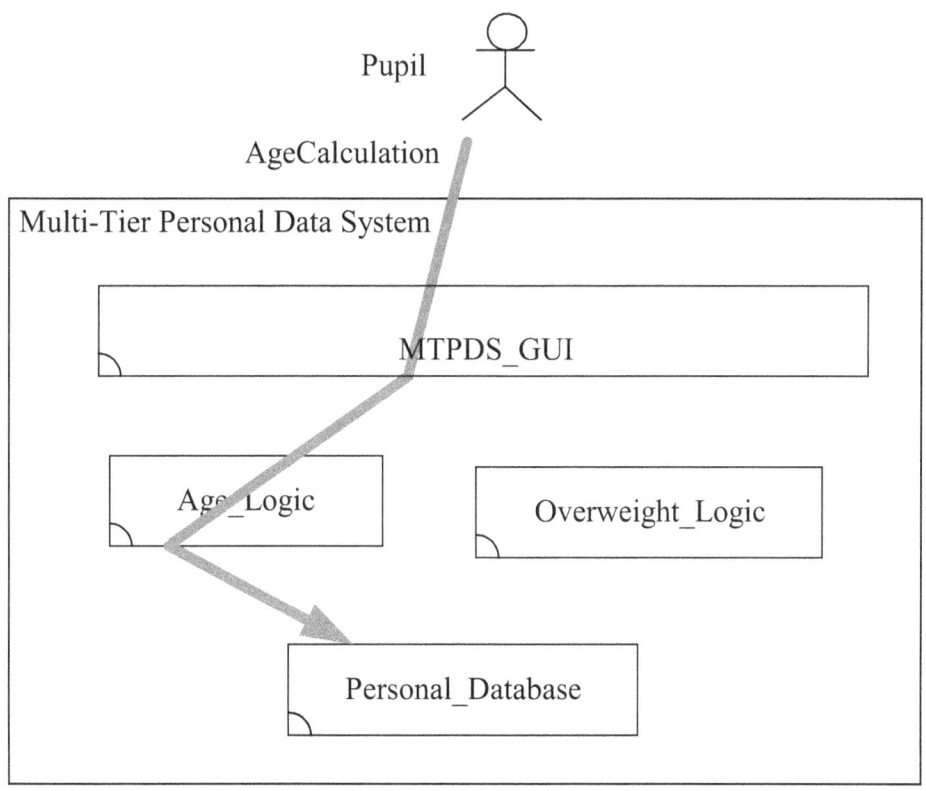

Figure 5-2   All Components, Actors, and the *AgeCalculation* Behavior

Adding the *OverweightCalculation* behavior to Figure 5-2, we then obtain the graphical representation shown in Figure 5-3. In this Figure, the *OverweightCalculation* behavior indicates that actor *Pupil* interacts with the *MTPDS_GUI* component first, then component *MTPDS_GUI* interacts with the *Overweight_Logic* component later, then component *Overweight_Logic* interacts with the *Personal_Database* component finally.

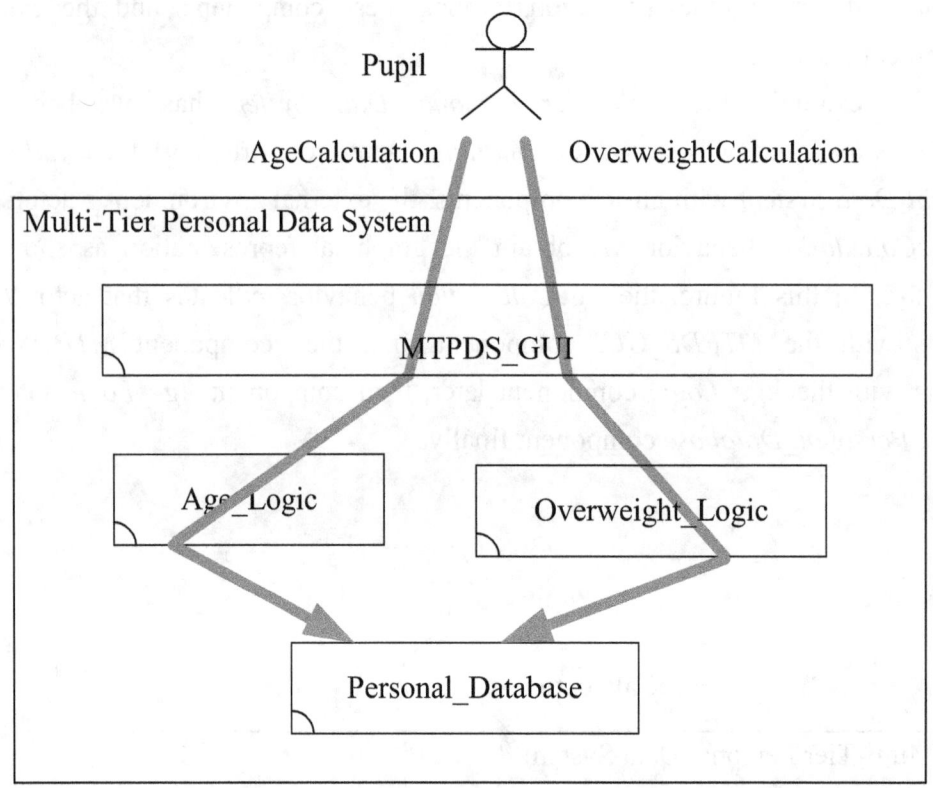

Figure 5-3　　Adding the *OverweightCalculation* Behavior to Figure 5-2

After finishing Figure 5-3, we actually have accomplished all the works needed to draw an entire SBCD of the *Multi-Tier Personal Data System*. As a matter of fact, Figure 5-3 shows exactly the *Multi-Tier Personal Data System*'s SBCD.

5-2 Interaction Flow Diagram

An interaction flow diagram (IFD) is utilized to describe each individual behavior of the overall behavior of a system. IFD is the sixth fundamental diagram to achieve structure-behavior coalescence.

5-2-1 Individual Systems Behavior Represented by Interaction Flow Diagram

The overall behavior of a system consists of many individual behaviors. Each individual behavior represents an execution path. An IFD is utilized to represent such an individual behavior.

Figure 5-4 demonstrates that the *Automobile* hardware system has two behaviors; thus, it has two IFDs.

Hardware System	IFD
Automobile	Accelerate_the_Car
	Stop_the_Car

Figure 5-4   *Automobile* has Two IFDs

Figure 5-5 demonstrates that the *Multi-Tier Personal Data System* has two behaviors; thus, it has two IFDs.

Software System	IFD
Multi-Tier Personal Data System	AgeCalculation
	OverweightCalculation

Figure 5-5   *Multi-Tier Personal Data System* has Two IFDs

Figure 5-6 demonstrates that the *Stanford University* has eight behaviors; thus it has eight IFDs.

Enterprise System	IFD
Stanford University	Study_Calculus_Course
	Study_Algebra_Course
	Study_Mechanics_Course
	Study_Quantum_Course
	Study_Database_Course
	Study_Networking_Course
	Study_Economics_Course
	Study_Accounting_Course

Figure 5-6   *Stanford University* has Eight IFDs

Figure 5-7 demonstrates that the *Robot* knowledge system has two behaviors; thus it has two IFDs.

Knowledge System	IFD
Robot	Writing
	Walking

Figure 5-7   *Robot* has Two IFDs

Figure 5-8 demonstrates that the *Strategic Thinking of an Airline* has two behaviors; thus it has two IFDs.

Thinking System	IFD
Strategic Thinking of an Airline	Productivity_Improvement
	Revenue_Growth

Figure 5-8 *Strategic Thinking of an Airline* has Two IFDs

5-2-2 Drawing the Interaction Flow Diagram

    Let us now explain the usage of interaction flow diagram (IFD) by drawing an IFD step by step. Figure 5-9 demonstrates an IFD of the *SaleInput* behavior. The X-axis direction is from the left side to right side and the Y-axis direction is from the above to the below. Inside an IFD, there are four elements: a) external environment's actor, b) components, c) interactions and d) input/output parameters. Participants of the interaction, such as the external environment's actor and each component, are laid aside along the X-axis direction on the top of the diagram. The external environment's actor which initiates the sequential interactions is always placed on the most left side of the X-axis. Then, interactions among the external environment's actor and components successively in turn decorate along the Y-axis direction. The first interaction is placed on the top of the Y-axis position. The last interaction is placed on the bottom of the Y-axis position. Each interaction may carry several input and/or output parameters.

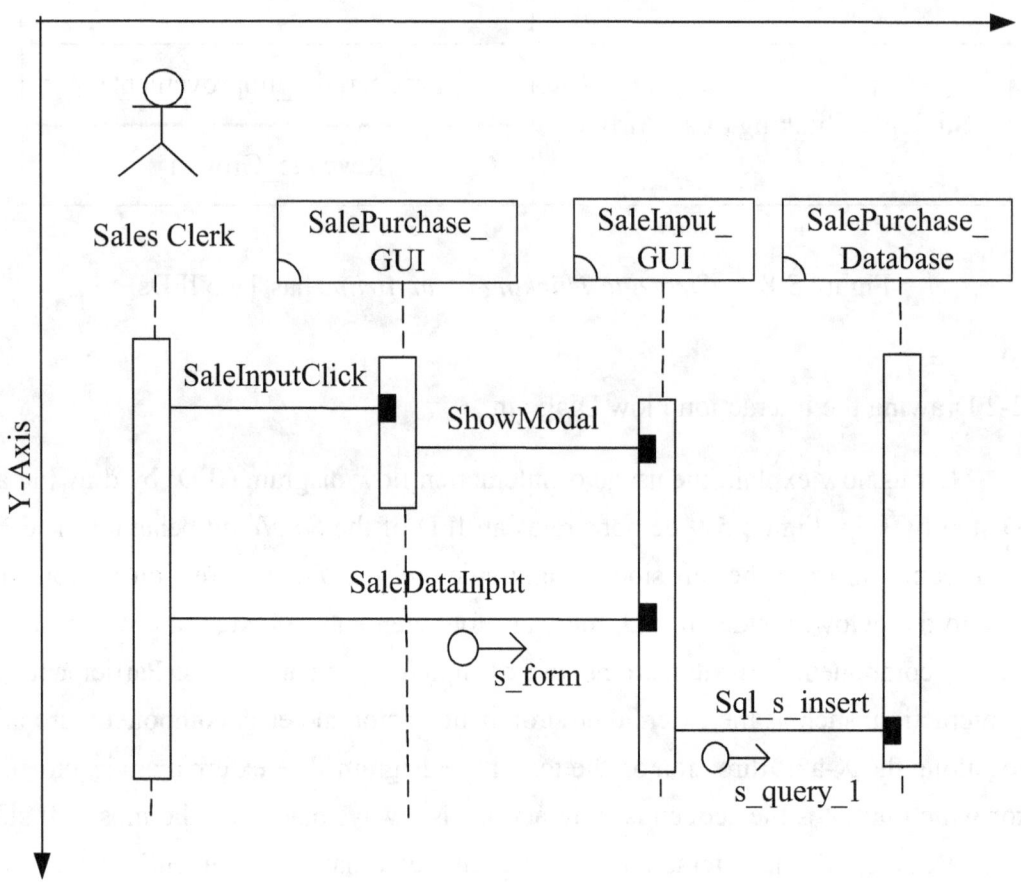

Figure 5-9   IFD of the *SaleInput* Behavior

In Figure 5-9, *Sales Clerk* is an external environment's actor. *SalePurchase_GUI*, *SaleInput_GUI* and *SalePurchase_Database* are components. *SaleInputClick* is an operation which is provided by the *SalePurchase_GUI* component. *ShowModal* is an operation which is provided by the *SaleInput_GUI* component. *SaleDataInput* is an operation, carrying the *s_form* input parameter, which is also provided by the *SaleInput_GUI* component. *Sql_s_insert* is an operation, carrying the *s_query_1* input parameter, which is provided by the *SalePurchase_Database* component.

The execution path of Figure 5-9 is as follows. First, actor *Sales Clerk* interacts with the *SalePurchase_GUI* component through the *SaleInputClick* operation call interaction. Next, component *SalePurchase_GUI* interacts with the *SaleInput_GUI* component through the *ShowModal* operation call interaction. Continuingly, actor *Sales Clerk* interacts with the *SaleInput_GUI* component through the *SaleDataInput* operation call interaction, carrying the *s_form* input parameter. Finally, component *SaleInput_GUI* interacts with the *SalePurchase_Database* component through the *Sql_s_insert* operation call interaction, carrying the *s_query_1* input parameter.

For each interaction, the solid line stands for operation call while the dashed line stands for operation return. The operation call and operation return interactions, if using the same operation name, belong to the identical operation. Figure 5-10 exhibits two interactions (operation call interaction and operation return interaction) having the identical "*Request*" operation.

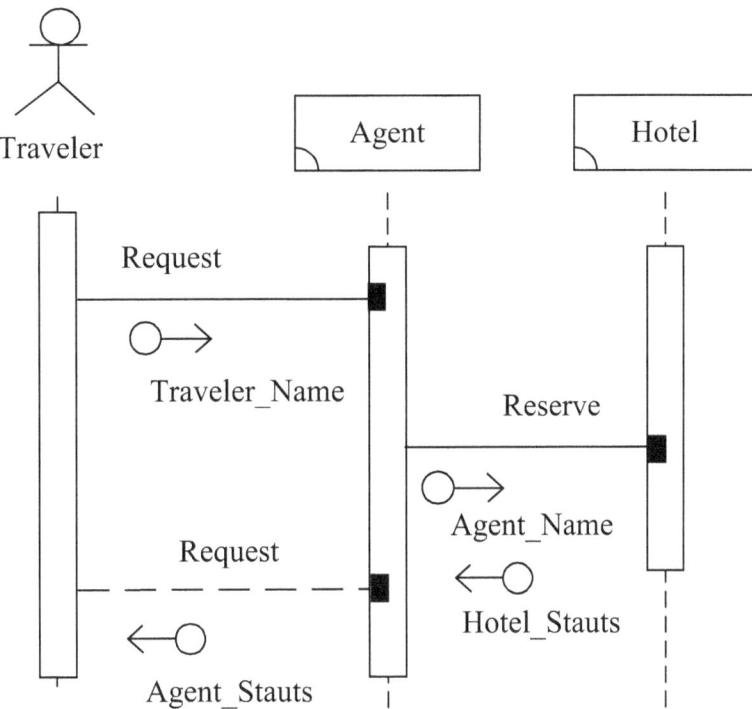

Figure 5-10 Two Interactions Have the Identical Operation

The execution path of Figure 5-10 is as follows. First, external environment's actor *Traveler* interacts with the *Agent* component through the *Request* operation call interaction, carrying the *Traveler_Name* input parameter. Next, component *Agent* interacts with the *Hotel* component through the *Reserve* operation call interaction, carrying the *Agent_Name* input parameter and *Hotel_Stauts* output parameter. Finally, external environment's actor *Traveler* interacts with the *Agent* component through the *Request* operation return interaction, carrying the *Agent_Stauts* output parameter.

An interaction flow diagram may contain a conditional expression. Figure 5-11 shows such an example which has the following execution path. First, external environment's actor *Employee* interacts with the *Computer* component through the *Open* operation call interaction, carrying the *Task_No* input parameter. Next, if the *var_1 < 4 & var_2 > 7* condition is true then component *Computer* shall interact with the *Skype* component through the *Op_1* operation call interaction and component *Skype* shall interact with the *Earphone* component through the *Op_4* operation call interaction, carrying the *Skype_Earphone* output parameter; else if the *var_3 = 99* condition is true then component *Computer* shall interact with the *Skype* component through the *Op_2* operation call interaction and component *Skype* shall interact with the *Speaker* component through the *Op_5* operation call interaction, carrying the *Skype_Speaker* output parameter; else component *Computer* shall interact with the *Youtube* component through the *Op_3* operation call interaction and component *Youtube* shall interact with the *Speaker* component through the *Op_6* operation call interaction, carrying the *Youtube_Speaker* output parameter. Continuingly, if the *var_1 < 4 & var_2 > 7* condition is true then component *Computer* shall interact with the *Skype* component through the *Op_1* operation return interaction, carrying the *Status_1* output parameter; else if the *var_3 = 99* condition is true then component *Computer* shall interact with the *Skype* component through the *Op_2* operation return interaction, carrying the *Status_2* output parameter; else component *Computer* shall interact with the *Youtube* component through the *Op_3* operation return interaction, carrying the *Status_3* output parameter. Finally, external environment's actor *Employee* interacts with the *Computer* component through the *Open* operation return interaction, carrying the *Status* output parameter.

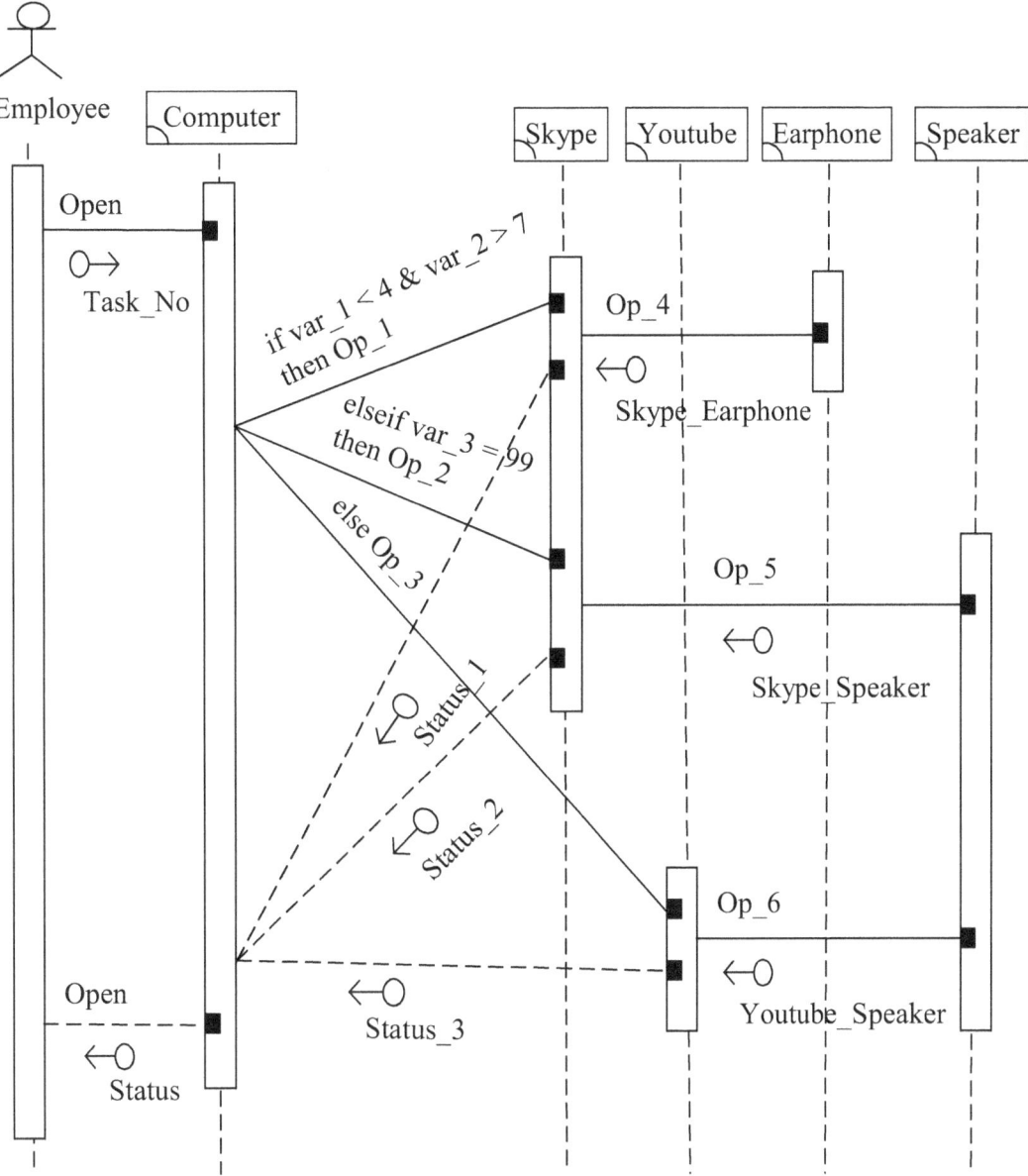

Figure 5-11   Conditional Interaction

Several Boolean conditions are shown in Figure 5-11. They are "*var_1* < *4 &
var_2* > *7*" and "*var_3* = *99*". Variables, such as *var_1*, *var_2* and *var_3*, appearing
in the Boolean condition can be local or global variables [Prat00, Seth96].

100

PART III: CONCEPT VIEW

Chapter 6: Principle of the Concept View

Concept view corresponds to an executive summary for an administrator who wants an overview or estimate of the scope of the system, what it would cost, and how it would relate to the general environment in which it will operate.

6-1 Architecting the Concept View

According to the SBC multi-level (hierarchical) view, systems architects construct the concept's systems architecture for the administrator to view. This concept's systems architecture is called the concept view as shown in Figure 6-1.

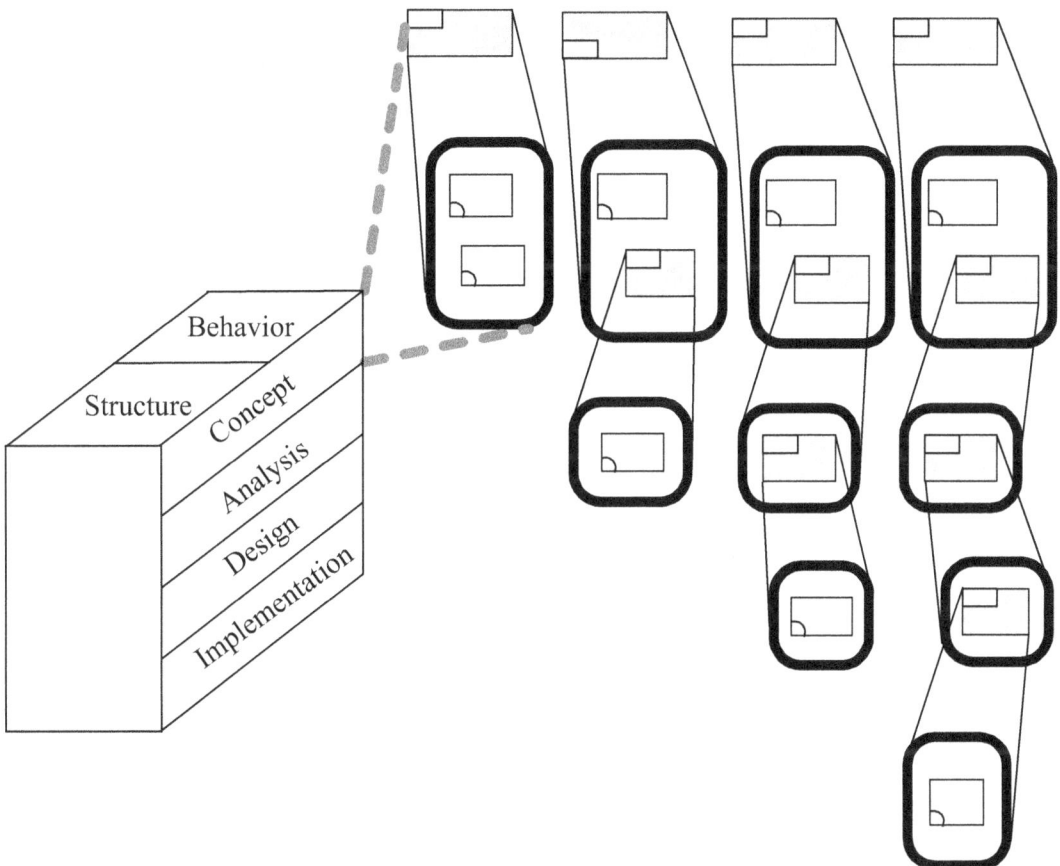

Figure 6-1   Concept View

The concept view consists of: a) concept's systems structure and b) concept's systems behavior.

6-2 Concept's Systems Structure

The entire SBC concept's systems structure includes: a) *Concept's AHD*, b) *Concept's FD*, c) *Concept's COD* and d) *Concept's CCD*, as shown in Figure 6-2.

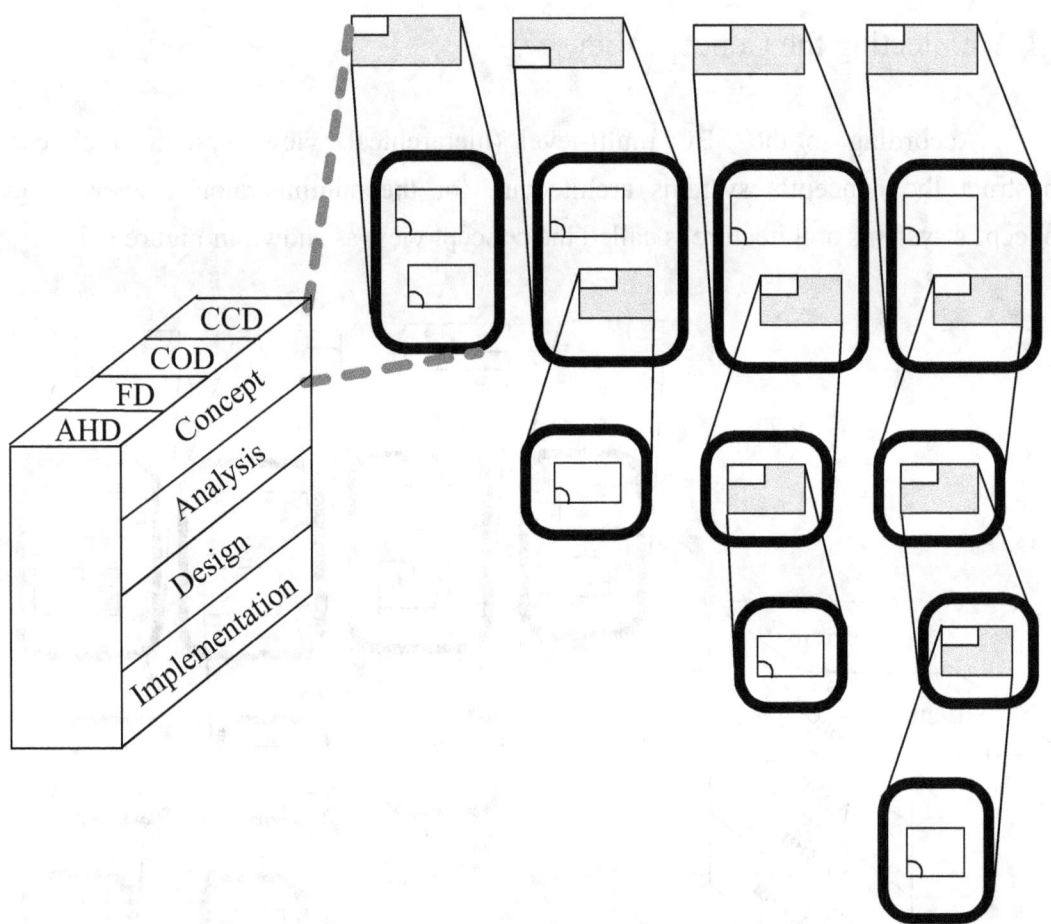

Figure 6-2   SBC Concept's Systems Structure

Another drawing can also be used to illustrate the SBC concept's systems structure as shown in Figure 6-3.

	Systems Structure			
	Architecture Hierarchy Diagram	Framework Diagram	Component Operation Diagram	Component Connection Diagram
Concept	Concept's AHD	Concept's FD	Concept's COD	Concept's CCD

Figure 6-3   SBC Concept's Systems Structure

More complete discussions on the concept's systems structure will be elaborated in later chapters which will provide a detailed explanation of its purpose.

6-3 Concept's Systems Behavior

The entire SBC concept's systems behavior includes: a) *Concept's SBCD* and b) *Concept's IFD* as shown in Figure 6-4.

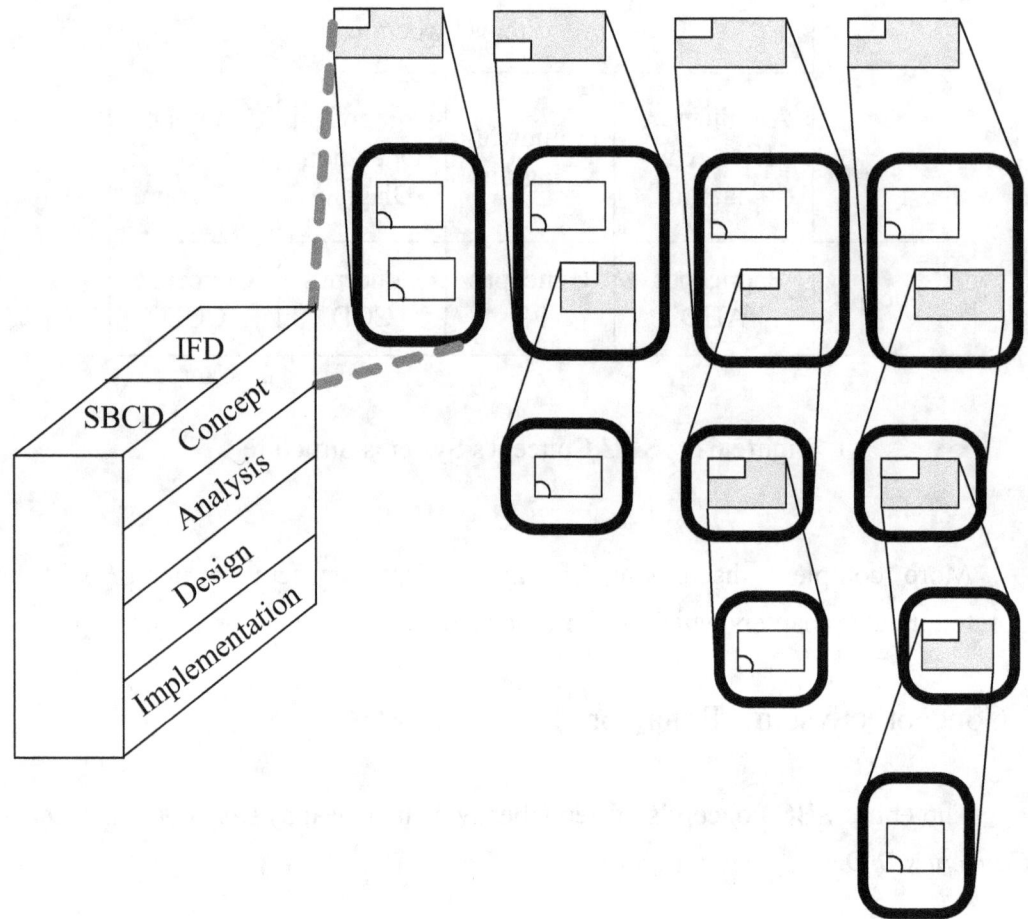

Figure 6-4 SBC Concept's Systems Behavior

Let us use another drawing to illustrate the SBC concept's systems behavior as shown in Figure 6-5.

	Systems Behavior	
	Structure-Behavior Coalescence Diagram	Interaction Flow Diagram
Concept	Concept's SBCD	Concept's IFD

Figure 6-5 SBC Concept's Systems Behavior

More complete discussions on the concept's systems behavior will be elaborated in later chapters which will provide a detailed explanation of its purpose.

Chapter 7: Concept's Systems Structure

The concept's systems structure includes: a) *Concept's AHD*, b) *Concept's FD*, c) *Concept's COD* and d) *Concept's CCD* as shown in Figure 7-1.

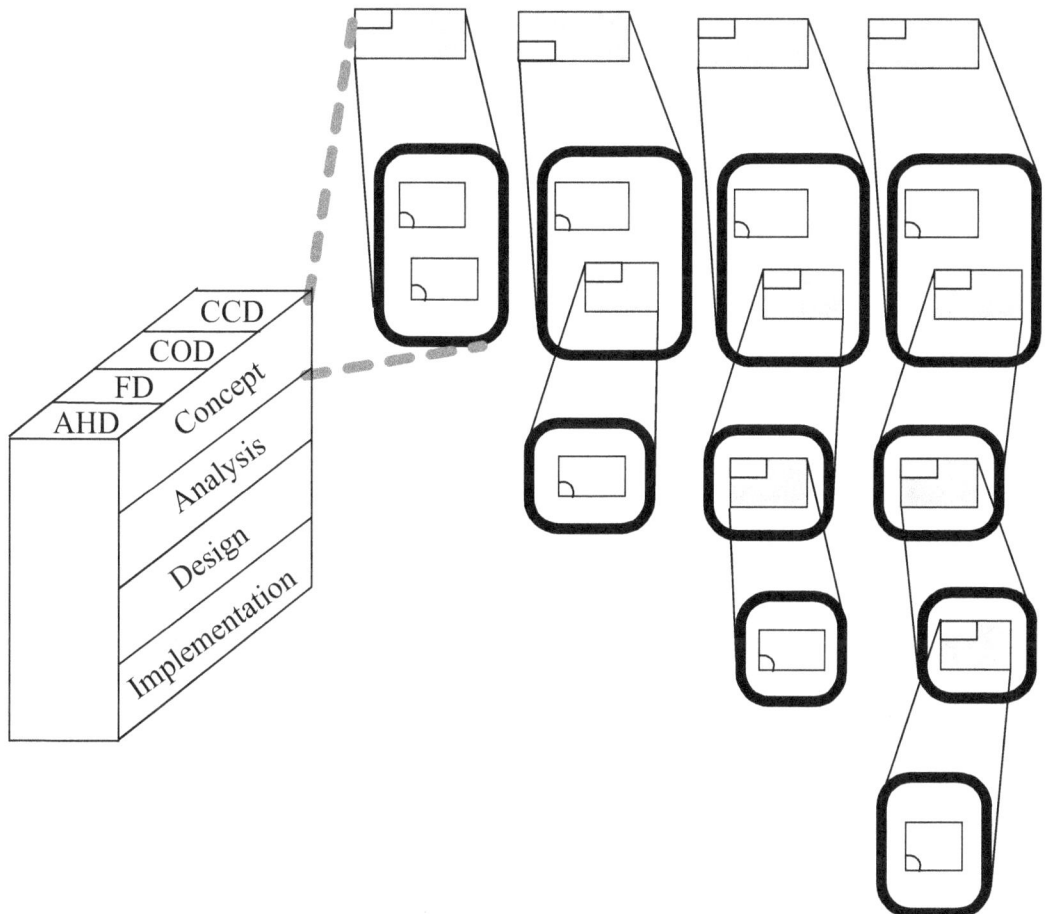

Figure 7-1   SBC Concept's Systems Structure

7-1 Concept's AHD

Concept's AHD is the architecture hierarchy diagram we obtain after the concept phase is finished. Figure 7-2 shows the concept's AHD of the *QQQ* system. In the figure, *QQQ* is composed of *B1*, *B2*, *A1*, *D1*, *D2* and *T1*.

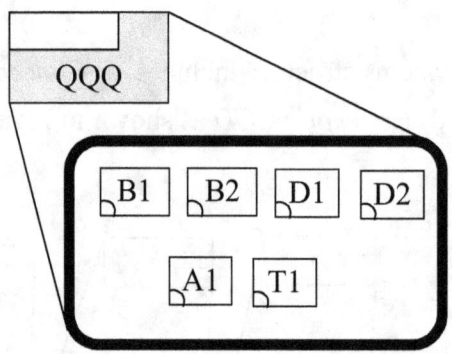

Figure 7-2   Concept's AHD of the *QQQ* System

In Figure 7-2, *QQQ* is an aggregated system while *B1*, *B2*, *A1*, *D1*, *D2* and *T1* are non-aggregated systems.

7-2 Concept's FD

Concept's FD is the framework diagram we obtain after the concept phase is finished. Figure 7-3 shows the concept's FD of the *QQQ* system. In the figure, *Business_Layer* contains the *B1* and *B2* components; *Application_Layer* contains the *A1* component; *Data_Layer* contains the *D1* and *D2* components; *Technology_Layer* contains the *T1* component.

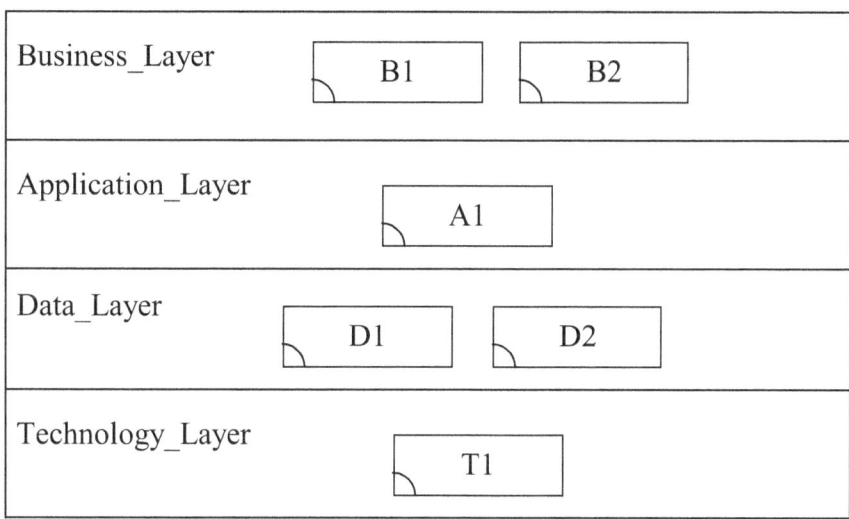

Figure 7-3 Concept's FD of the *QQQ* System

7-3 Concept's COD

Concept's COD is the component operation diagram we obtain after the concept phase is finished. Figure 7-4 shows the concept's COD of the *QQQ* system. In the figure, component *B1* has two operations: *op_01* and *op_02*; component *B2* has one operation: *op_03*; component *A1* has three operations: *op_04*, *op_05* and *op_06*; component *D1* has two operations: *op_07* and *op_08*; component *D2* has one operation: *op_9*; component *T1* has two operations: *op_10* and *op_11*.

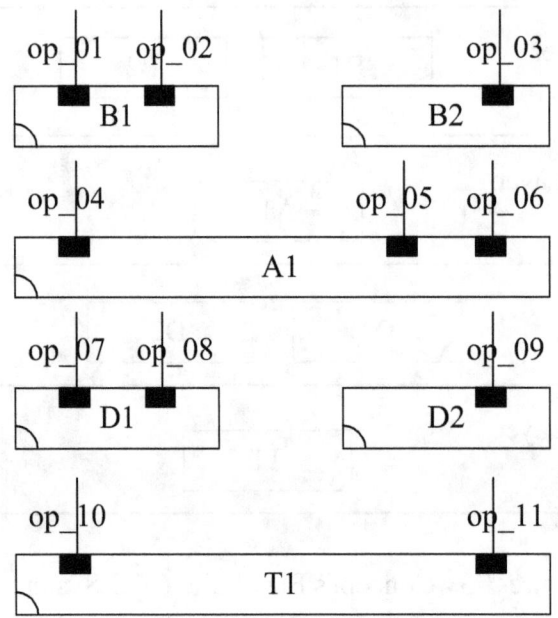

Figure 7-4　Concept's COD of the QQQ System

7-4 Concept's CCD

Concept's CCD is the component connection diagram we obtain after the concept phase is finished. Figure 7-5 shows the concept's CCD of the QQQ system.

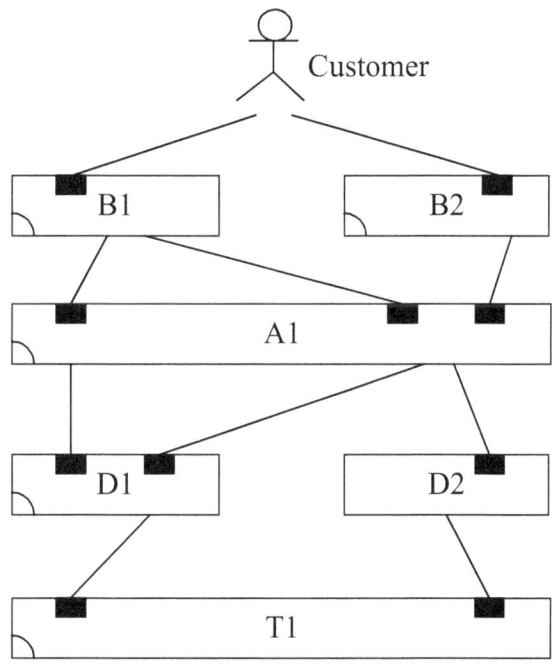

Figure 7-5   Concept's CCD of the *QQQ* System

In Figure 7-5, we see that actor *Customer* has a connection with each of the *B1* and *B2* components; component *B1* has two connections with the *A1* component; component *B2* has a connection with the *A1* component; component *A1* has two connections with the *D1* component; component *A1* has a connection with the *D2* components; component *D1* has a connection with the *T1* component; component *D2* has a connection with the *T1* component.

Chapter 8: Concept's Systems Behavior

The concept's systems behavior includes: a) *Concept's SBCD* and b) *Concept's IFD* as shown in Figure 8-1.

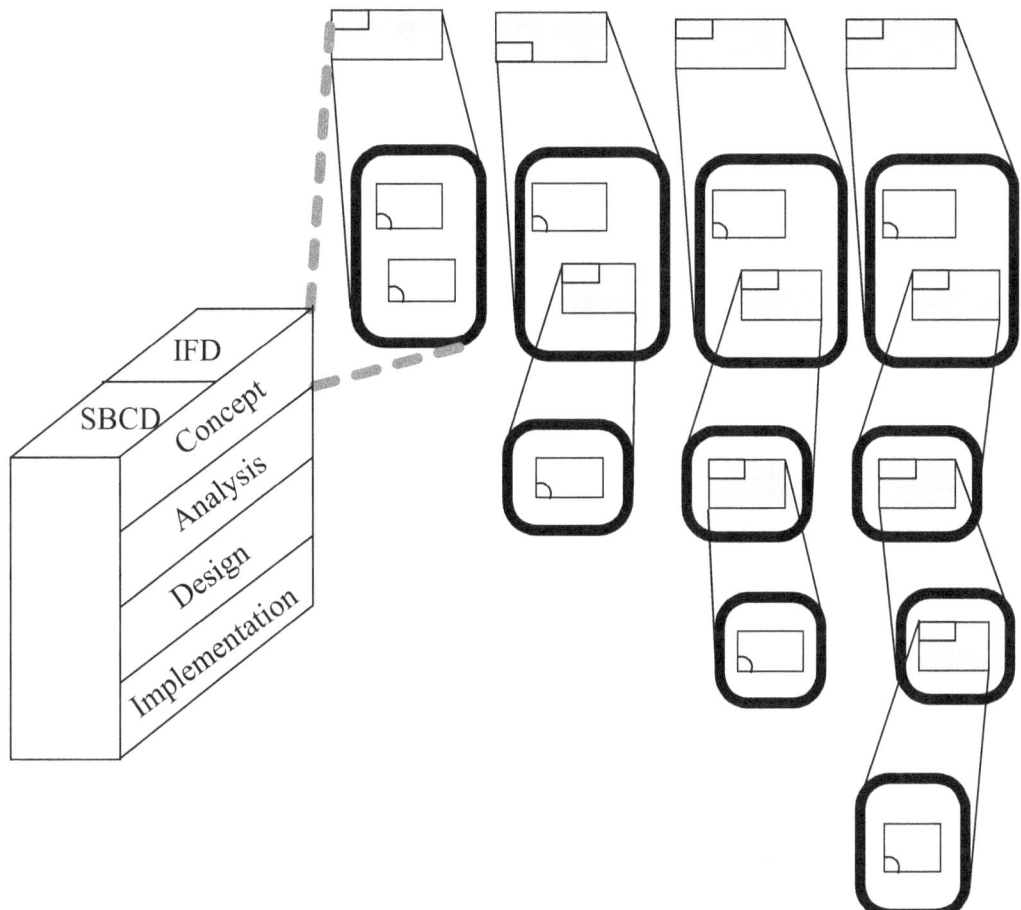

Figure 8-1   SBC Concept's Systems Behavior

8-1 Concept's SBCD

Concept's SBCD is the structure-behavior coalescence diagram we obtain after the concept phase is finished. Figure 8-2 shows the concept's SBCD of the *QQQ* system. In this example, an actor interacting with six components shall describe the overall concept's systems behavior. Interactions among the *Customer* actor and the *B1*, *A1*, *D1* components draw forth the *qqq_1* behavior. Interactions among the *Customer* actor and the *B1*, *A1*, *D1*, *T1* components draw forth the *qqq_2* behavior. Interactions among the *Customer* actor and the *B2, A1, D2, T1* components draw forth the *qqq_3* behavior.

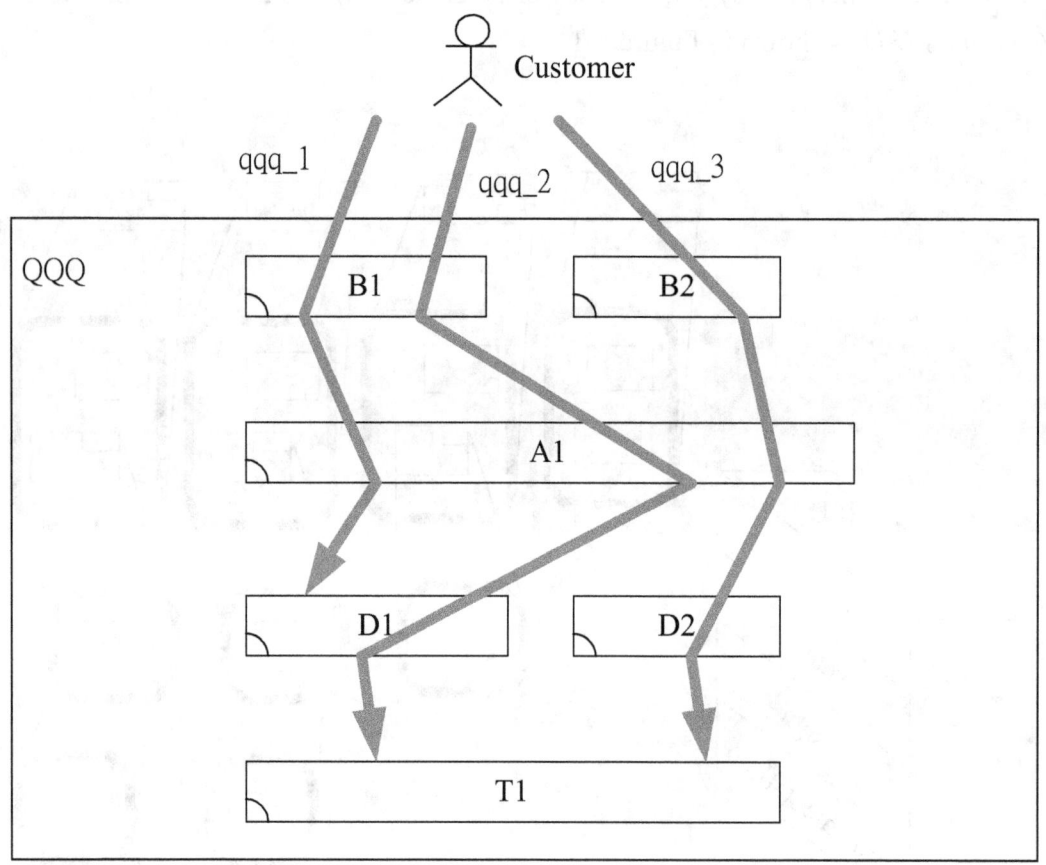

Figure 8-2   Concept's SBCD of the *QQQ* System

The overall behavior of a system is a collection of all its individual behaviors. All individual behaviors are mutually independent of each other. They tend to be executed concurrently [Hoar85, Miln89, Miln99]. For example, the overall concept's behavior of the *QQQ* system includes the *qqq_1*, *qqq_2* and *qqq_3* behaviors. In other words, the *qqq_1*, *qqq_2* and *qqq_3* behaviors are combined to produce the overall concept's behavior of the *QQQ* system.

The major purpose of using the architectural approach, instead of separating the structure model from the behavior model, is to achieve one single coalesced model. In Figure 8-2, systems architects are able to see that the systems structure and systems behavior coexist in the concept's SBCD. That is, in the concept's SBCD of the *QQQ* system, systems architects not only see its systems structure but also see (at the same time) its systems behavior.

8-2 Concept's IFD

Concept's IFD are the interaction flow diagrams we obtain after the concept phase is finished. The overall concept's behavior of the *QQQ* system includes three behaviors: *qqq_1*, *qqq_2* and *qqq_3*. Each of them is described by an individual IFD. Figure 8-3 shows the concept's IFD of the *qqq_1* behavior. First, actor *Customer* interacts with the *B1* component through the *op_01* operation call interaction. Next, component *B1* interacts with the *A1* component through the *op_04* operation call interaction. Finally, component *A1* interacts with the *D1* component through the *op_07* operation call interaction.

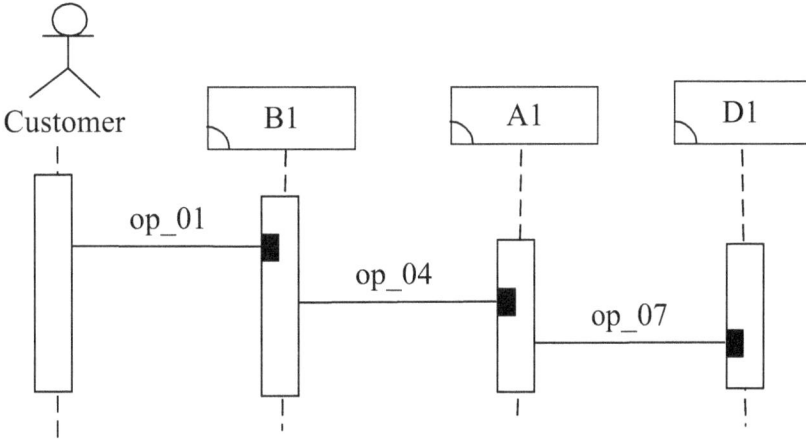

Figure 8-3   Concept's IFD of the *qqq_1 Behavior*

Figure 8-4 shows the concept's IFD of the *qqq_2* behavior. First, actor *Customer* interacts with the *B1* component through the *op_02* operation call interaction. Next, component *B1* interacts with the *A1* component through the *op_05* operation call interaction. Continuingly, component *A1* interacts with the *D1* component through the *op_08* operation call interaction. Finally, component *D1* interacts with the *T1* component through the *op_10* operation call interaction.

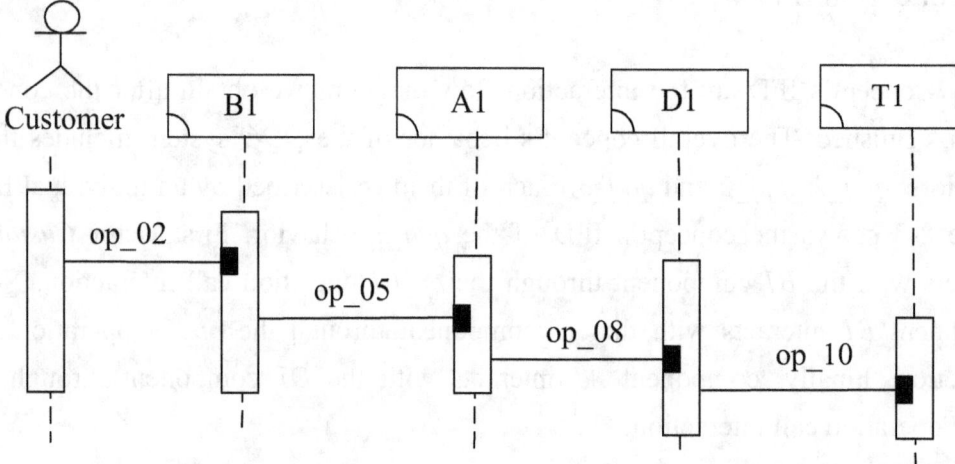

Figure 8-4   Concept's IFD of the *qqq_2* Behavior

Figure 8-5 shows the concept's IFD of the *qqq_3* behavior. First, actor *Customer* interacts with the *B2* component through the *op_03* operation call interaction. Next, component *B2* interacts with the *A1* component through the *op_06* operation call interaction. Continuingly, component *A1* interacts with the *D2* component through the *op_09* operation call interaction. Finally, component *D2* interacts with the *T1* component through the *op_11* operation call interaction.

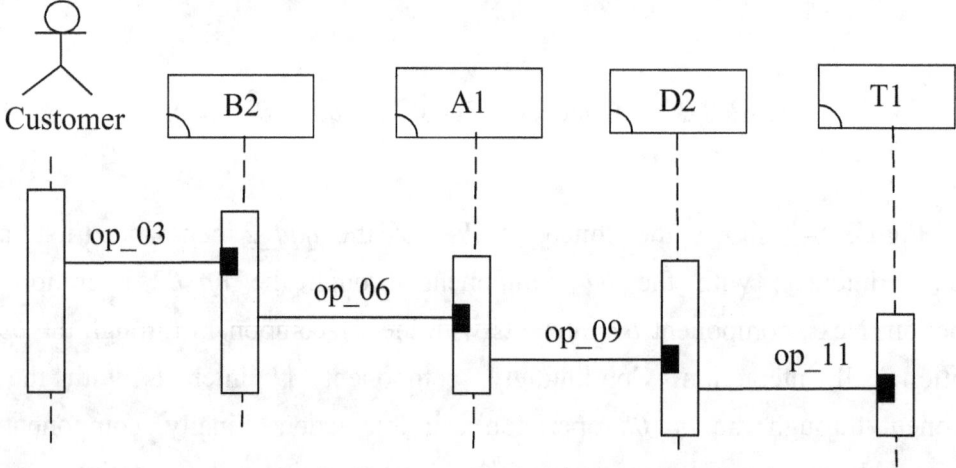

Figure 8-5   Concept's IFD of the *qqq_3* Behavior

PART IV: ANALYSIS VIEW

Chapter 9: Principle of the Analysis View

Analysis view corresponds to a summary for an analyzer who works on the analysis of a system. The analysis is mainly to find out what the system is. When working on the analysis, we only ask what this system is about, but may not provide sufficient focus on how the system is actually designed.

Analysis view is one level down structural decomposition (with observation congruence verification) of the concept view [Chao15a, Chao15b, Chao15c, Chao15d, Chao15e]. That is, we shall not create the analysis view from the scratch. Instead, we will architect the analysis view by decomposing the concept view.

9-1 Architecting the Analysis View

According to the SBC multi-level (hierarchical) view, systems architects construct the analysis' systems architecture for the analyzer to view. This analysis' systems architecture is called the analysis view as shown in Figure 9-1.

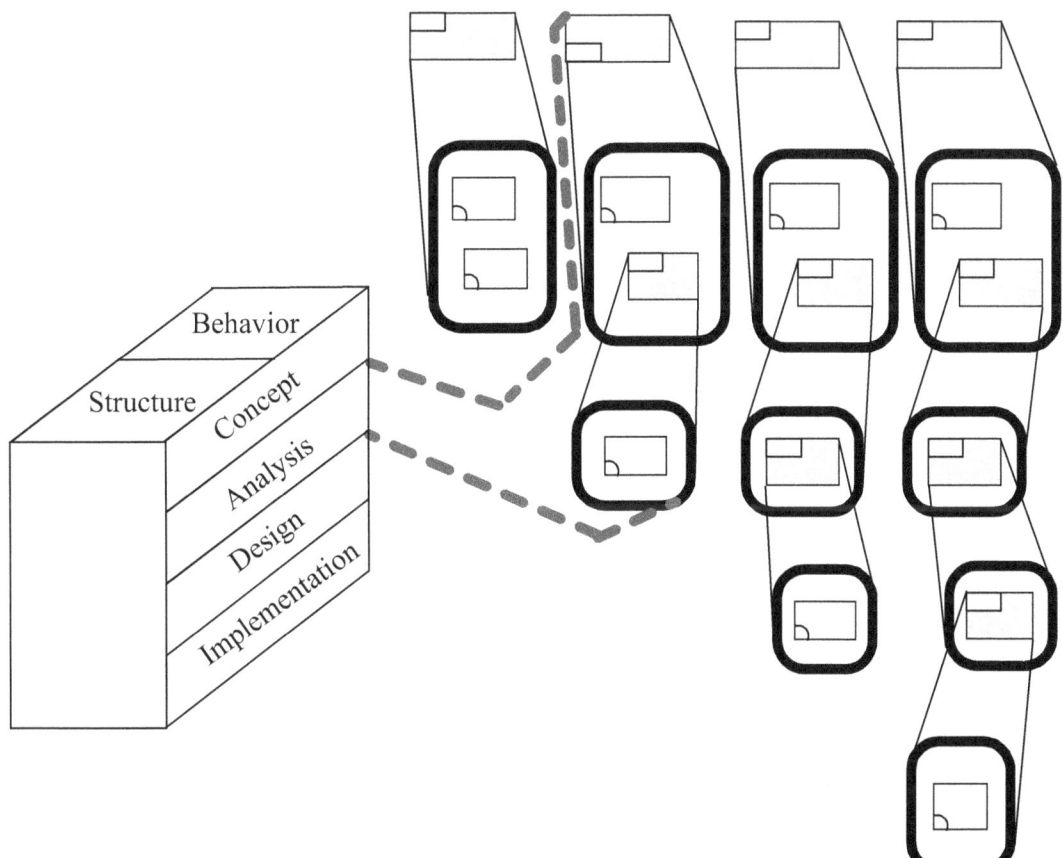

Figure 9-1   Analysis View

The analysis view consists of: a) analysis' systems structure and b) analysis' systems behavior.

9-2 Analysis' Systems Structure

The entire SBC analysis' systems structure includes: a) *Analysis' AHD*, b) *Analysis' FD*, c) *Analysis' COD* and d) *Analysis' CCD*, as shown in Figure 9-2.

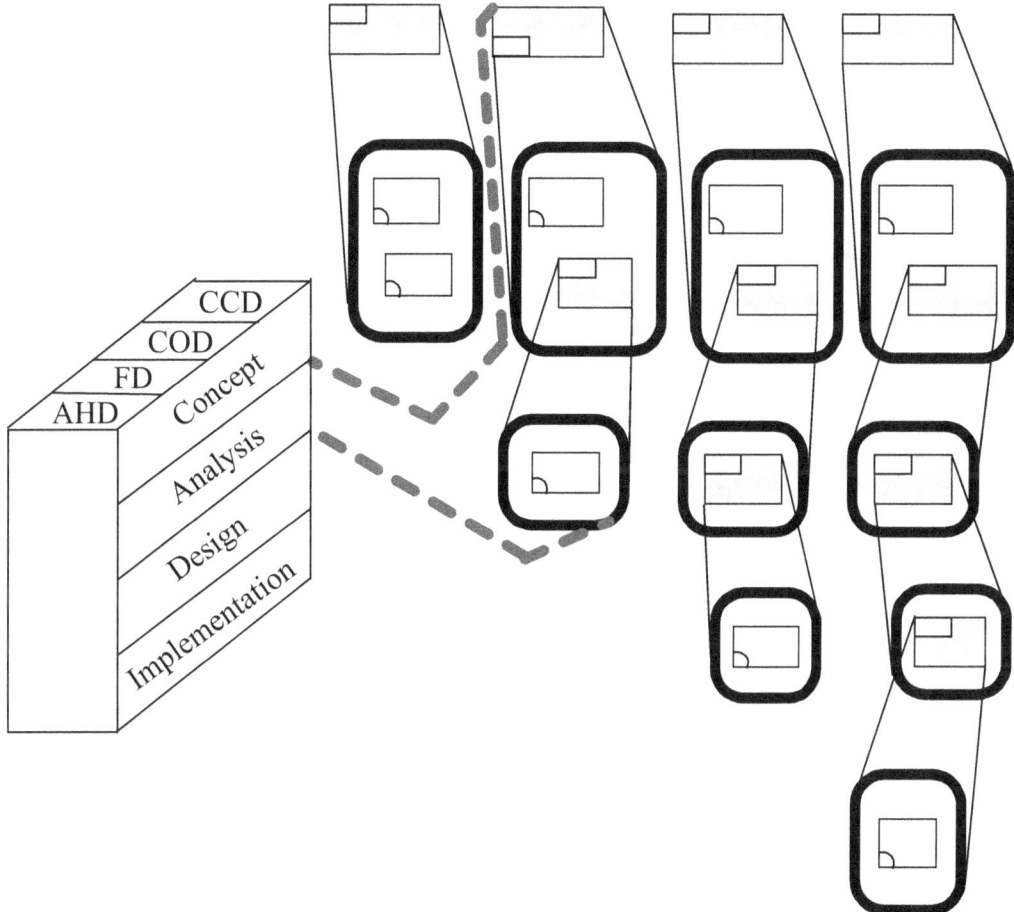

Figure 9-2   SBC Analysis' Systems Structure

Another drawing can also be used to illustrate the SBC analysis' systems structure as shown in Figure 9-3.

	Systems Structure			
	Architecture Hierarchy Diagram	Framework Diagram	Component Operation Diagram	Component Connection Diagram
Analysis	Analysis' AHD	Analysis' FD	Analysis' COD	Analysis' CCD

Figure 9-3   SBC Analysis' Systems Structure

More complete discussions on the analysis' systems structure will be elaborated in later chapters which will provide a detailed explanation of its purpose.

9-3 Analysis' Systems Behavior

The entire SBC analysis' systems behavior includes: a) *Analysis' SBCD* and b) *Analysis' IFD* as shown in Figure 9-4.

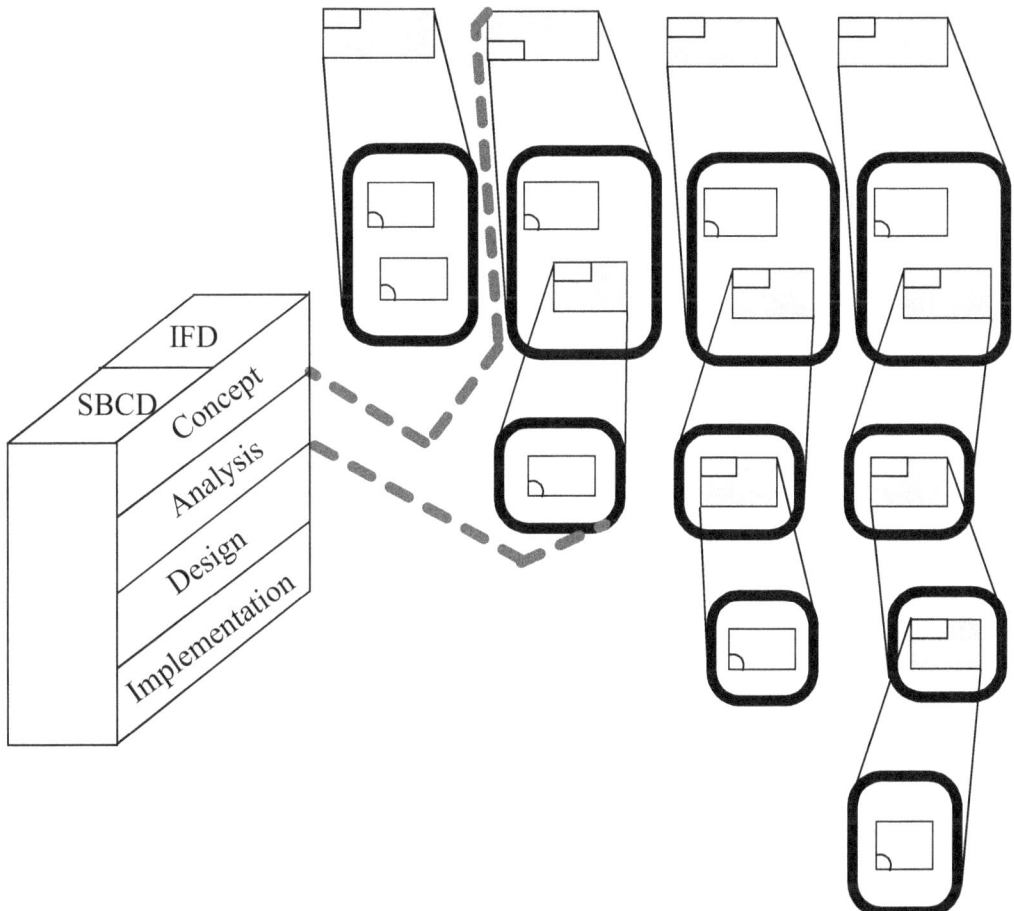

Figure 9-4   SBC Analysis' Systems Behavior

Let us use another drawing to illustrate the SBC analysis' systems behavior as shown in Figure 9-5.

	Systems Behavior	
	Structure-Behavior Coalescence Diagram	Interaction Flow Diagram
Analysis	Analysis' SBCD	Analysis' IFD

Figure 9-5   SBC Analysis' Systems Behavior

More complete discussions on the analysis' systems behavior will be elaborated in later chapters which will provide a detailed explanation of its purpose.

Chapter 10: Analysis' Systems Structure

The analysis' systems structure includes: a) *Analysis' AHD*, b) *Analysis' FD*, c) *Analysis' COD* and d) *Analysis' CCD* as shown in Figure 10-1.

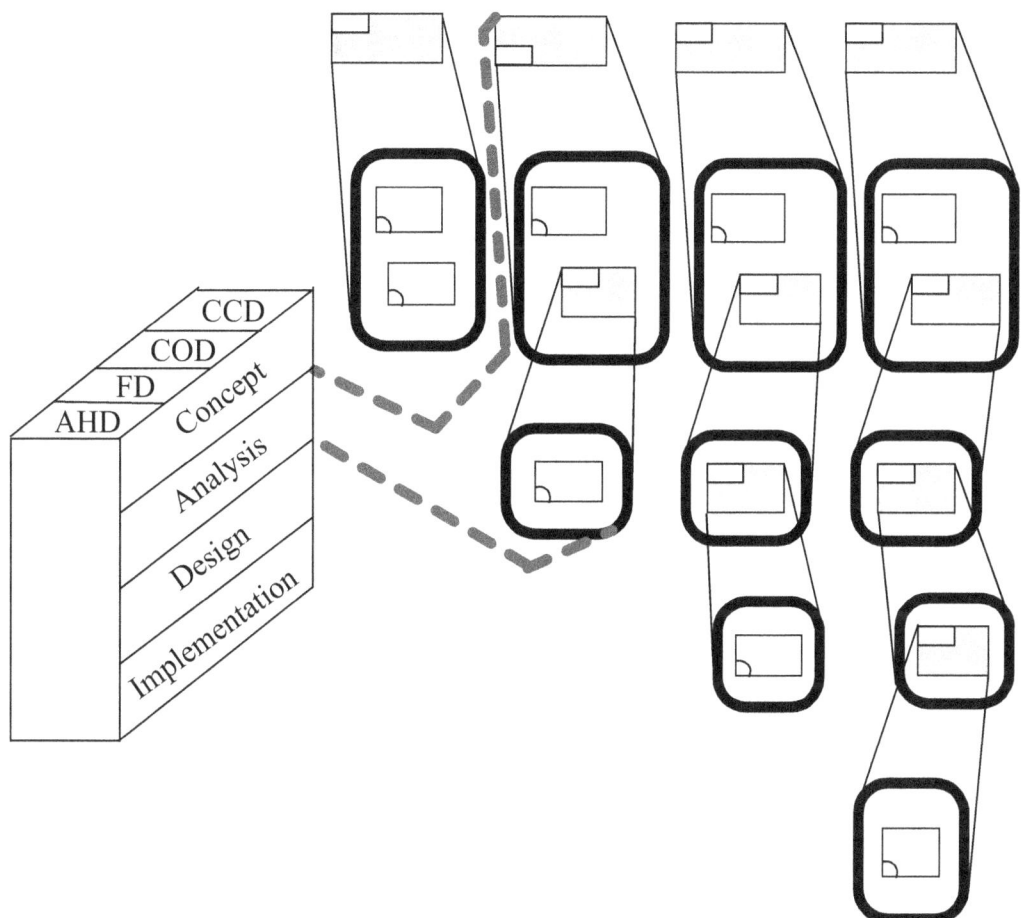

Figure 10-1   SBC Analysis' Systems Structure

10-1 Analysis' AHD

Analysis' AHD is the architecture hierarchy diagram we obtain after the analysis phase is finished. Figure 10-2 shows the analysis' AHD of the *QQQ* system. In the figure, *QQQ* is composed of *B1*, *B2*, *A1*, *D1*, *D2* and *T1*; *A1* is composed of *A11*; *T1* is composed of *T11* and *T12*.

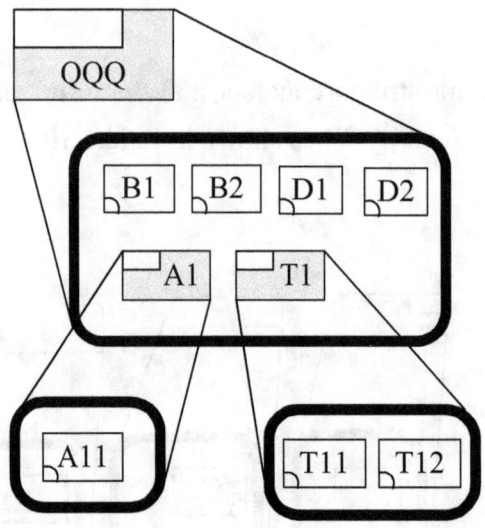

Figure 10-2   Analysis' AHD of the *QQQ* System

In Figure 10-2, *QQQ*, *A1* and *T1* are aggregated systems while *B1*, *B2*, *A11*, *D1*, *D2*, *T11* and *T12* are non-aggregated systems.

We validate our claim that the analysis view is one level down structural decomposition of the concept view by comparing Figure 7-2 with Figure 10-2. In Figure 7-2, *A1* and *T1* are non-aggregated systems. As an interesting contrast, in Figure 10-2, *A1* becomes an aggregated system and is composed of *A11*; *T1* also becomes an aggregated system and is composed of *T11* and *T12*.

10-2 Analysis' FD

Analysis' FD is the framework diagram we obtain after the analysis phase is finished. Figure 10-3 shows the analysis' FD of the *QQQ* system. In the figure, *Business_Layer* contains the *B1* and *B2* components; *Application_Layer* contains the *A11* component; *Data_Layer* contains the *D1* and *D2* components; *Technology_Layer* contains the *T11* and *T12* components.

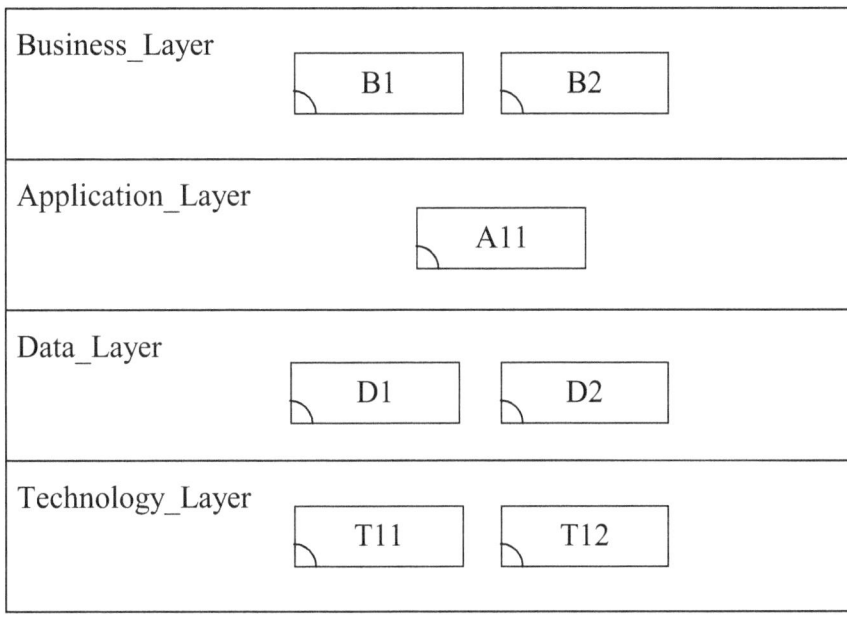

Figure 10-3 Analysis' FD of the *QQQ* System

10-3 Analysis' COD

Analysis' COD is the component operation diagram we obtain after the analysis phase is finished. Figure 10-4 shows the analysis' COD of the *QQQ* system. In the figure, component *B1* has two operations: *op_01* and *op_02*; component *B2* has one operation: *op_03*; component *A11* has three operations: *op_04*, *op_05 and op_06*; component *D1* has two operations: *op_07* and *op_08*; component *D2* has one operation: *op_9*; component *T11* has one operation: *op_10*; component *T12* has one operation: *op_11*.

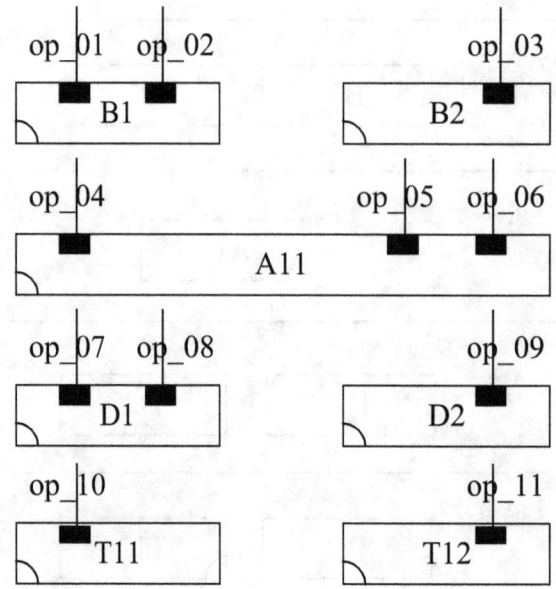

Figure 10-4 Analysis' COD of the *QQQ* System

10-4 Analysis' CCD

Analysis' CCD is the component connection diagram we obtain after the analysis phase is finished. Figure 10-5 shows the analysis' CCD of the *QQQ* system.

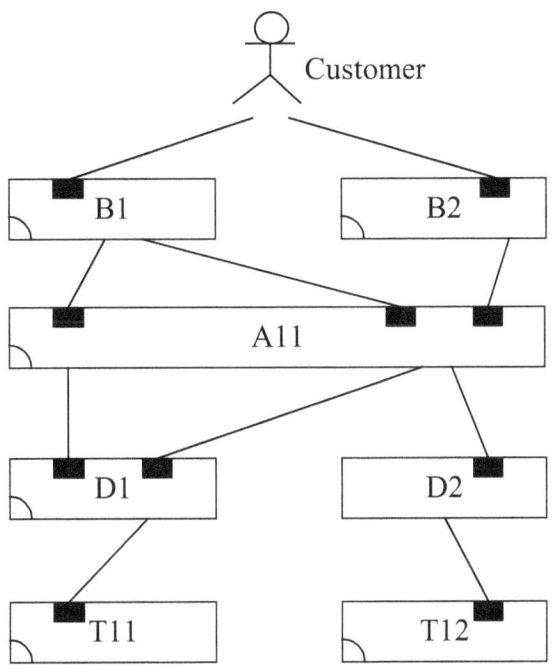

Figure 10-5 Analysis' CCD of the *QQQ* System

In Figure 10-5, we see that actor *Customer* has a connection with each one of the *B1* and *B2* components; component *B1* has two connections with the *A11* component; component *B2* has a connection with the *A11* component; component *A11* has two connections with the *D1* component; component *A11* has a connection with the *D2* component; component *D1* has a connection with the *T11* component; component *D2* has a connection with the *T12* component.

Chapter 11: Analysis' Systems Behavior

The analysis' systems behavior includes: a) *Analysis' SBCD* and *Analysis' IFD* as shown in Figure 11-1.

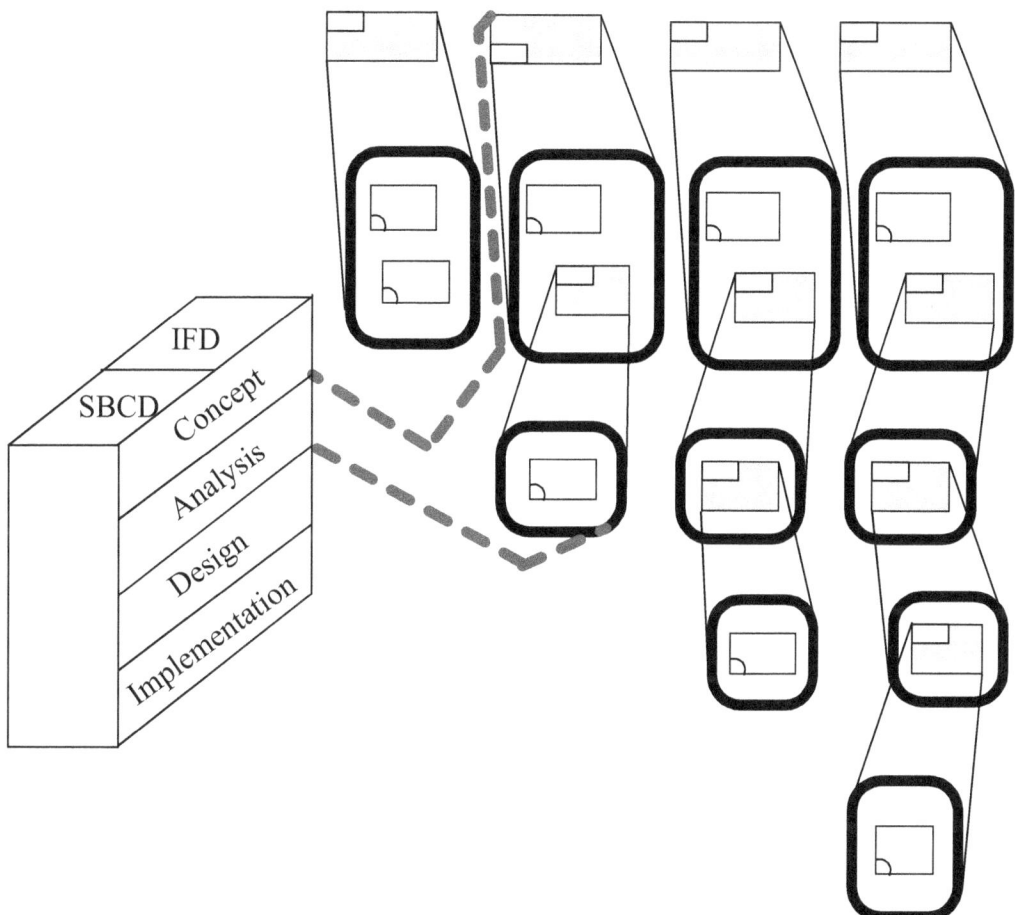

Figure 11-1   SBC Analysis' Systems Behavior

11-1 Analysis' SBCD

Analysis' SBCD is the structure-behavior coalescence diagram we obtain after the analysis phase is finished. Figure 11-2 shows the analysis' SBCD of the *QQQ* system. In this example, an actor interacting with seven components shall describe the overall analysis' systems behavior. Interactions among the *Customer* actor and the *B1*, *A11*, *D1* components draw forth the *qqq_1* behavior. Interactions among the *Customer* actor and the *B1*, *A11*, *D1*, *T11* components draw forth the *qqq_2* behavior.

Interactions among the *Customer* actor and the *B2*, *A11*, *D2*, *T12* components draw forth the *qqq_3* behavior.

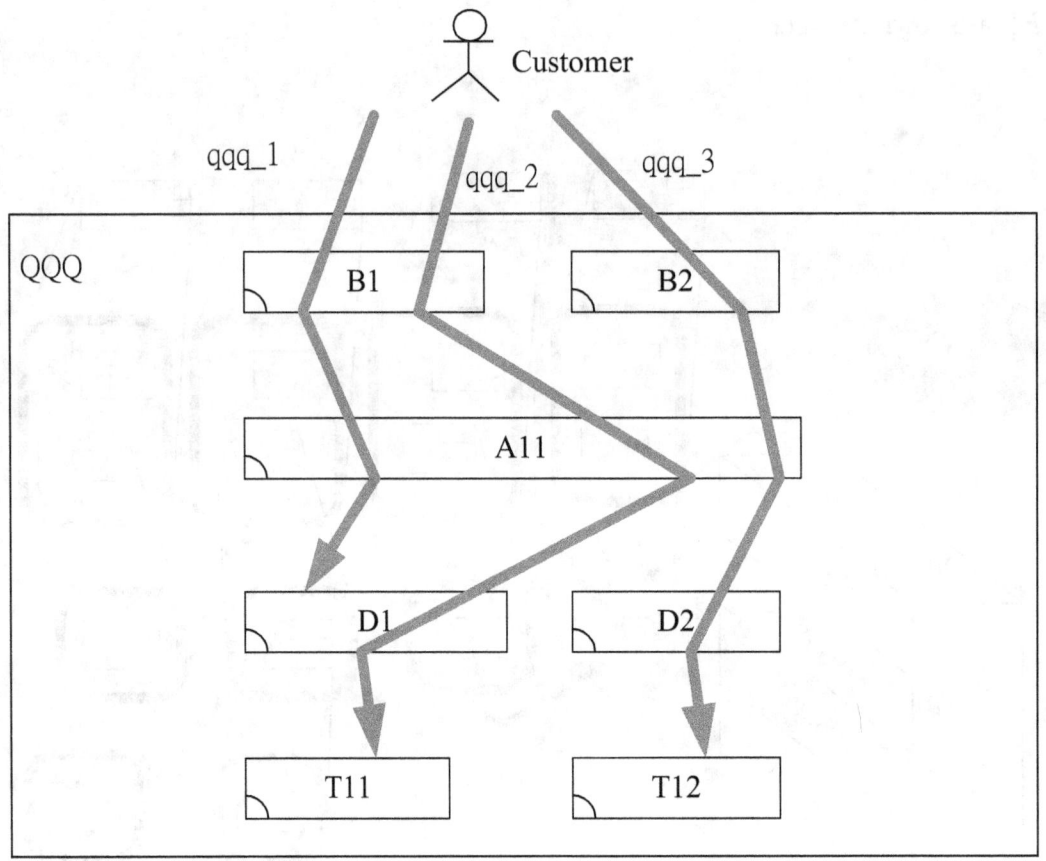

Figure 11-2 Analysis' SBCD of the *QQQ* System

The overall behavior of a system is a collection of all its individual behaviors. All individual behaviors are mutually independent of each other. They tend to be executed concurrently. For example, the overall analysis' behavior of the *QQQ* system includes the *qqq_1*, *qqq_2* and *qqq_3* behaviors. In other words, the *qqq_1*, *qqq_2* and *qqq_3* behaviors are combined to produce the overall analysis' behavior of the *QQQ* system.

The major purpose of using the architectural approach, instead of separating the structure model from the behavior model, is to achieve one single coalesced model. In Figure 11-2, systems architects are able to see that the systems structure and systems behavior coexist in the analysis' SBCD. That is, in the analysis' SBCD of the *QQQ* system, systems architects not only see its systems structure but also see (at the same time) its systems behavior.

11-2 Analysis' IFD

Analysis' IFDs are the interaction flow diagrams we obtain after the analysis phase is finished. The overall analysis' behavior of the *QQQ* system includes three behaviors: *qqq_1*, *qqq_2* and *qqq_3*. Each of them is described by an individual IFD. Figure 11-3 shows the analysis' IFD of the *qqq_1* behavior. First, actor *Customer* interacts with the *B1* component through the *op_01* operation call interaction. Next, component *B1* interacts with the *A11* component through the *op_04* operation call interaction. Finally, component *A11* interacts with the *D1* component through the *op_07* operation call interaction.

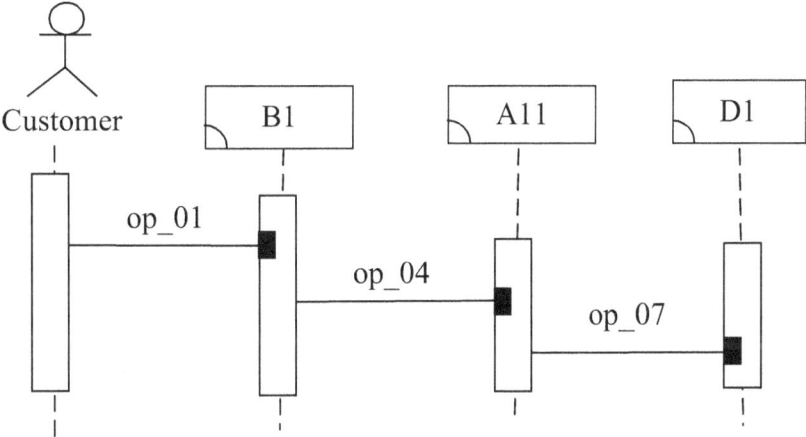

Figure 11-3   Analysis' IFD of the *qqq_1* Behavior

Figure 11-4 shows the analysis' IFD of the *qqq_2* behavior. First, actor *Customer* interacts with the *B1* component through the *op_02* operation call interaction. Next, component *B1* interacts with the *A11* component through the *op_05* operation call interaction. Continuingly, component *A11* interacts with the *D1* component through the *op_08* operation call interaction. Finally, component *D1* interacts with the *T11* component through the *op_10* operation call interaction.

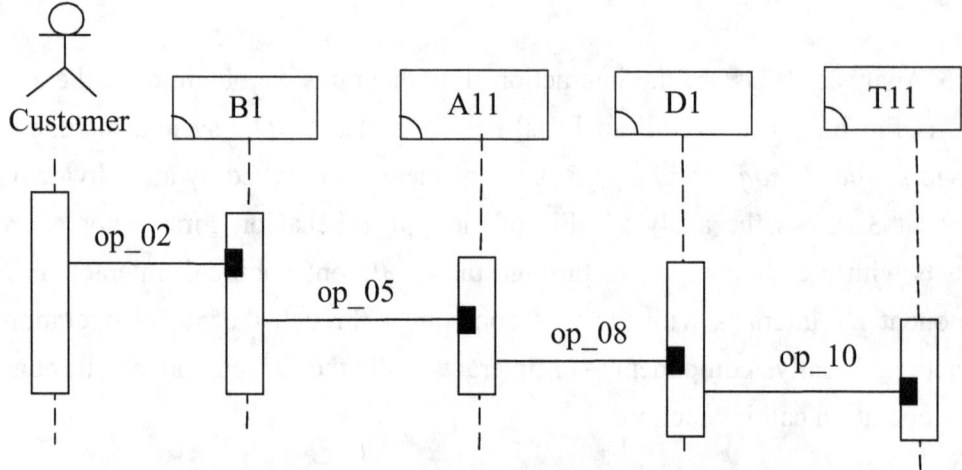

Figure 11-4   Analysis' IFD of the *qqq_2* Behavior

Figure 11-5 shows the analysis' IFD of the *qqq_3* behavior. First, actor *Customer* interacts with the *B2* component through the *op_03* operation call interaction. Next, component *B2* interacts with the *A11* component through the *op_06* operation call interaction. Continuingly, component *A11* interacts with the *D2* component through the *op_09* operation call interaction. Finally, component *D2* interacts with the *T12* component through the *op_11* operation call interaction.

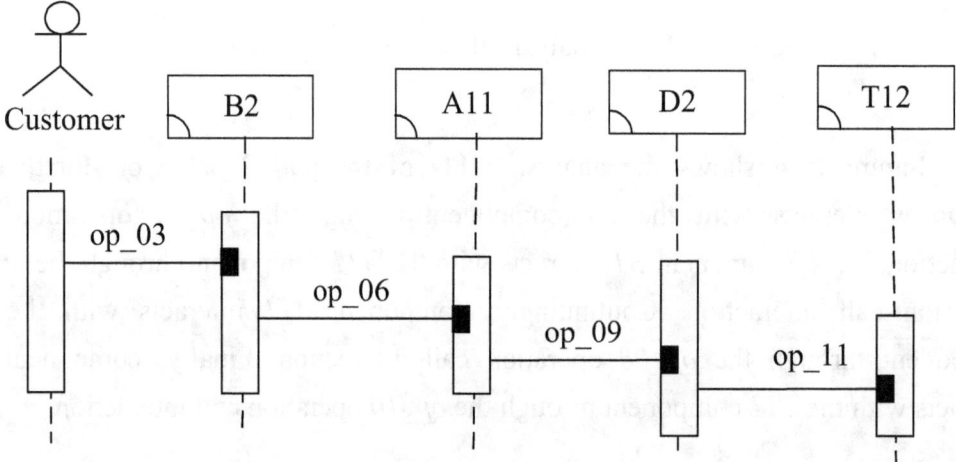

Figure 11-5   Analysis' IFD of the *qqq_3* Behavior

PART V: DESIGN VIEW

Chapter 12: Principle of the Design View

Design view describes what a designer has accomplished for his work. Design view is one level down structural decomposition (with observation congruence verification) of the analysis view [Chao15a, Chao15b, Chao15c, Chao15d, Chao15e]. That is, we shall not create the design view from the scratch. Instead, we will architect the design view by decomposing the analysis view.

12-1 Architecting the Design View

According to the SBC multi-level (hierarchical) view, systems architects construct the design's systems architecture for the designer to view. This design's systems architecture is called the design view as shown in Figure12-1.

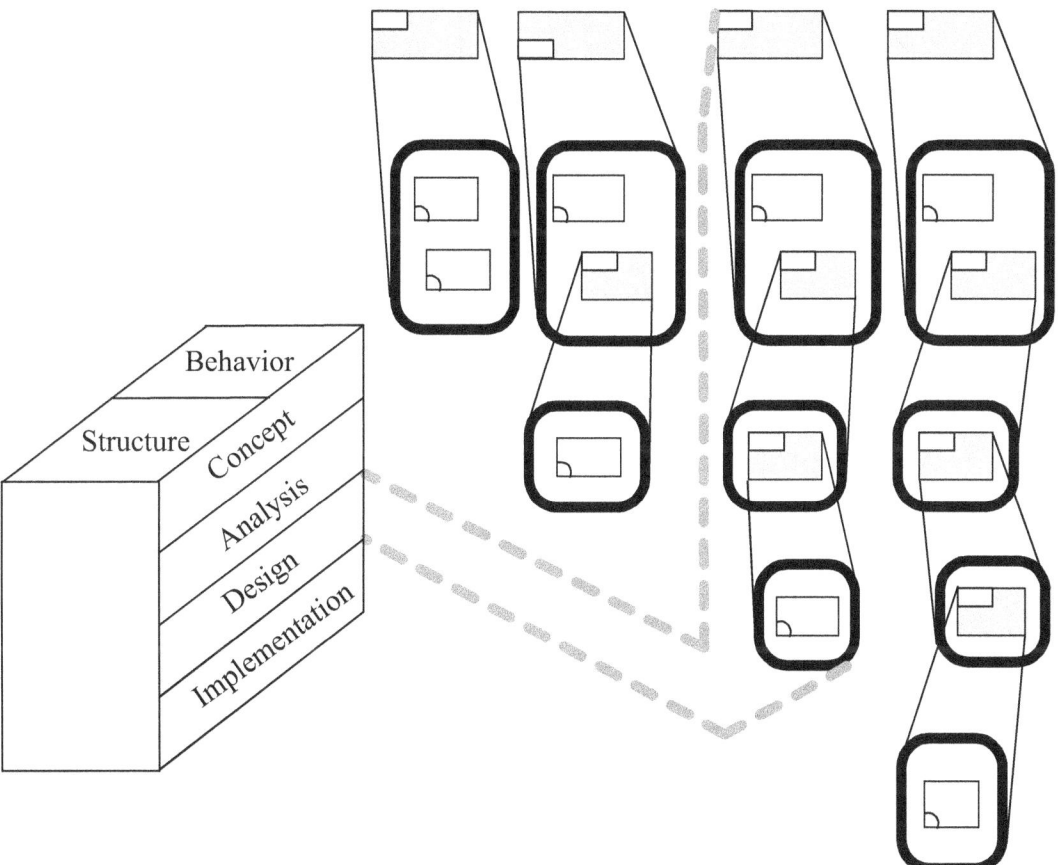

Figure 12-1 Design View

The design view consists of: a) design's systems structure and b) design's systems behavior.

12-2 Design's Systems Structure

The entire SBC design's systems structure includes: a) *Design's AHD*, b) *Design's FD*, c) *Design's COD* and d) *Design's CCD*, as shown in Figure 12-2.

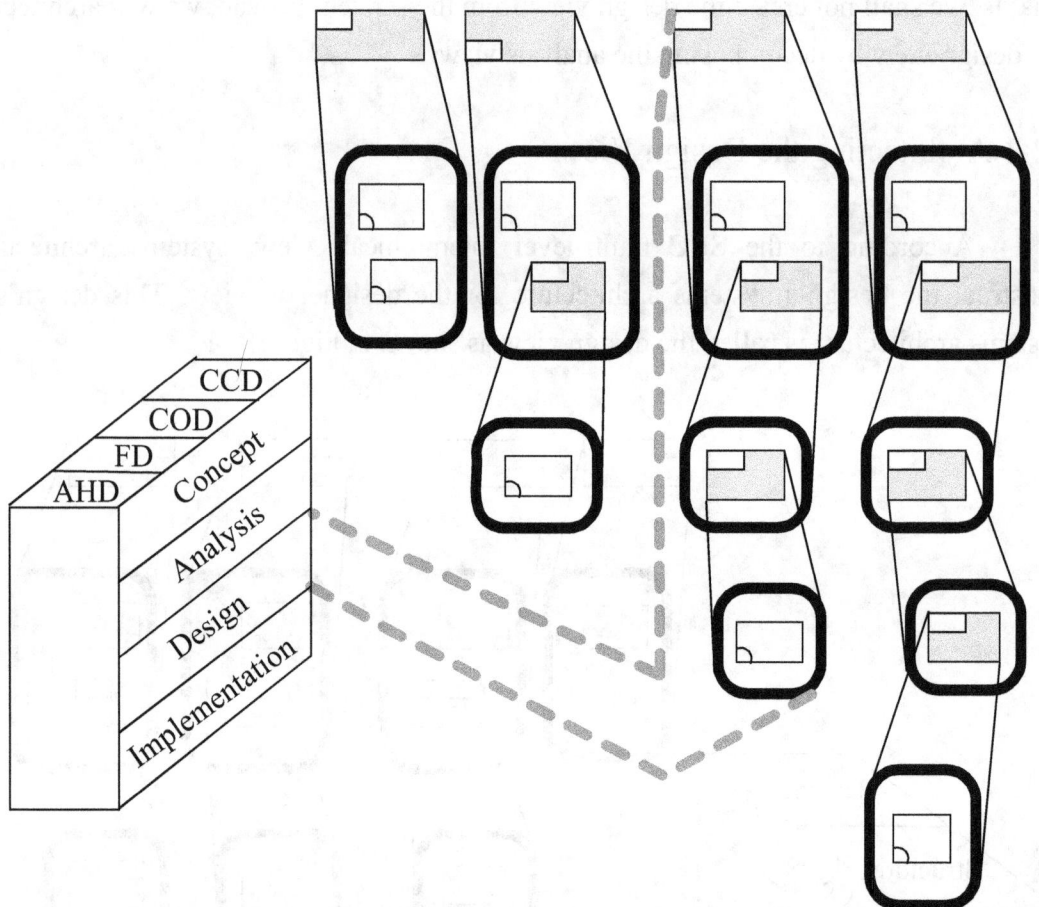

Figure 12-2　SBC Design's Systems Structure

Another drawing can also be used to illustrate the SBC design's systems structure as shown in Figure 12-3.

	Systems Structure			
	Architecture Hierarchy Diagram	Framework Diagram	Component Operation Diagram	Component Connection Diagram
Design	Design's AHD	Design's FD	Design's COD	Design's CCD

Figure 12-3 SBC Design's Systems Structure

More complete discussions on the design's systems structure will be elaborated in later chapters which will provide a detailed explanation of its purpose.

12-3 Design's Systems Behavior

The entire SBC design's systems behavior includes: a) *Design's SBCD* and b) *Design's IFD* as shown in Figure 12-4.

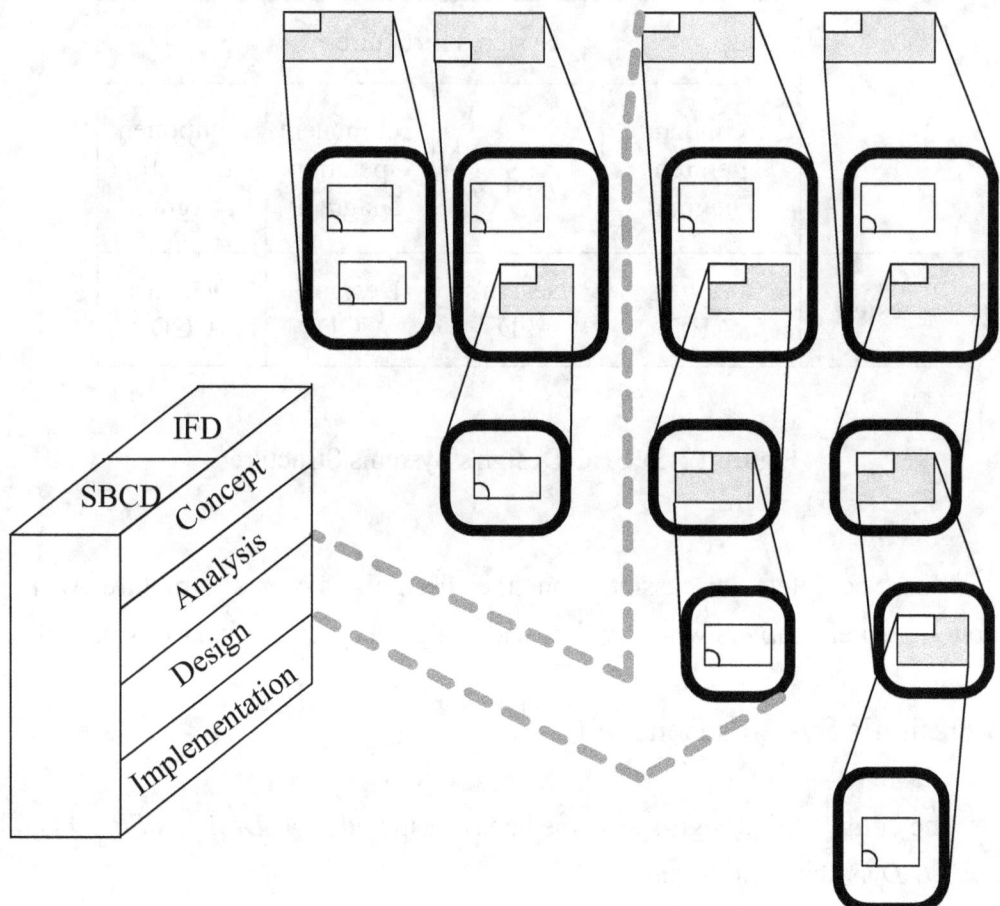

Figure 12-4   SBC Design's Systems Behavior

Let us use another drawing to illustrate the SBC design's systems behavior as shown in Figure 12-5.

	Systems Behavior	
	Structure-Behavior Coalescence Diagram	Interaction Flow Diagram
Design	Design's SBCD	Design's IFD

Figure 12-5 SBC Design's Systems Behavior

More complete discussions on the design's systems behavior will be elaborated in later chapters which will provide a detailed explanation of its purpose.

Chapter 13: Design's Systems Structure

The design's systems structure includes: a) *Design's AHD*, b) *Design's FD*, c) *Design's COD* and d) *Design's CCD* as shown in Figure 13-1.

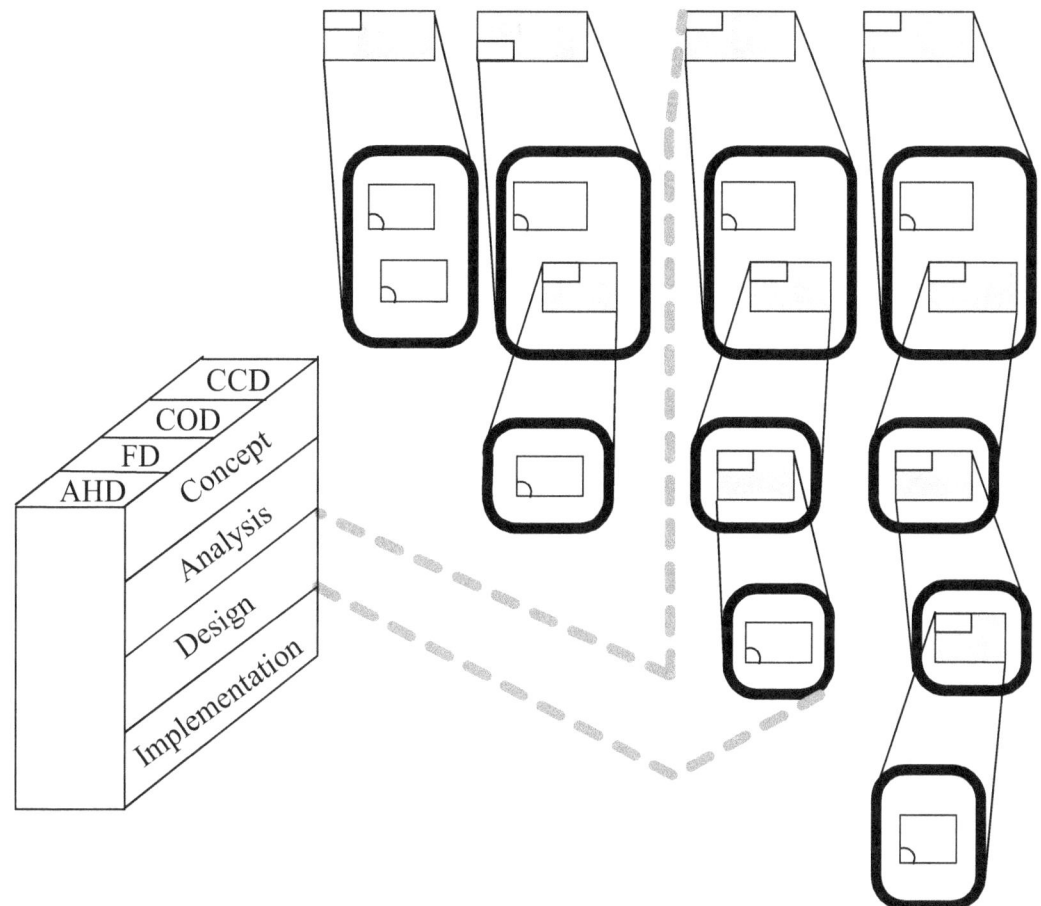

Figure 13-1   SBC Design's Systems Structure

13-1 Design's AHD

Design's AHD is the architecture hierarchy diagram we obtain after the designing phase is finished. Figure 13-2 shows the design's AHD of the *QQQ* system. In the figure, *QQQ* is composed of *B1*, *B2*, *A1*, *D1*, *D2* and *T1*; *A1* is composed of *A11*; *T1* is composed of *T11* and *T12*; *A11* is composed of *A111*.

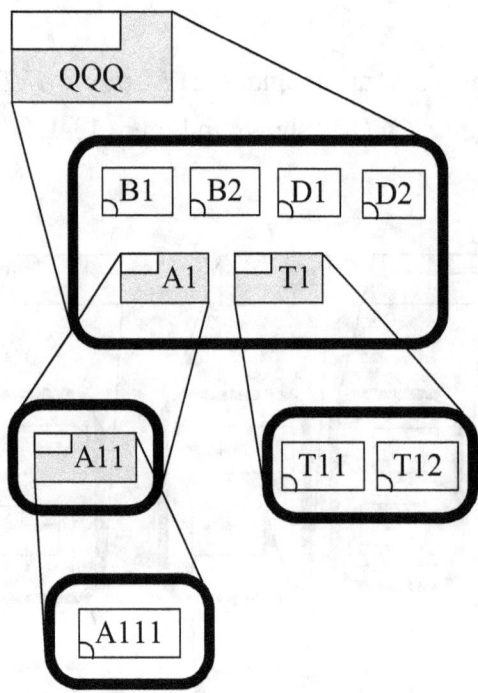

Figure 13-2   Design's AHD of the *QQQ* System

In Figure 13-2, *QQQ*, *A1*, *T1* and *A11* are aggregated systems while *B1*, *B2*, *A111*, *D1*, *D2*, *T11* and *T12* are non-aggregated systems.

We validate our claim that the design view is one level down structural decomposition of the analysis view by comparing Figure 10-2 with Figure 13-2. In Figure 10-2, *A11* is a non-aggregated system. As an interesting contrast, in Figure 13-2, *A11* becomes an aggregated system and is composed of *A111*.

13-2 Design's FD

Design's FD is the framework diagram we obtain after the designing phase is finished. Figure 13-3 shows the design's FD of the *QQQ* system. In the figure, *Business_Layer* contains the *B1* and *B2* components; *Application_Layer* contains the *A111* component; *Data_Layer* contains the *D1* and *D2* components; *Technology_Layer* contains the *T11* and *T12* components.

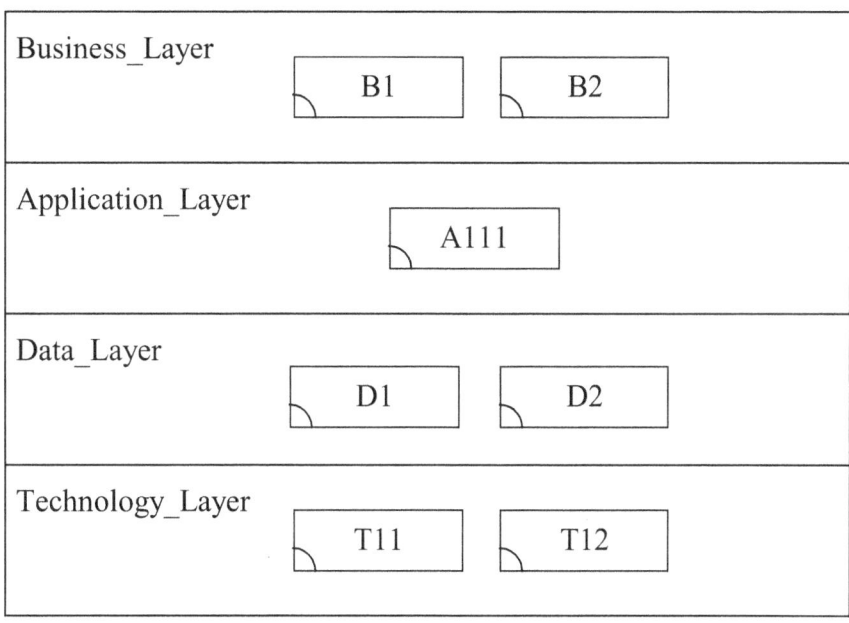

Figure 13-3 Design's FD of the *QQQ* System

13-3 Design's COD

Design's COD is the component operation diagram we obtain after the designing phase is finished. Figure 13-4 shows the design's COD of the *QQQ* system. In the figure, component *B1* has two operations: *op_01* and *op_02*; component *B2* has one operation: *op_03*; component *A111* has three operations: *op_04*, *op_05 and op_06*; component *D1* has two operations: *op_07* and *op_08*; component *D2* has one operation: *op_9*; component *T11* has one operation: *op_10*; component *T12* has one operation: *op_11*.

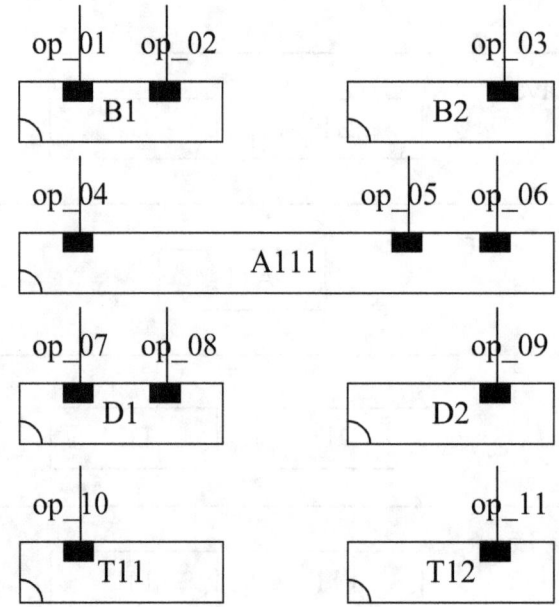

Figure 13-4   Design's COD of the *QQQ* System

13-4 Design's CCD

Design's CCD is the component connection diagram we obtain after the designing phase is finished. Figure 13-5 shows the design's CCD of the *QQQ* system.

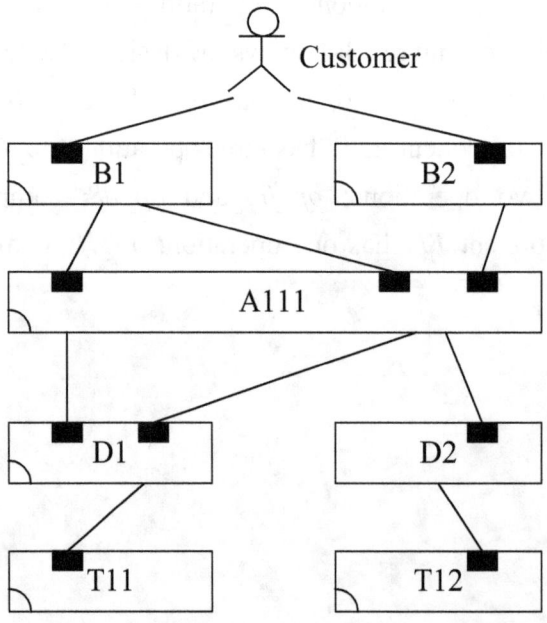

Figure 13-5   Design's CCD of the *QQQ* System

In Figure 13-5, we see that actor *Customer* has a connection with each one of the *B1* and *B2* components; component *B1* has two connections with the *A111* component; component *B2* has a connection with the *A111* component; component *A111* has two connections with the *D1* component; component *A111* has a connection with the *D2* component; component *D1* has a connection with the *T11* component; component *D2* has a connection with the *T12* component.

Chapter 14: Design's Systems Behavior

The design's systems behavior includes: a) *Design's SBCD* and b) *Design's IFD* as shown in Figure 14-1.

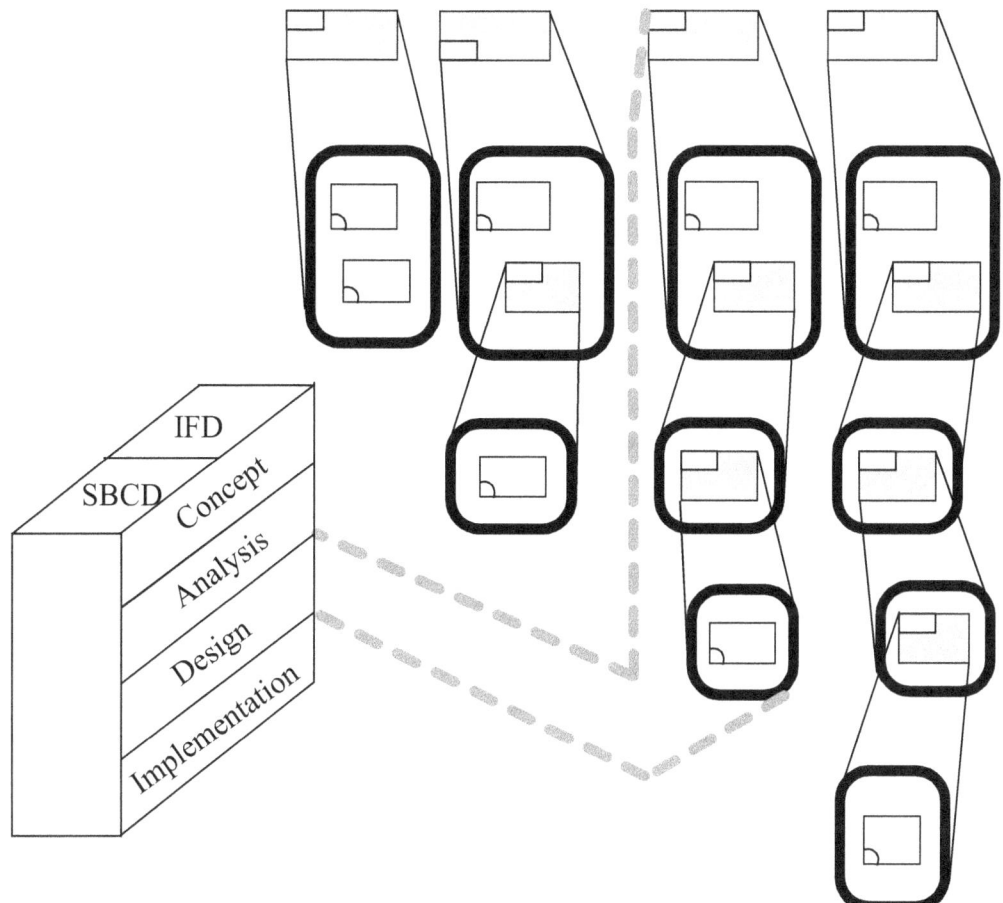

Figure 14-1   SBC Design's Systems Behavior

14-1 Design's SBCD

Design's SBCD is the structure-behavior coalescence diagram we obtain after the designing phase is finished. Figure 14-2 shows the design's SBCD of the *QQQ* system. In this example, an actor interacting with seven components shall describe the overall design's systems behavior. Interactions among the *Customer* actor and the *B1*, *A111*, *D1* components draw forth the *qqq_1* behavior. Interactions among the *Customer* actor and the *B1*, *A111*, *D1*, *T11* components draw forth the *qqq_2* behavior.

Interactions among the *Customer* actor and the *B2, A111, D2, T12* components draw forth the *qqq_3* behavior.

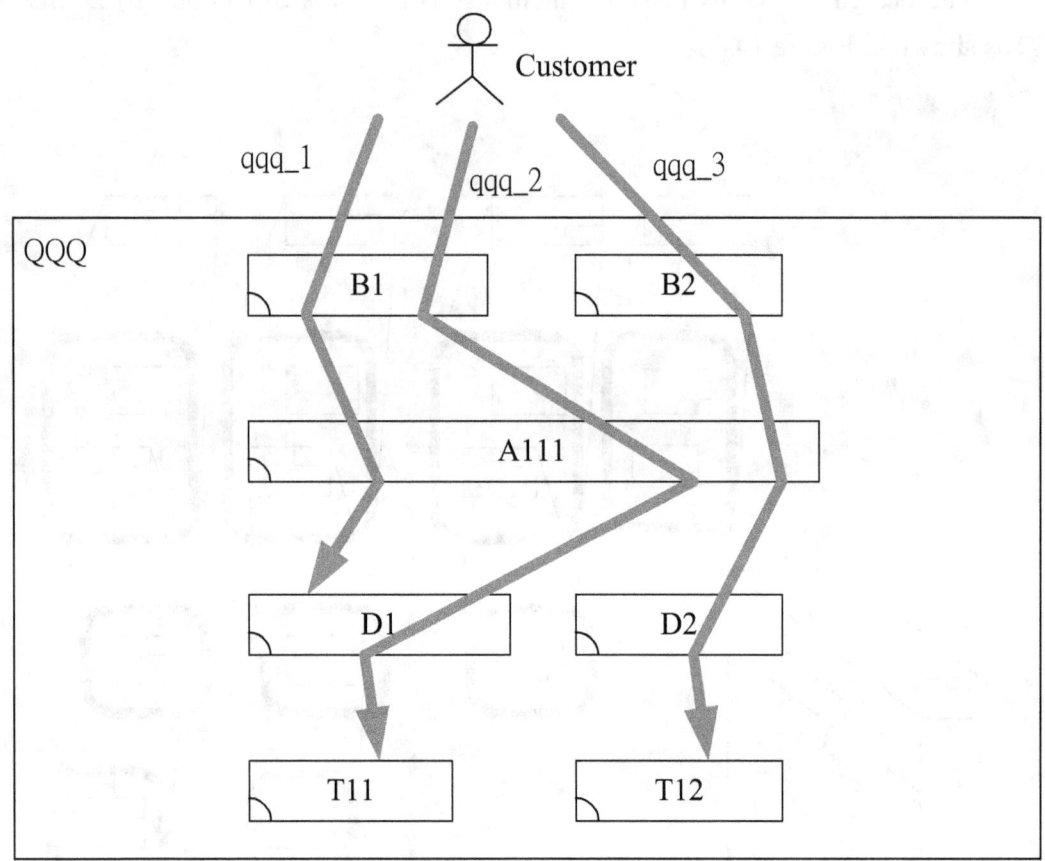

Figure 14-2 Design's SBCD of the *QQQ* System

The overall behavior of a system is a collection of all its individual behaviors. All individual behaviors are mutually independent of each other. They tend to be executed concurrently. For example, the overall design's behavior of the *QQQ* system includes the *qqq_1*, *qqq_2* and *qqq_3* behaviors. In other words, the *qqq_1*, *qqq_2* and *qqq_3* behaviors are combined to produce the overall design's behavior of the *QQQ* system.

The major purpose of using the architectural approach, instead of separating the structure model from the behavior model, is to achieve one single coalesced model. In Figure 14-2, systems architects are able to see that the systems structure and systems behavior coexist in the design's SBCD. That is, in the design's SBCD of the *QQQ* system, systems architects not only see its systems structure but also see (at the same time) its systems behavior.

14-2 Design's IFD

Design's IFDs are the interaction flow diagrams we obtain after the designing phase is finished. The overall design's behavior of the *QQQ* system includes three behaviors: *qqq_1*, *qqq_2* and *qqq_3*. Each of them is described by an individual IFD. Figure 14-3 shows the design's IFD of the *qqq_1* behavior. First, actor *Customer* interacts with the *B1* component through the *op_01* operation call interaction. Next, component *B1* interacts with the *A111* component through the *op_04* operation call interaction. Finally, component *A111* interacts with the *D1* component through the *op_07* operation call interaction.

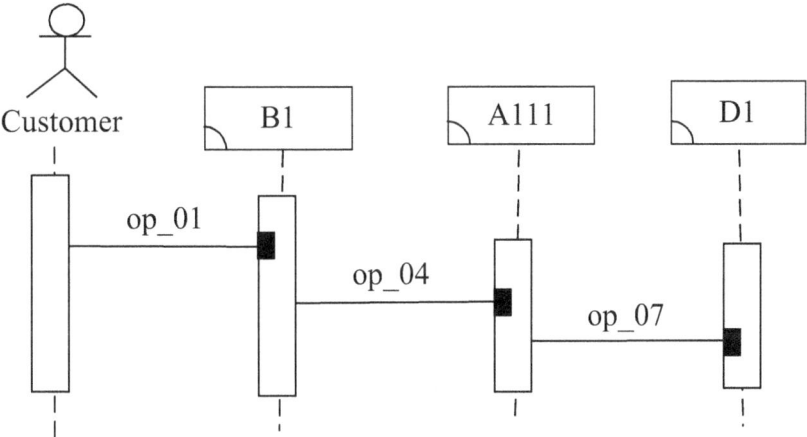

Figure 14-3   Design's IFD of the *qqq_1* Behavior

Figure 14-4 shows the design's IFD of the *qqq_2* behavior. First, actor *Customer* interacts with the *B1* component through the *op_02* operation call interaction. Next, component *B1* interacts with the *A111* component through the *op_05* operation call interaction. Continuingly, component *A111* interacts with the *D1* component through the *op_08* operation call interaction. Finally, component *D1* interacts with the *T11* component through the *op_10* operation call interaction.

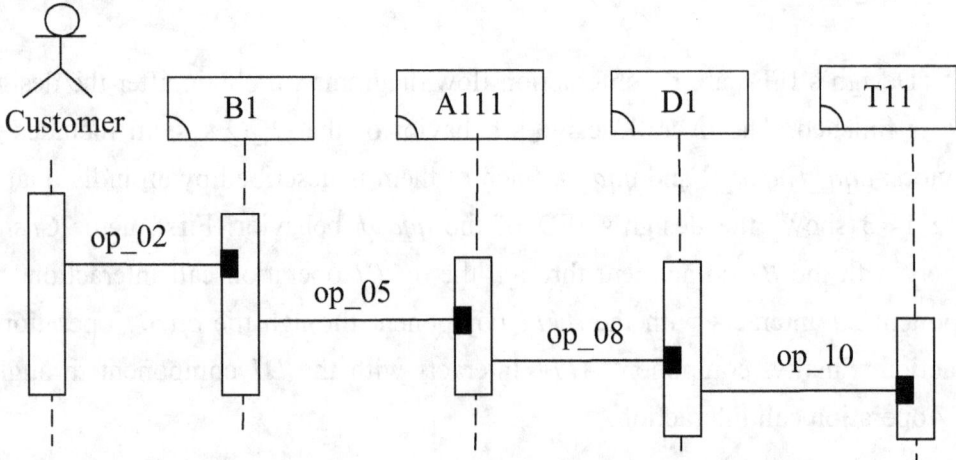

Figure 14-4   Design's IFD of the *qqq_2* Behavior

Figure 14-5 shows the design's IFD of the *qqq_3* behavior. First, actor *Customer* interacts with the *B2* component through the *op_03* operation call interaction. Next, component *B2* interacts with the *A111* component through the *op_06* operation call interaction. Continuingly, component *A111* interacts with the *D2* component through the *op_09* operation call interaction. Finally, component *D2* interacts with the *T12* component through the *op_11* operation call interaction.

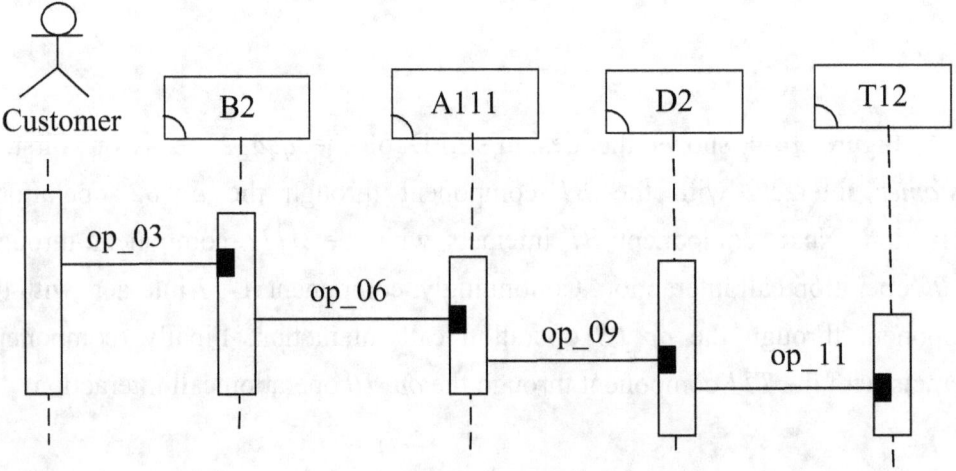

Figure 14-5   Design's IFD of the *qqq_3* Behavior

PART VI: IMPLEMENTATION VIEW

Chapter 15: Principle of the Implementation View

Implementation view describes what an implementer has done for his work. Implementation view is one level down structural decomposition (with observation congruence verification) of the design view [Chao15a, Chao15b, Chao15c, Chao15d, Chao15e]. That is, we shall not create the implementation view from the scratch. Instead, we will architect the implementation view by decomposing the design view.

15-1 Architecting the Implementation View

According to the SBC multi-level (hierarchical) view, systems architects construct the implementation's systems architecture for the implementer to view. This implementation's systems architecture is called the implementation view as shown in Figure 15-1.

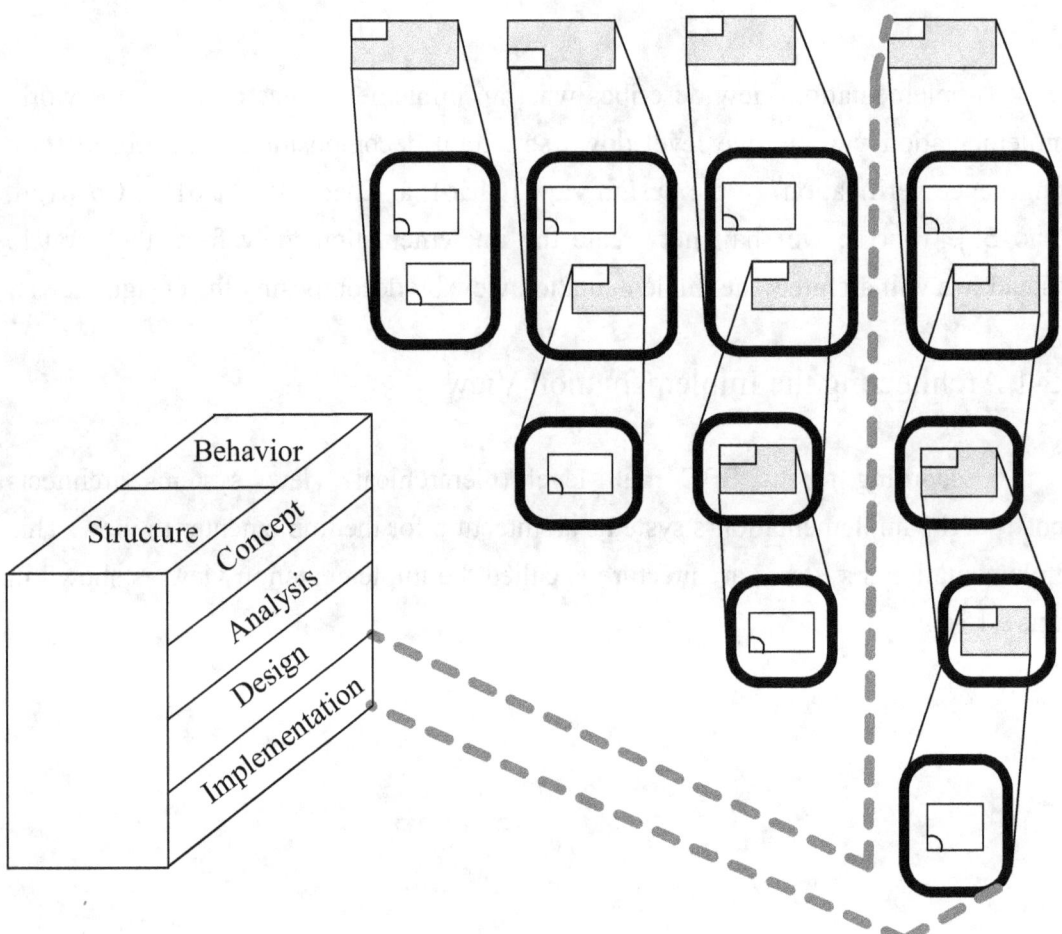

Figure 15-1   Implementation View

The implementation view consists of: a) implementation's systems structure and b) implementation's systems behavior.

15-2 Implementation's Systems Structure

The entire SBC implementation's systems structure includes: a) *Implementation's AHD*, b) *Implementation's FD*, c) *Implementation's COD* and d) *Implementation's CCD*, as shown in Figure 15-2.

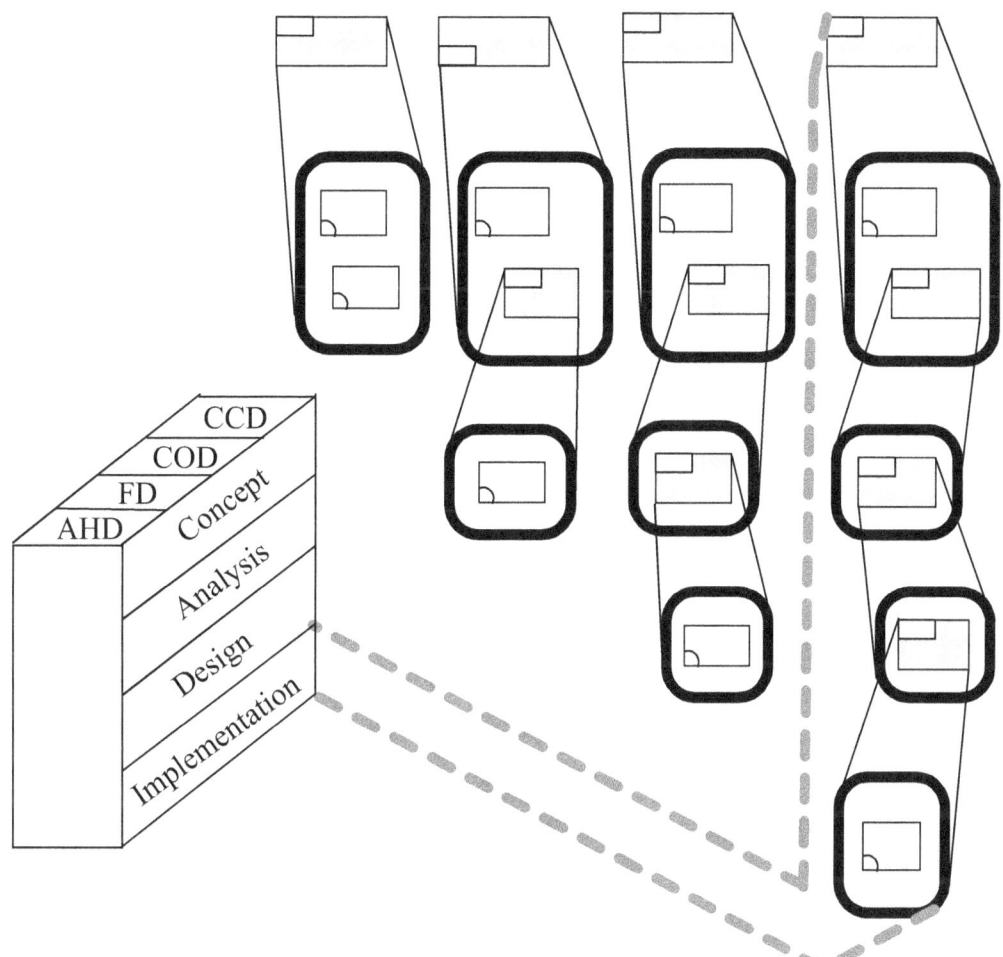

Figure 15-2   SBC Implementation's Systems Structure

Another drawing can also be used to illustrate the SBC implementation's systems structure as shown in Figure 15-3.

	Systems Structure			
	Architecture Hierarchy Diagram	Framework Diagram	Component Operation Diagram	Component Connection Diagram
Implementation	Implementation's AHD	Implementation's FD	Implementation's COD	Implementation's CCD

Figure 15-3 SBC Implementation's Systems Structure

More complete discussions on the implementation's systems structure will be elaborated in later chapters which will provide a detailed explanation of its purpose.

15-3 Implementation's Systems Behavior

The entire SBC implementation's systems behavior includes: a) *Implementation's SBCD* and b) *Implementation's IFD* as shown in Figure 15-4.

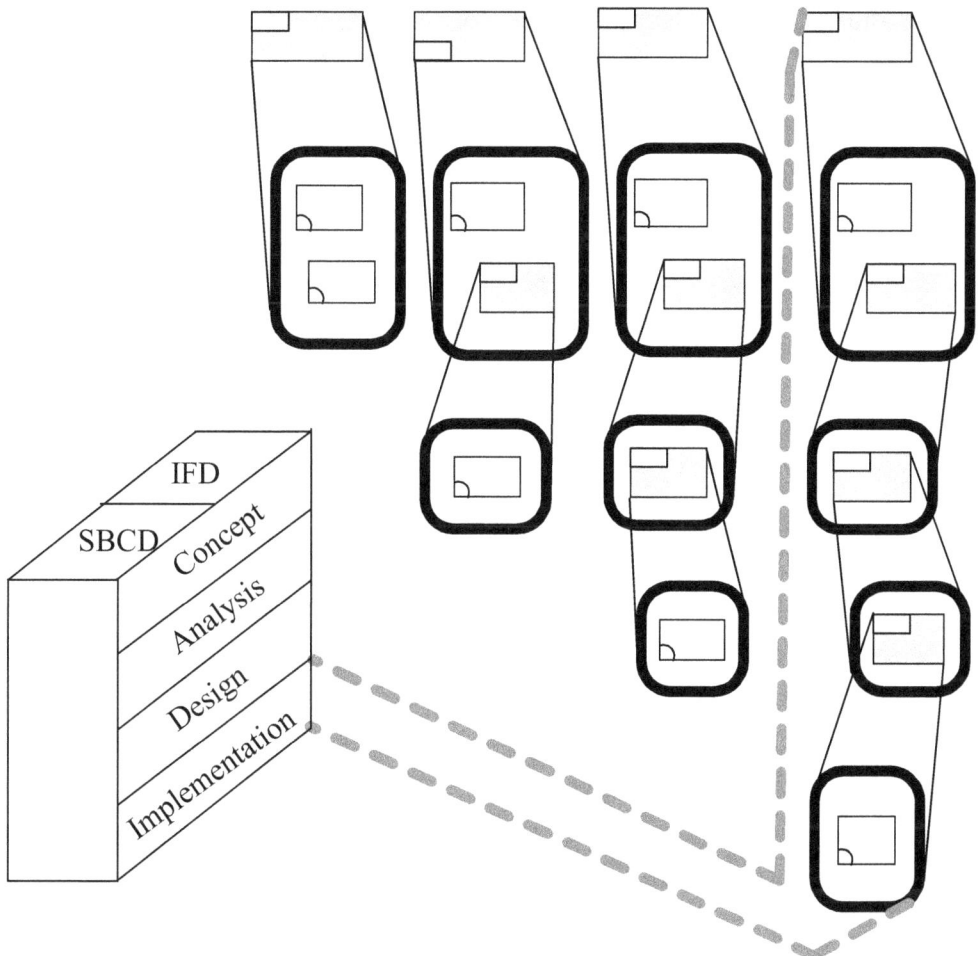

Figure 15-4   SBC Implementation's Systems Behavior

Let us use another drawing to illustrate the SBC implementation's systems behavior as shown in Figure 15-5.

	Systems Behavior	
	Structure-Behavior Coalescence Diagram	Interaction Flow Diagram
Implementation	Implementation's SBCD	Implementation's IFD

Figure 15-5   SBC Implementation's Systems Behavior

More complete discussions on the SBC implementation's systems behavior will be elaborated in later chapters which will provide a detailed explanation of its purpose.

Chapter 16: Implementation's Systems Structure

The implementation's systems structure includes: a) *Implementation's AHD*, b) *Implementation's FD*, c) *Implementation's COD* and d) *Implementation's CCD* as shown in Figure 16-1.

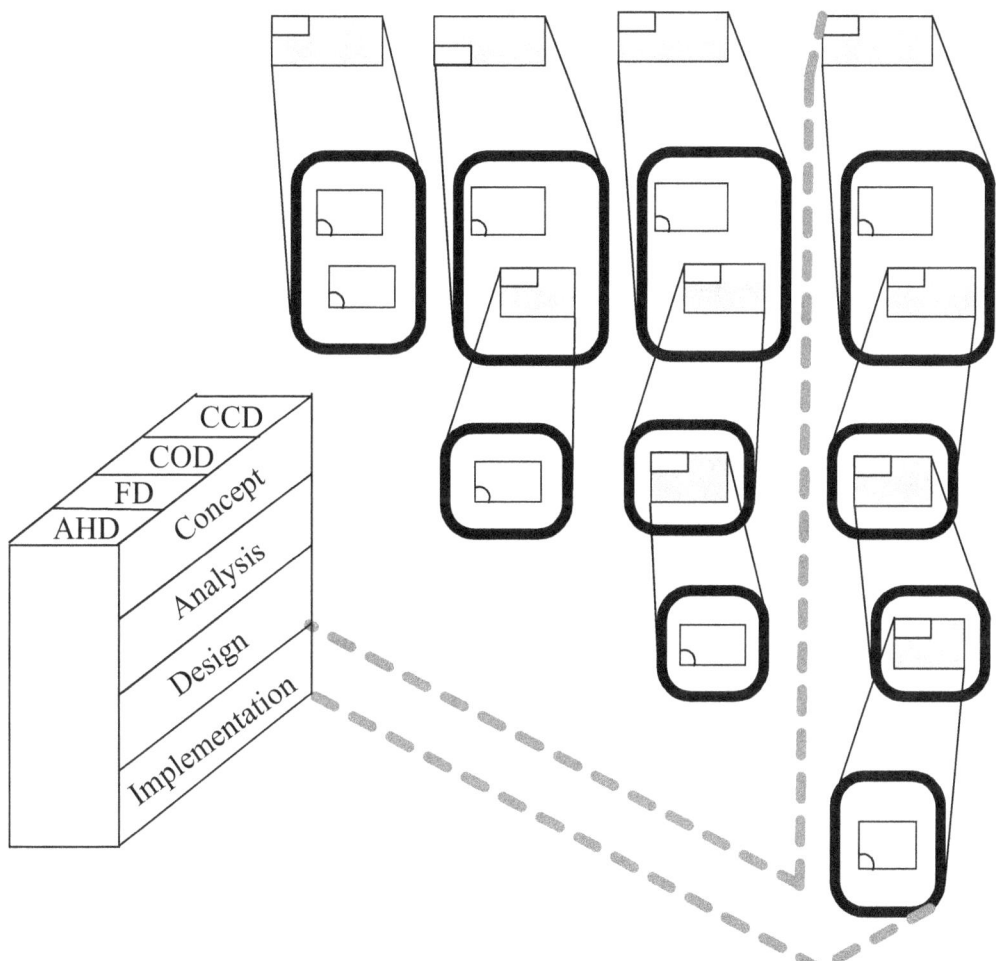

Figure 16-1   SBC   Implementation's Systems Structure

16-1 Implementation's AHD

Implementation's AHD is the architecture hierarchy diagram we obtain after the implementation phase is finished. Figure 16-2 shows the implementation's AHD of the *QQQ* system. In the figure, *QQQ* is composed of *B1*, *B2*, *A1*, *D1*, *D2* and *T1*;

A1 is composed of *A11*; *T1* is composed of *T11* and *T12*; *A11* is composed of *A111*; *A111* is composed of *A1111* and *A1112*.

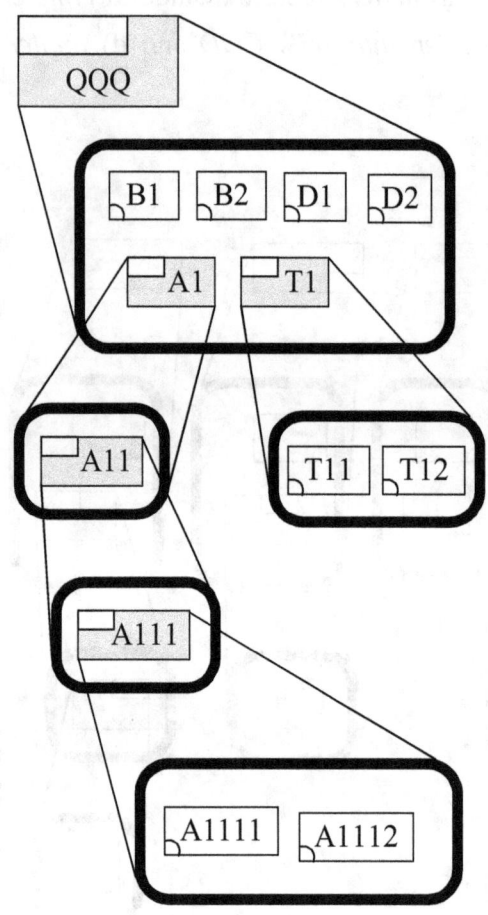

Figure 16-2   Implementation's AHD of the *QQQ* System

In Figure 16-2, *QQQ*, *A1*, *T1*, *A11* and *A111* are aggregated systems while *B1*, *B2*, *A1111*, *A1112*, *D1*, *D2*, *T11* and *T12* are non-aggregated systems.

We validate our claim that the implementation view is one level down structural decomposition of the design view by comparing Figure 13-2 with Figure 16-2. In Figure 13-2, *A111* is a non-aggregated system. As an interesting contrast, in Figure 16-2, *A111* becomes an aggregated system and is composed of *A1111* and *A1112*.

16-2 Implementation's FD

Implementation's FD is the framework diagram we obtain after the implementation phase is finished. Figure 16-3 shows the implementation's FD of the

QQQ system. In the figure, *Business_Layer* contains the *B1* and *B2* components; *Application_Layer* contains the *A1111* and *A1112* components; *Data_Layer* contains the *D1* and *D2* components; *Technology_Layer* contains the *T11* and *T12* components.

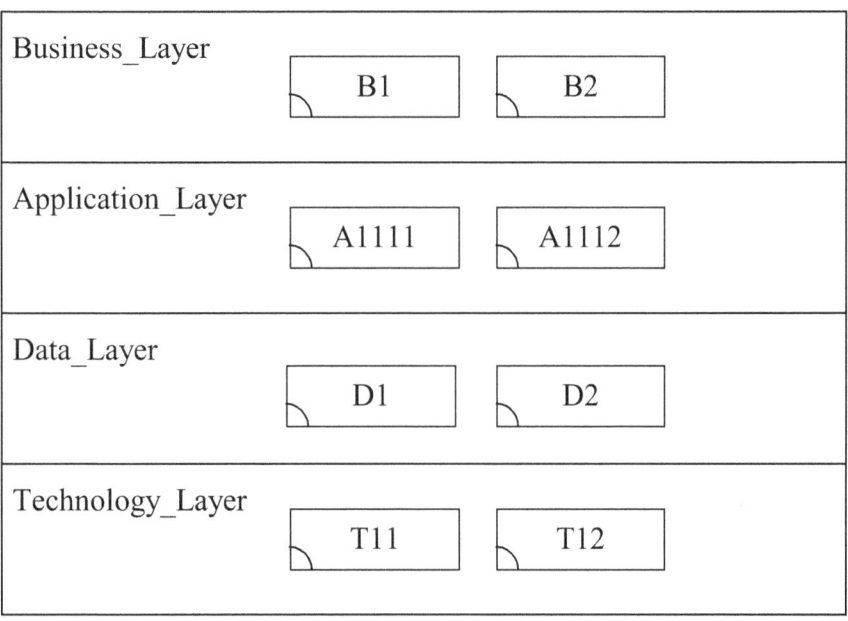

Figure 16-3 Implementation's FD of the *QQQ* System

16-3 Implementation's COD

Implementation's COD is the component operation diagram we obtain after the implementation phase is finished. Figure 16-4 shows the implementation's COD of the *QQQ* system. In the figure, component *B1* has two operations: *op_01* and *op_02*; component *B2* has one operation: *op_03*; component *A1111* has one operation: *op_04*; component *A1112* has two operations: *op_05* and *op_06*; component *D1* has two operations: *op_07* and *op_08*; component *D2* has one operation: *op_9*; component *T11* has one operation: *op_10*; component *T12* has one operation: *op_11*.

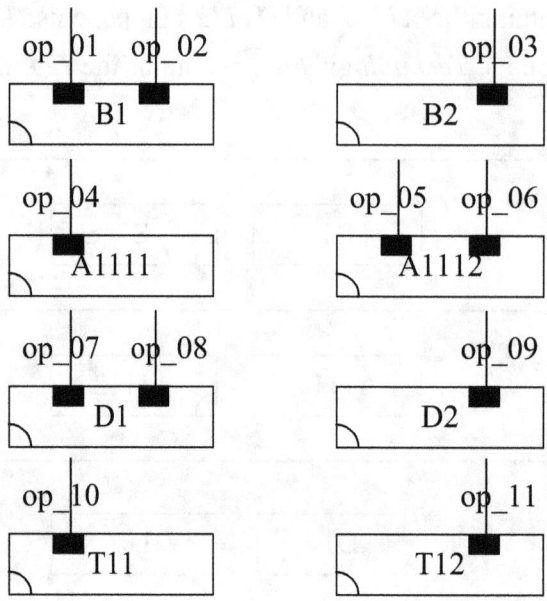

Figure 16-4   Implementation's COD of the *QQQ* System

16-4 Implementation's CCD

Implementation's CCD is the component connection diagram we obtain after the implementation phase is finished. Figure 16-5 shows the implementation's CCD of the *QQQ* system.

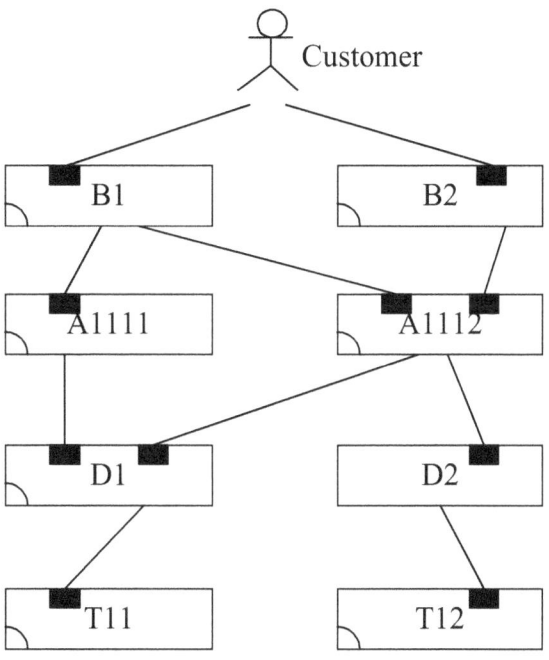

Figure 16-5   Implementation's CCD of the *QQQ* System

In Figure 16-5, we see that actor *Customer* has a connection with each one of the *B1* and *B2* components; component *B1* has a connection with each one of the *A1111* and *A1112* components; component *B2* has a connection with the *A1112* component; component *A1111* has a connection with the *D1* component; component *A1112* has a connection with each one of the *D1* and *D2* components; component *D1* has a connection with the *T11* component; component *D2* has a connection with the *T12* component.

Chapter 17: Implementation's Systems Behavior

The implementation's systems behavior includes: a) *Implementation's SBCD* and b) *Implementation's IFD* as shown in Figure 17-1.

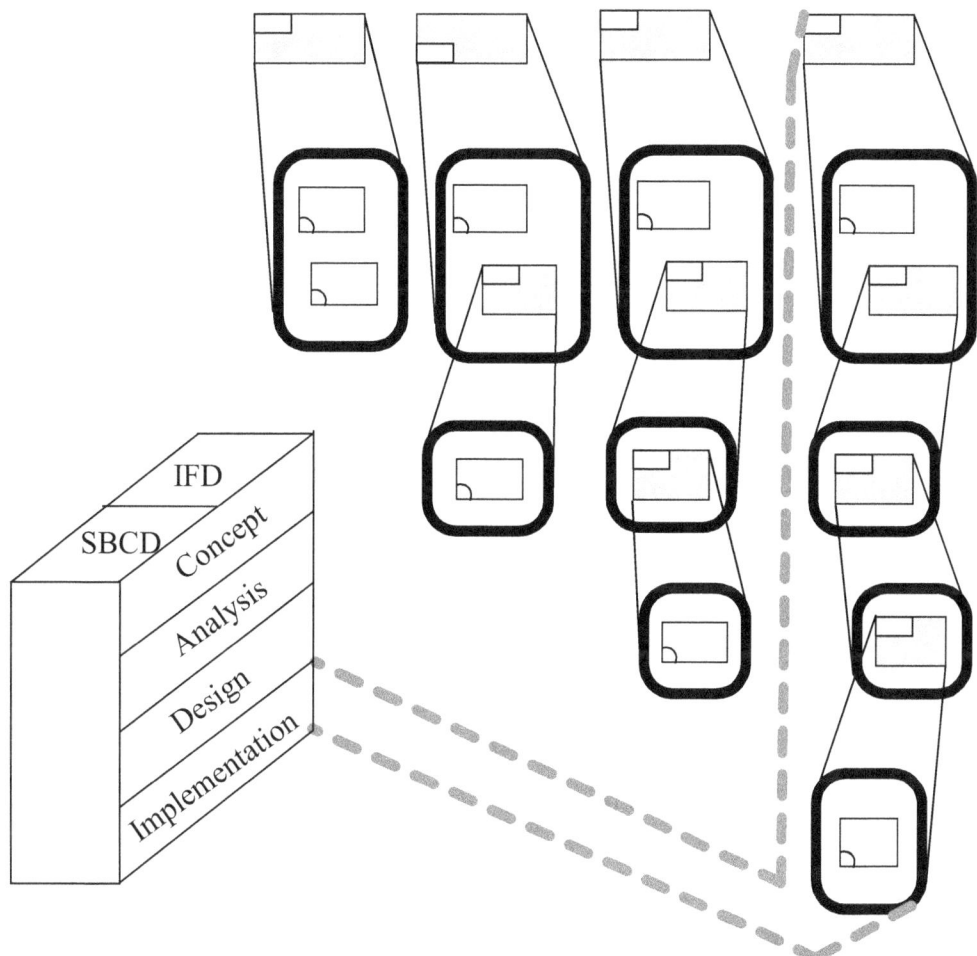

Figure 17-1   SBC Implementation's Systems Behavior

17-1 Implementation's SBCD

Implementation's SBCD is the structure-behavior coalescence diagram we obtain after the implementation phase is finished. Figure 17-2 shows the implementation's SBCD of the *QQQ* system. In this example, an actor interacting with eight components shall describe the overall implementation's systems behavior. Interactions among the *Customer* actor and the *B1*, *A1111*, *D1* components draw forth the *qqq_1* behavior. Interactions among the *Customer* actor and the *B1*, *A1112*, *D1*,

T11 components draw forth the *qqq_2* behavior. Interactions among the *Customer* actor and the *B2, A1112, D2, T12* components draw forth the *qqq_3* behavior.

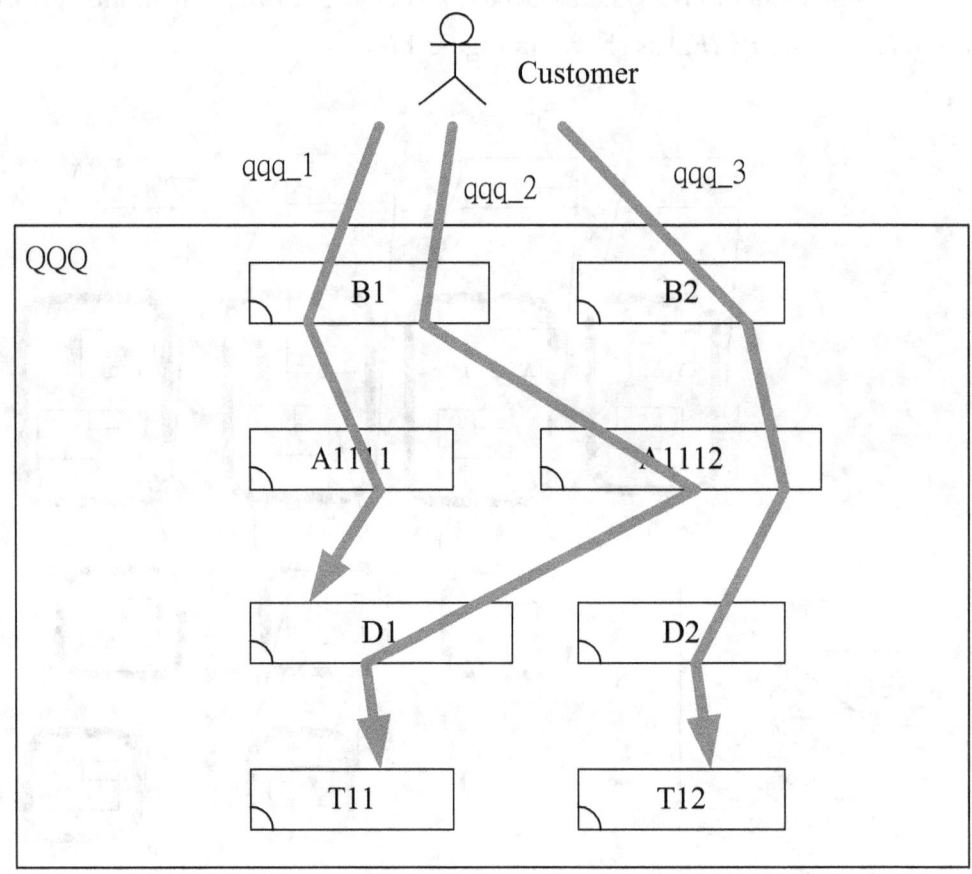

Figure 17-2   Implementation's SBCD of the *QQQ* System

The overall behavior of a system is a collection of all its individual behaviors. All individual behaviors are mutually independent of each other. They tend to be executed concurrently. For example, the overall implementation's behavior of the *QQQ* system includes the *qqq_1*, *qqq_2* and *qqq_3* behaviors. In other words, the *qqq_1*, *qqq_2* and *qqq_3* behaviors are combined to produce the overall implementation's behavior of the *QQQ* system.

The major purpose of using the architectural approach, instead of separating the structure model from the behavior model, is to achieve one single coalesced model. In Figure 17-2, we are able to see that the systems structure and systems behavior coexist in the implementation's SBCD. That is, in the implementation's SBCD of the *QQQ* system, a systems architect not only sees its systems structure but also sees (at the same time) its systems behavior.

17-2 Implementation's IFD

Implementation's IFDs are the interaction flow diagrams we obtain after the implementation phase is finished. The overall implementation's behavior of the *QQQ* system includes three behaviors: *qqq_1*, *qqq_2* and *qqq_3*. Each of them is described by an individual IFD. Figure 17-3 shows the implementation's IFD of the *qqq_1* behavior. First, actor *Customer* interacts with the *B1* component through the *op_01* operation call interaction. Next, component *B1* interacts with the *A1111* component through the *op_04* operation call interaction. Finally, component *A1111* interacts with the *D1* component through the *op_07* operation call interaction.

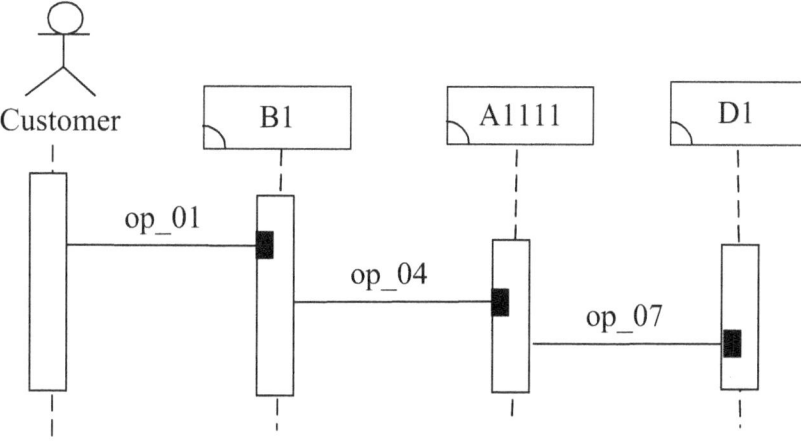

Figure 17-3   Implementation's IFD of the *qqq_1* Behavior

Figure 17-4 shows the implementation's IFD of the *qqq_2* behavior. First, actor *Customer* interacts with the *B1* component through the *op_02* operation call interaction. Next, component *B1* interacts with the *A1112* component through the *op_05* operation call interaction. Continuingly, component *A1112* interacts with the *D1* component through the *op_08* operation call interaction. Finally, component *D1* interacts with the *T11* component through the *op_10* operation call interaction.

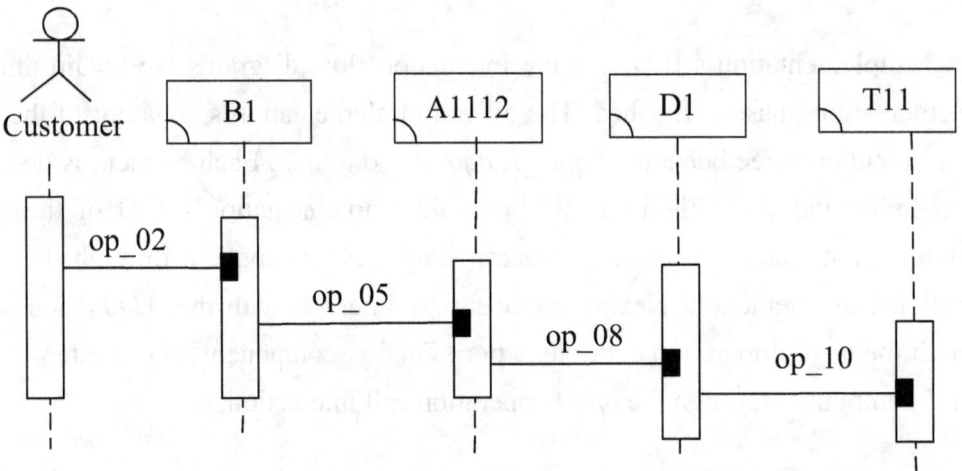

Figure 17-4   Implementation's IFD of the *qqq_2* Behavior

Figure 17-5 shows the implementation's IFD of the *qqq_3* behavior. First, actor *Customer* interacts with the *B2* component through the *op_03* operation call interaction. Next, component *B2* interacts with the *A1112* component through the *op_06* operation call interaction. Continuingly, component *A1112* interacts with the *D2* component through the *op_09* operation call interaction. Finally, component *D2* interacts with the *T12* component through the *op_11* operation call interaction.

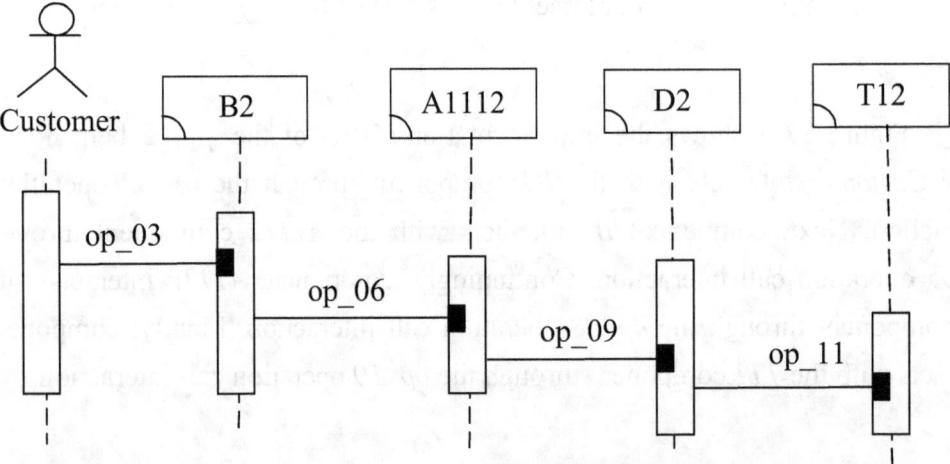

Figure 17-5   Implementation's IFD of the *qqq_3* Behavior

PART VII: CASE STUDIES

Chapter 18: Automobile --Hardware Architecture

This chapter examines the *Automobile* which represents a case study of hardware architecture [Engl09] for hardware systems modeling and architecting. An automobile, either motor car or car, is a wheeled motor vehicle used for transporting passengers. A driver may depress the gas or brake pedal to accelerate or stop the car. The overall behavior of the *Automobile* is eminently represented by two behaviors: *Accelerate_the_Car* and *Stop_the_Car* as shown in Figure 18-1.

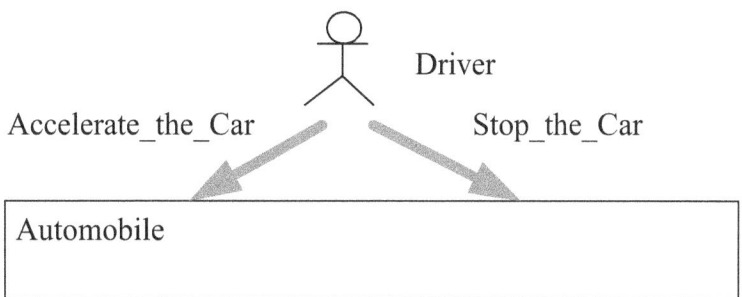

Figure 18-1 Two Behaviors of the *Automobile*

Using the SBC hardware multi-level (hierarchical) view, an architect goes through: a) analysis view and b) design view for the *Automobile* hardware systems architecture as shown in Figure 18-2.

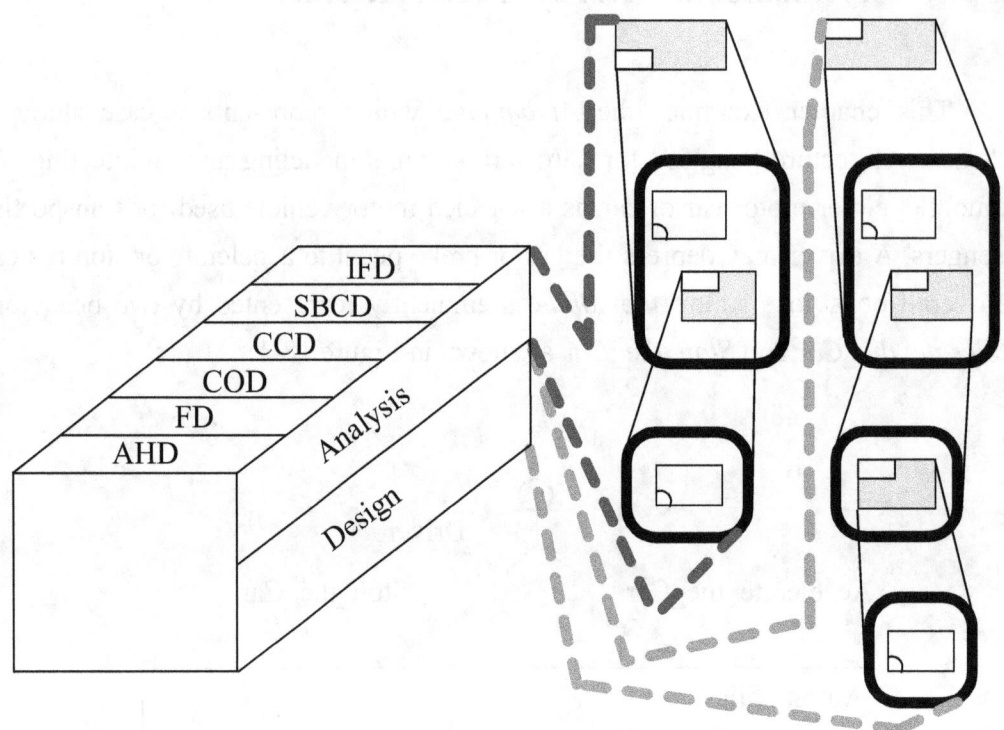

Figure 18-2   SBC Hardware Multi-Level (Hierarchical) View

18-1 Analysis View of the Automobile

In the SBC hardware multi-level (hierarchical) view, an architect constructs the analysis' systems architecture for the analyzer to view. This analysis' systems architecture is called the analysis view as shown in Figure 18-3.

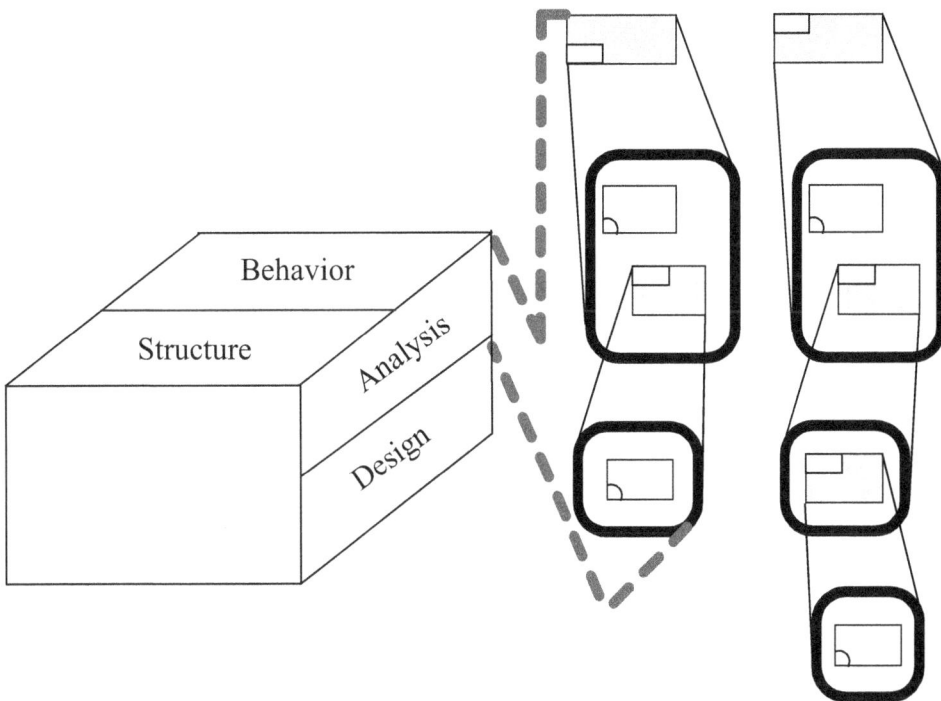

Figure 18-3　Analysis View

The analysis view consists of: a) analysis' systems structure and b) analysis' systems behavior.

18-1-1 Analysis' Systems Structure

The entire SBC analysis' systems structure includes: a) *Analysis' AHD*, b) *Analysis' FD*, c) *Analysis' COD* and d) *Analysis' CCD*.

We first draw the analysis' AHD of the *Automobile*. As shown in Figure 18-4, *Automobile* is composed of *Acceleration_Subsystem* and *Brake_Subsystem*.

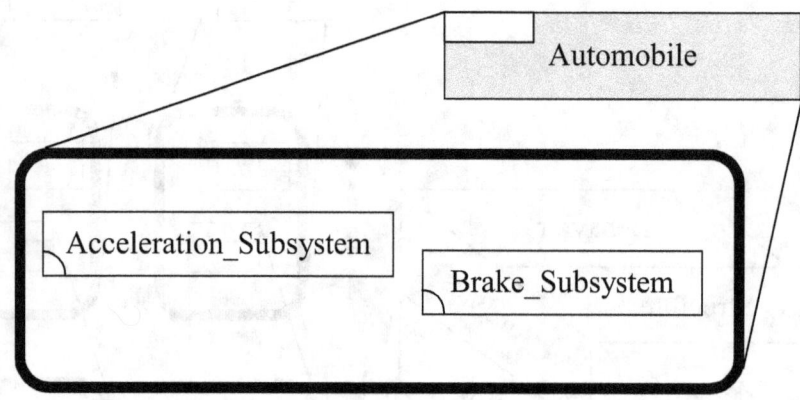

Figure 18-4 Analysis' AHD of the *Automobile*

In Figure 18-4, *Automobile* is an aggregated system while *Acceleration_Subsystem* and *Brake_Subsystem* are non-aggregated systems.

Analysis' FD is the framework diagram we obtain after the analysis phase is finished. Figure 18-5 shows the analysis' FD of the *Automobile*. In the figure, *Technology_SubLayer_1* contains the *Acceleration_Subsystem* and *Brake_Subsystem* components.

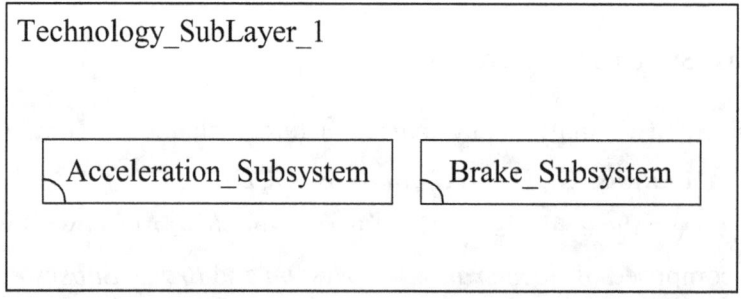

Figure 18-5 Analysis' FD of the *Automobile*

Analysis' COD is the component operation diagram we obtain after the analysis phase is finished. Figure 18-6 shows the analysis' COD of the *Automobile*. In the figure, component *Acceleration_Subsystem* has one operation: *Depress_Gas_Pedal*; component *Brake_Subsystem* has one operation: *Depress_Brake_Pedal*.

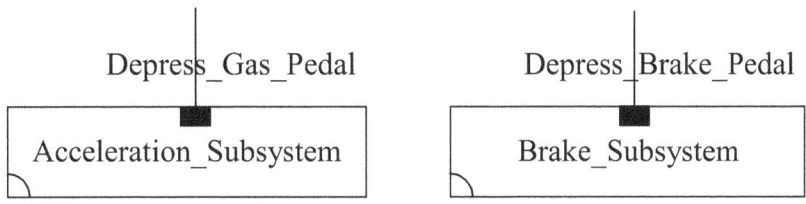

Figure 18-6 Analysis' COD of the *Automobile*

Analysis' CCD is the component connection diagram we obtain after the analysis phase is finished. Figure 18-7 shows the analysis' CCD of the *Automobile*.

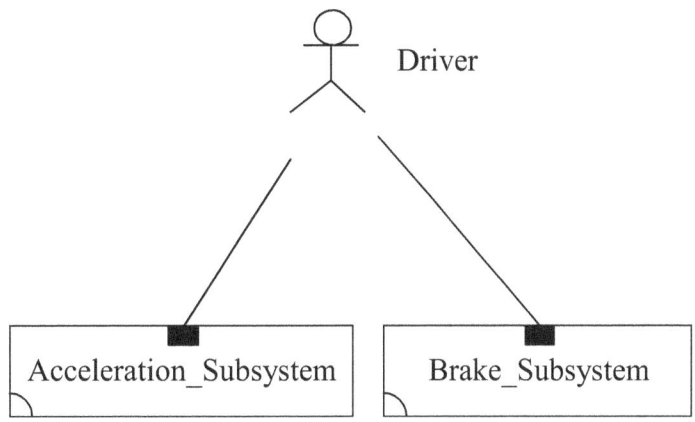

Figure 18-7 Analysis' CCD of the *Automobile*

In Figure 18-7, actor *Driver* has a connection with each one of the *Acceleration_Subsystem* and *Brake_Subsystem* components.

18-1-2 Analysis' Systems Behavior

The entire SBC analysis' systems behavior includes: a) *Analysis' SBCD* and b) *Analysis' IFD*.

Analysis' SBCD is the structure-behavior coalescence diagram we obtain after the analysis phase is finished. Figure 18-8 shows the analysis' SBCD of the *Automobile* in which interactions among the *Driver* actor and the *Acceleration_Subsystem*, *Brake_Subsystem* components shall draw forth the *Accelerate_the_Car* and *Stop_the_Car* behaviors.

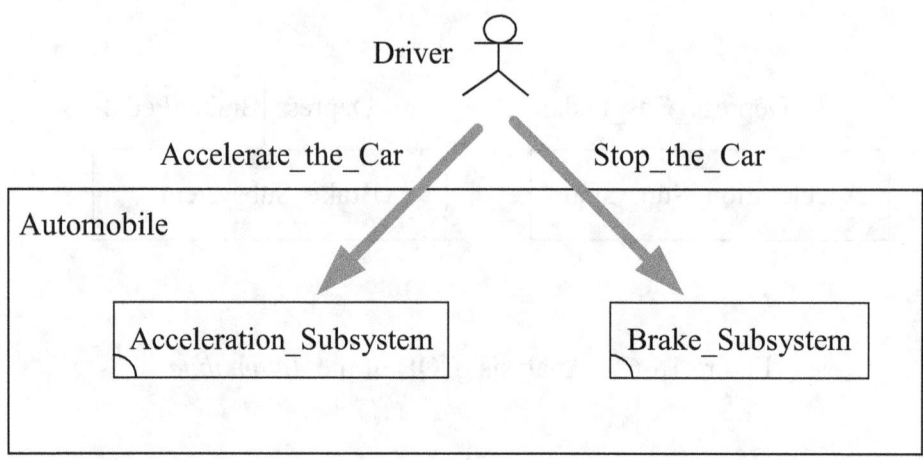

Figure 18-8   Analysis' SBCD of the *Automobile*

The overall behavior of a hardware system is contained in its individual behaviors. After the analysis phase is finished, the overall analysis' behavior of the *Automobile* includes the *Accelerate_the_Car* and *Stop_the_Car* behaviors. In other words, the *Accelerate_the_Car* and *Stop_the_Car* behaviors together provide the overall analysis' behavior of the *Automobile* after the analysis phase is finished.

Be noticed that the *Accelerate_the_Car* behavior and the *Stop_the_Car* behavior are mutually independent of each other. They shall be executed concurrently [Hoar85, Miln89, Miln99].

The major purpose of using the architectural approach, instead of separating the structure model from the behavior model, is to achieve a coalesced model. In Figure 18-8, we not only see its hardware systems structure, also see at the same time its hardware systems behavior in the analysis' SBCD of the *Automobile*.

After the analysis phase is finished, the overall analysis' *Automobile* behavior includes two behaviors: *Accelerate_the_Car* and *Stop_the_Car*. Each of them is described by an individual IFD. Figure 18-9 shows the analysis' IFD of the *Accelerate_the_Car* behavior. First, actor *Driver* interacts with the *Acceleration_Subsystem* component through the *Depress_Gas_Pedal* operation call interaction.

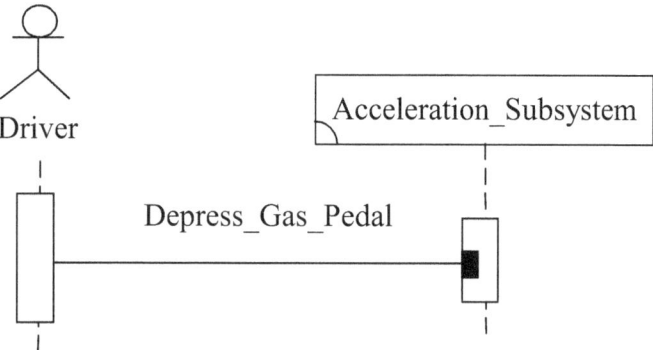

Figure 18-9   Analysis' IFD of the *Accelerate_the_Car* Behavior

Figure 18-10 shows the analysis' IFD of the *Stop_the_Car* behavior. First, actor *Driver* interacts with the *Brake_Subsystem* component through the *Depress_Brake_Pedal* operation call interaction.

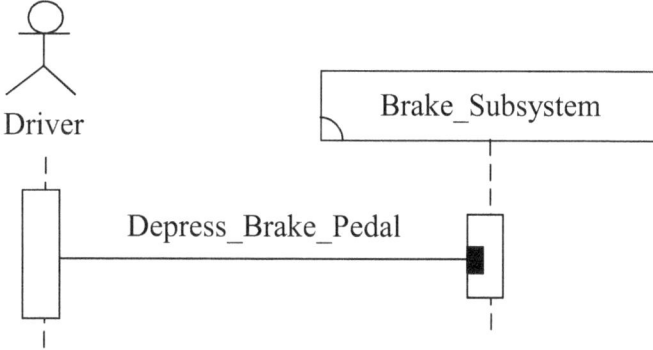

Figure 18-10   Analysis' IFD of the *Stop_the_Car* Behavior

18-2 Design View of the Automobile

In the SBC hardware multi-level (hierarchical) view, an architect constructs the design's systems architecture for the designer to view. This design's systems architecture is called the design view as shown in Figure 18-11. Design view is one level down structural decomposition (with observation congruence verification) of the analysis view [Chao15a, Chao15b, Chao15c, Chao15d, Chao15e].

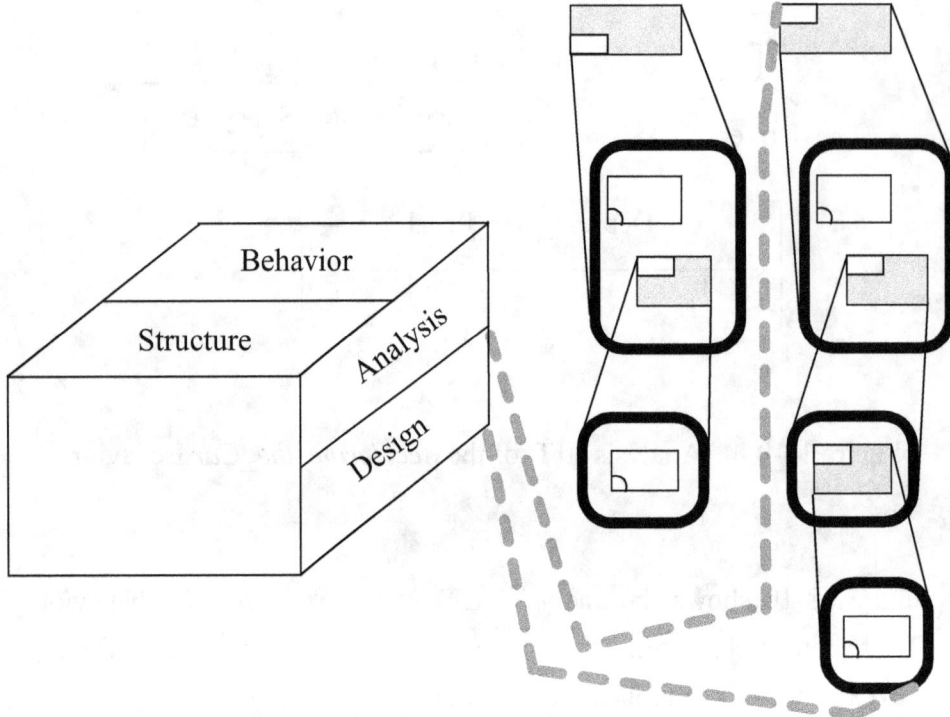

Figure 18-11   Design View

The design view consists of: a) design's systems structure and b) design's systems behavior.

18-2-1 Design's Systems Structure

The entire SBC design's systems structure includes: a) *Design's AHD*, b) *Design's FD*, c) *Design's COD* and d) *Design's CCD*.

We first draw the design's AHD of the *Automobile*. As shown in Figure 18-12, *Automobile* is composed of *Acceleration_Subsystem* and *Brake_Subsystem*; *Acceleration_Subsystem* is composed of *Gas_Pedal* and *Engine*; *Brake_Subsystem* is composed of *Brake_Pedal* and *Brake_Pad*.

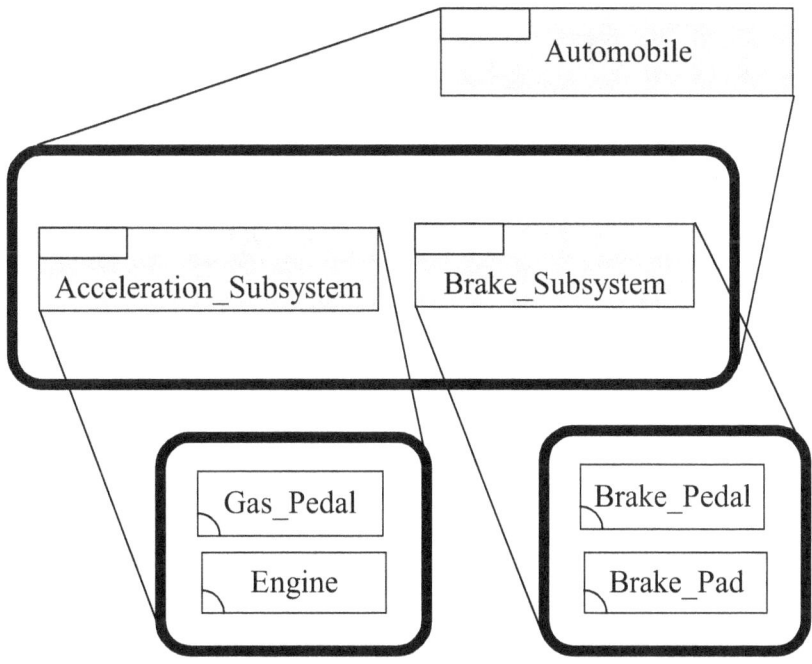

Figure 18-12 Design's AHD of the *Automobile*

In Figure 18-12, *Automobile*, *Acceleration_Subsystem* and *Brake_Subsystem* are aggregated systems while *Gas_Pedal*, *Engine*, *Brake_Pedal* and *Brake_Pad* are non-aggregated systems.

Design's FD is the framework diagram we obtain after the design phase is finished. Figure 18-13 shows the design's FD of the *Automobile*. In the figure, *Technology_SubLayer_2* contains the *Gas_Pedal* and *Brake_Pedal* components; *Technology_SubLayer_1* contains the *Engine* and *Brake_Pad* components.

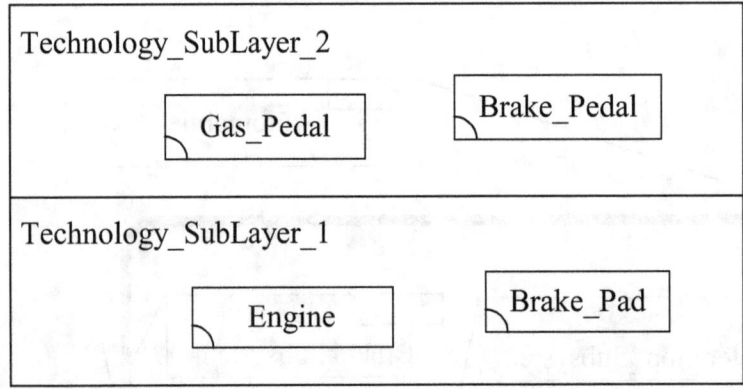

Figure 18-13   Design's FD of the *Automobile*

Design's COD is the component operation diagram we obtain after the design phase is finished. Figure 18-14 shows the design's COD of the *Automobile*. In the figure, component *Gas_Pedal* has one operation: *Depress_Gas_Pedal*; component *Brake_Pedal* has one operation: *Depress_Brake_Pedal*; component *Engine* has one operation: *Fuel_Supply*; component *Brake_Pad* has one operation: *Move_Inward*.

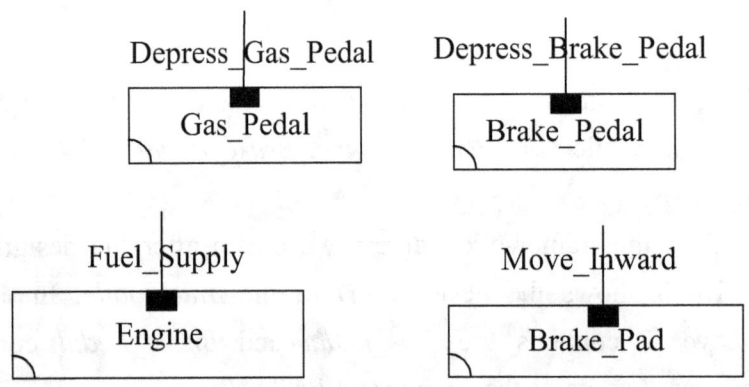

Figure 18-14   Design's COD of the *Automobile*

Design's CCD is the component connection diagram we obtain after the design phase is finished. Figure 18-15 shows the design's CCD of the *Automobile*.

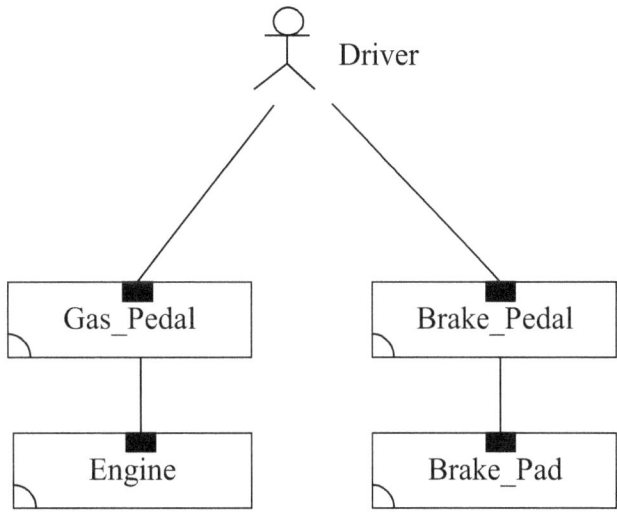

Figure 18-15　　Design's CCD of the *Automobile*

In Figure 18-15, actor *Driver* has a connection with each one of the *Gas_Pedal* and *Brake_Pedal* components; component *Gas_Pedal* has a connection with the *Engine* component; component *Brake_Pedal* has a connection with the *Brake_Pad* component.

18-2-2 Design's Systems Behavior

The entire SBC design's systems behavior includes: a) *Design's SBCD* and b) *Design's IFD*.

Design's SBCD is the structure-behavior coalescence diagram we obtain after the design phase is finished. Figure 18-16 shows the design's SBCD of the *Automobile* in which interactions among the *Driver* actor and the *Gas_Pedal*, *Engine*, *Brake_Pedal*, *Brake_Pad* components shall draw forth the *Accelerate_the_Car* and *Stop_the_Car* behaviors.

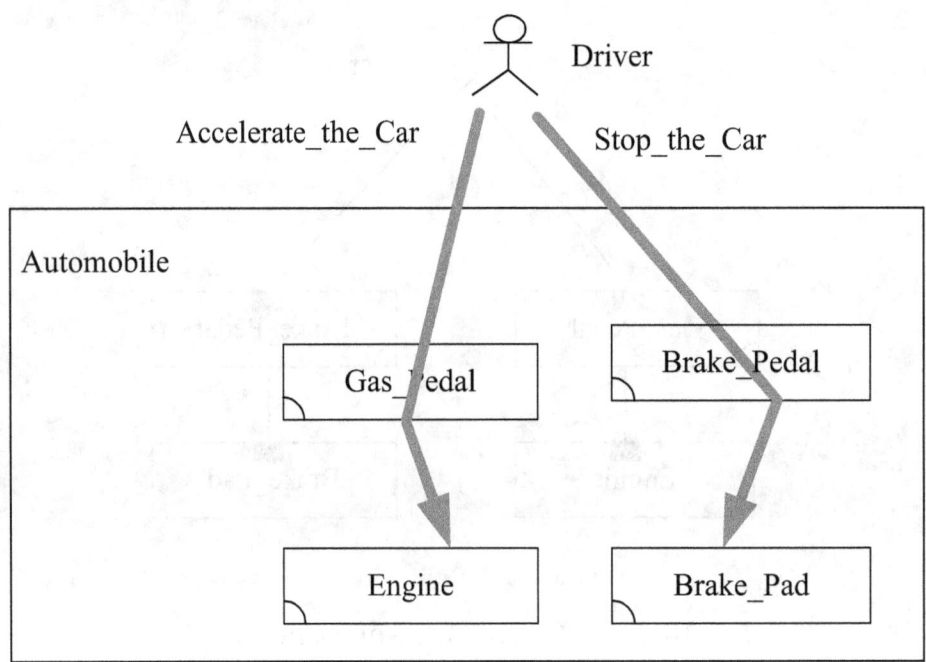

Figure 18-16   Design's SBCD of the *Automobile*

The overall behavior of a hardware system is contained in its individual behaviors. After the design phase is finished, the overall design's behavior of the *Automobile* includes the *Accelerate_the_Car* and *Stop_the_Car* behaviors. In other words, the *Accelerate_the_Car* and *Stop_the_Car* behaviors together provide the overall design's behavior of the *Automobile* after the design phase is finished.

Be noticed that the *Accelerate_the_Car* behavior and the *Stop_the_Car* behavior are mutually independent of each other. They shall be executed concurrently [Hoar85, Miln89, Miln99].

The major purpose of using the architectural approach, instead of separating the structure model from the behavior model, is to achieve a coalesced model. In Figure 18-16, we not only see its hardware systems structure, also see at the same time its hardware systems behavior in the design's SBCD of the *Automobile*.

After the design phase is finished, the overall design's behavior of the *Automobile* includes two behaviors: *Accelerate_the_Car* and *Stop_the_Car*. Each of them is described by an individual IFD. Figure 18-17 shows the design's IFD of the *Accelerate_the_Car* behavior. First, actor *Driver* interacts with the *Gas_Pedal* component through the *Depress_Gas_Pedal* operation call interaction. Finally, component *Gas_Pedal* interacts with the *Engine* component through the *Fuel_Supply* operation call interaction.

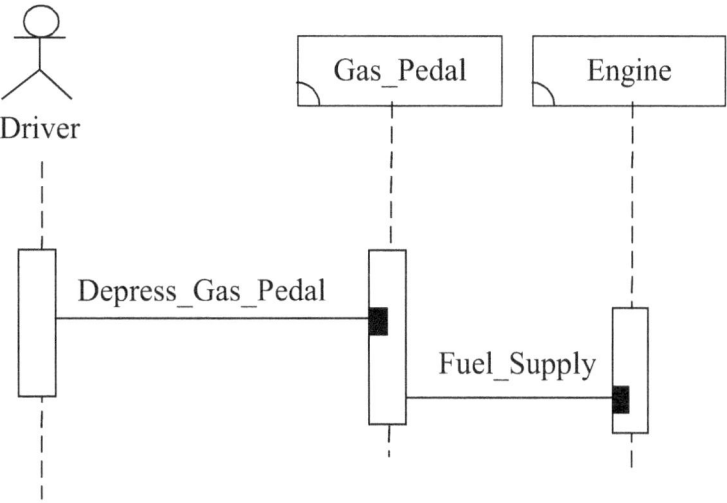

Figure 18-17   Design's IFD of the *Accelerate_the_Car* Behavior

Figure 18-18 shows the design's IFD of the *Stop_the_Car* behavior. First, actor *Driver* interacts with the *Brake_Pedal* component through the *Depress_Brake_Pedal* operation call interaction. Finally, component *Brake_Pedal* interacts with the *Brake_Pad* component through the *Move_Inward* operation call interaction.

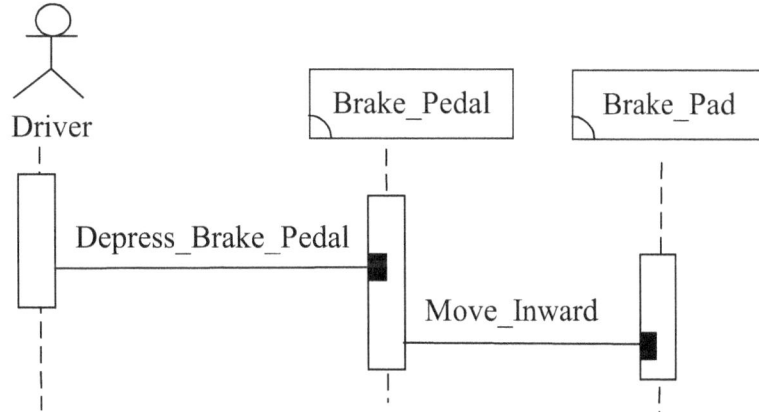

Figure 18-18   Design's IFD of the *Stop_the_Car* Behavior

Chapter 19: Multi-Tier Personal Data System -- Software Architecture

This chapter examines the *Multi-Tier Personal Data System* which represents a case study of software architecture [Bass03, Clem02, Clem10, Dike01, Roza05, Shaw96, Tayl09] for software systems modeling and architecting. After the systems development is finished, the *Multi-Tier Personal Data System* shall appear on a multi-tier platform [Wall04] as shown in Figure 19-1.

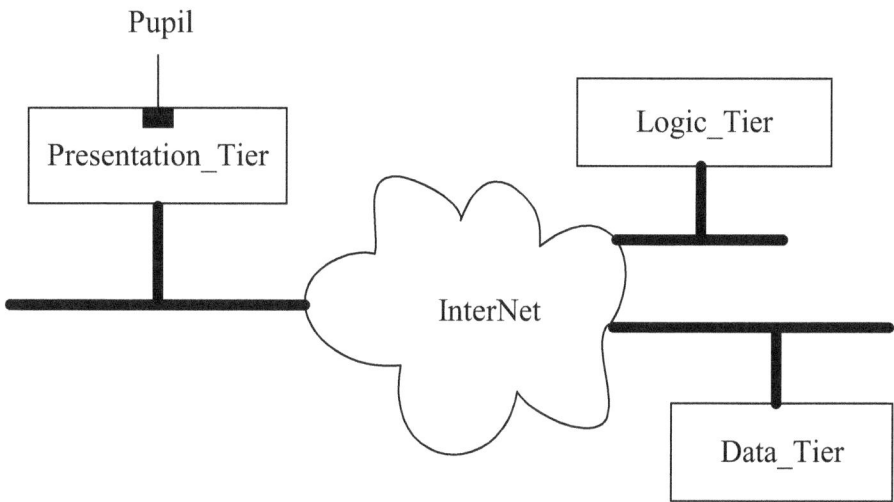

Figure 19-1   *Multi-Tier Personal Data System* on a Multi-Tier Platform

In the *Data_Layer*, there is a *Personal_Database* database [Date03, Elma10] which contains a *Personal_Data* table as shown in Figure 19-2.

Social_Secuirty_Number	Name	Date_of_Birth	Sex	Height (cm)	Weight (Kg)
318-49-2465	Mary R. Williams	June 17, 1976	Female	165	51
424-87-3651	Lee H. Wulf	July 24, 1982	Female	162	76
512-24-3722	John K. Bryant	May 12, 1954	Male	180	80

Figure 19-2   *Personal_Database* Contains *Personal_Data*

The overall behavior of the *Multi-Tier Personal Data System* is to provide a graphical user interface (GUI) [Gali07] for the *Pupil* actor to trigger two behaviors. The first behavior is *AgeCalculation* and the second behavior is *OverweightCalculation*, as shown in Figure 19-3.

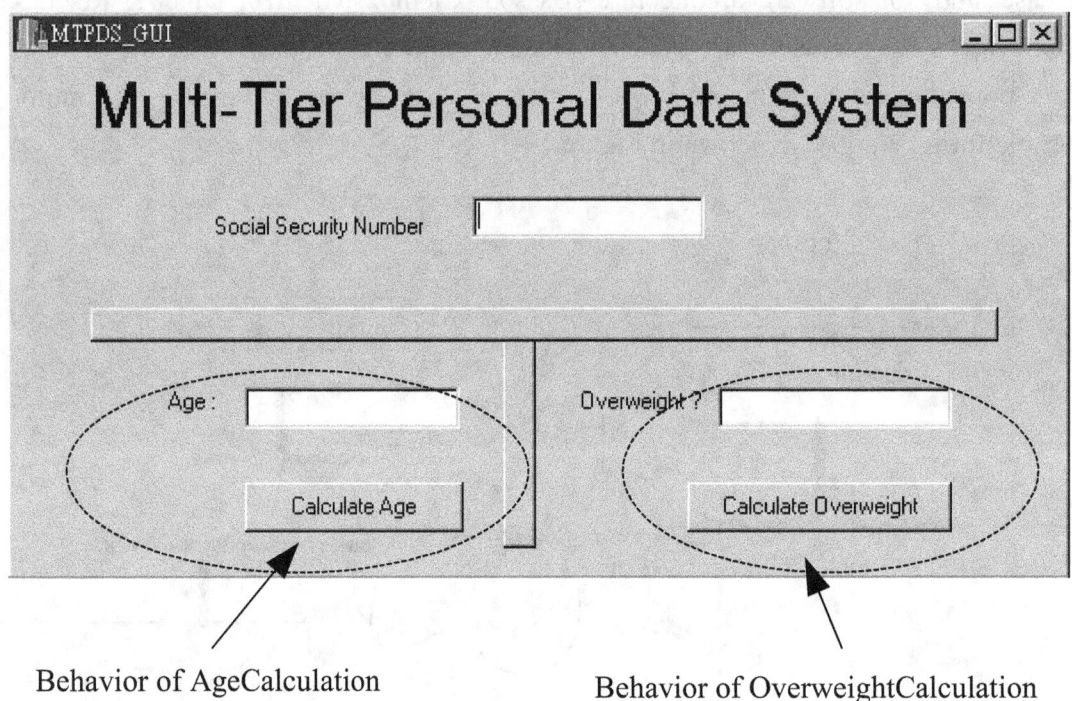

Figure 19-3   Two Behaviors

In the *AgeCalculation* behavior, actor *Pupil* inputs an integer *Social_Security_Number* value then presses down the *Calculate_Age* button. After that, the *Multi-Tier Personal Data System* retrieves the *Date_of_Birth* value from the database in line with the corresponding *Social_Security_Number* value. From the *Date_of_Birth* value, the *Multi-Tier Personal Data System* calculates the *Age* value and displays it on the screen. Figure 19-4 shows the *Social_Security_Number* value is 512-24-3722 and the retrieved *Date_of_Birth* value is May 12, 1954 and the calculated *Age* value, which is 60, is then displayed on the screen.

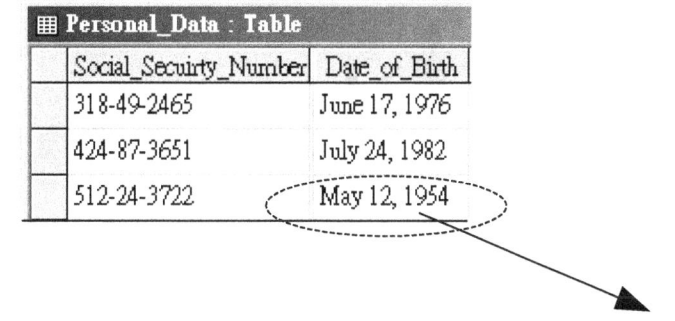

the calculated *Age* value when *Date_of_Birth* is May 12, 1954

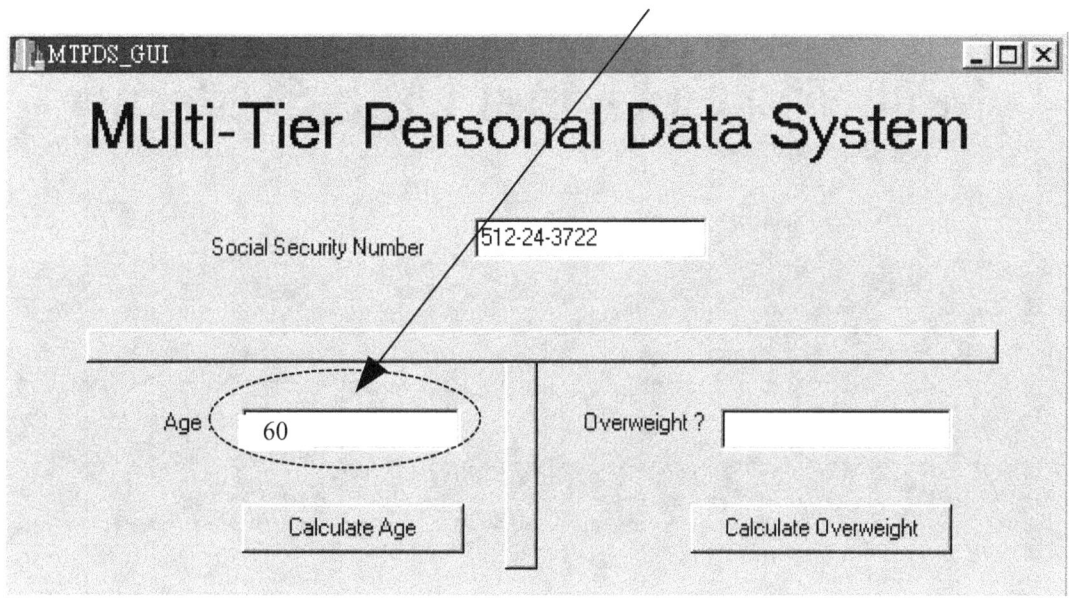

Figure 19-4   Behavior of *AgeCalculation*

In the *OverweightCalculation* behavior, actor *Pupil* inputs an integer *Social_Security_Number* value then presses down the *Calculate_Overweight* button. After that, the *Multi-Tier Personal Data System* retrieves the *Weight*, *Height* and *Sex* values from the database in line with the corresponding *Social_Security_Number* value. From the *Weight*, *Height* and *Sex* values, the *Multi-Tier Personal Data System* calculates the true-or-false *Overweight* value and displays it on the screen. Figure 19-5 shows the *Social_Security_Number* value is 318-49-2465 and the retrieved Sex, H*eight* and *Weight* values are Female, 165 and 51, respectively, the calculated *Overweight* value, which is *No*, is then displayed on the screen.

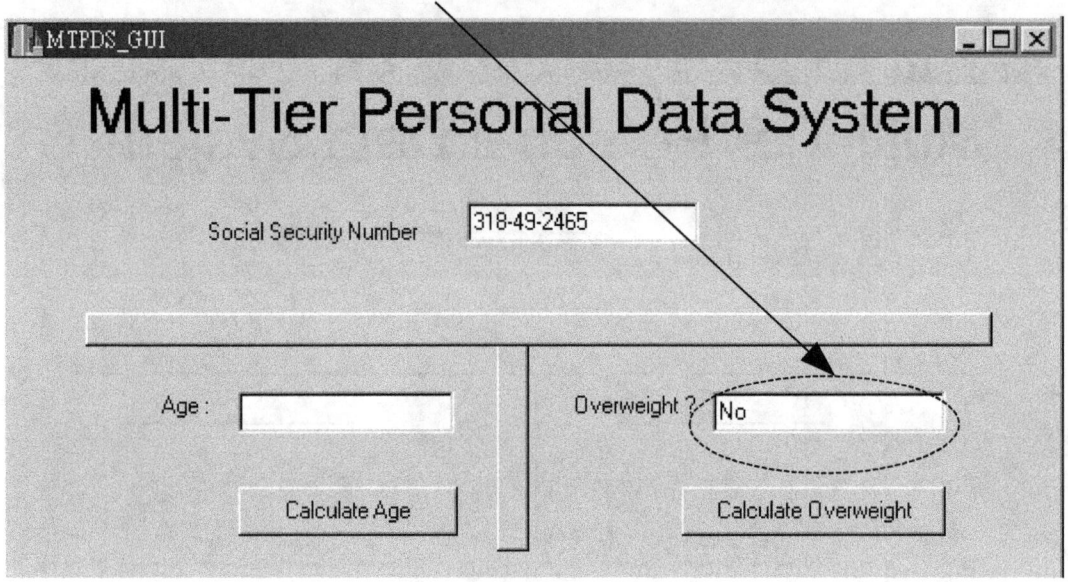

Figure 19-5   Behavior of *OverweightCalculation*

Using the SBC software multi-level (hierarchical) view an architect goes through: a) analysis view and b) design view for the *Multi-Tier Personal Data System* software systems architecture as shown in Figure 19-6.

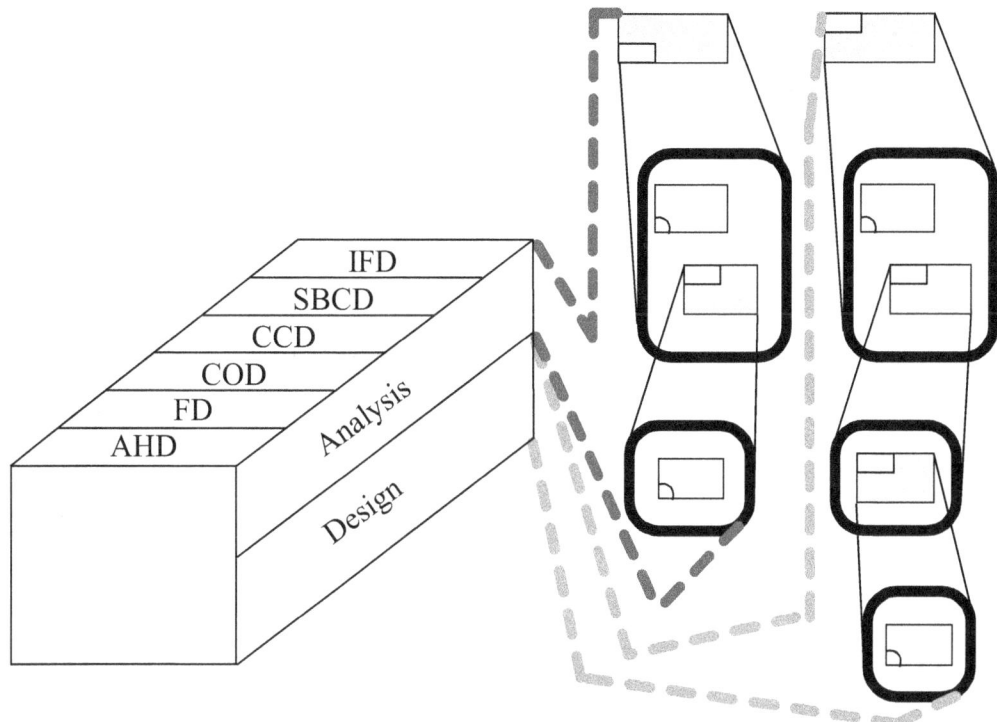

Figure 19-6 SBC Software Multi-Level (Hierarchical) View

19-1 Analysis View of the Multi-Tier Personal Data System

In the SBC software multi-level (hierarchical) view, an architect constructs the analysis' systems architecture for the analyzer to view. This analysis' systems architecture is called the analysis view as shown in Figure 19-7.

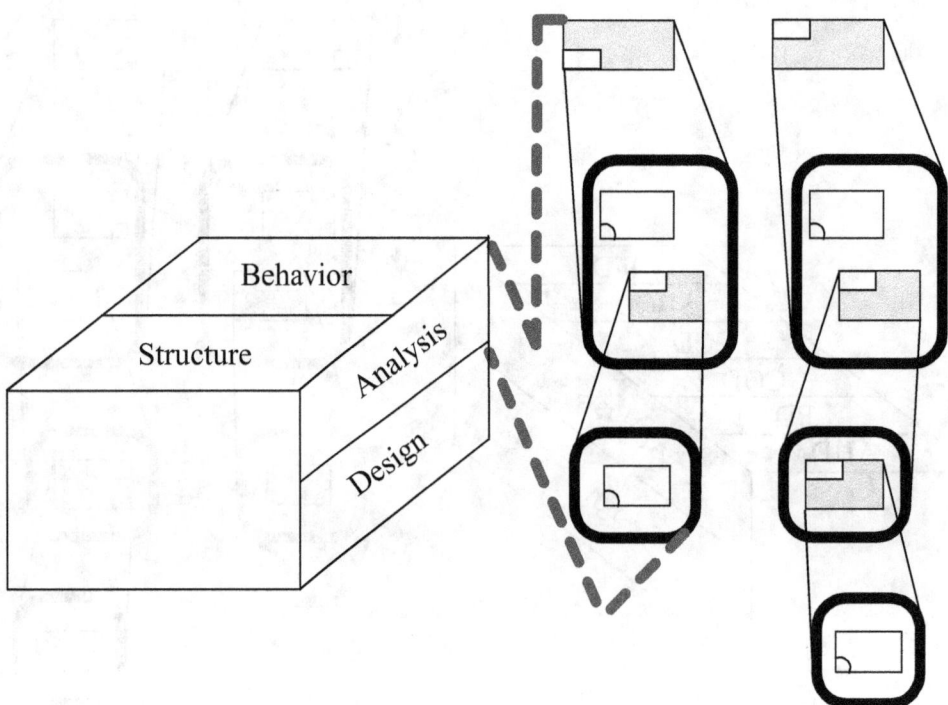

Figure 19-7 Analysis View

The analysis view consists of: a) analysis' systems structure and b) analysis' systems behavior.

19-1-1 Analysis' Systems Structure

The entire SBC analysis' systems structure includes: a) *Analysis' AHD*, b) *Analysis' FD*, c) *Analysis' COD* and d) *Analysis' CCD*.

We first draw the analysis' AHD of the *Multi-Tier Personal Data System*. As shown in Figure 19-8, *Multi-Tier Personal Data System* is composed of *MTPDS_GUI* and *M_Subsystem_2*.

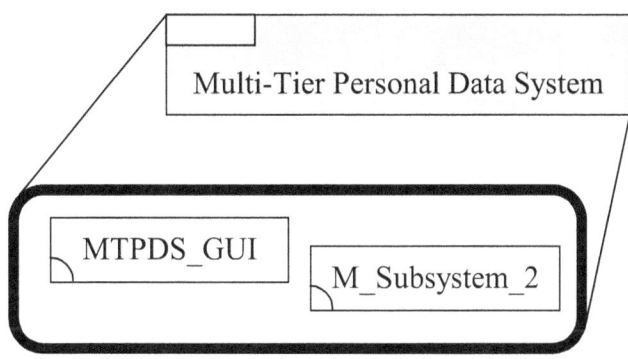

Figure 19-8　Analysis' AHD of the *Multi-Tier Personal Data System*

In Figure 19-8, *Multi-Tier Personal Data System* is an aggregated system while *MTPDS_GUI* and *M_Subsystem_2* are non-aggregated systems.

Analysis' FD is the framework diagram we obtain after the analysis phase is finished. Figure 19-9 shows the analysis' FD of the *Multi-Tier Personal Data System*. In the figure, *Presentation_Layer* contains the *MTPDS_GUI* and *M_Subsystem_2* components.

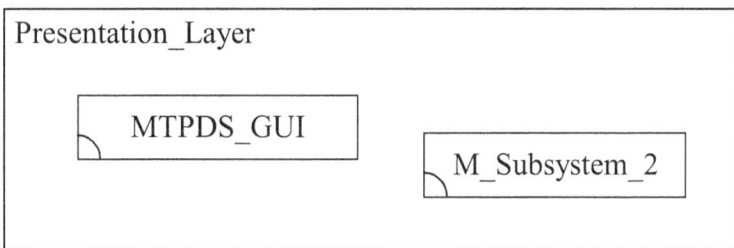

Figure 19-9　Analysis' FD of the *Multi-Tier Personal Data System*

Analysis' COD is the component operation diagram we obtain after the analysis phase is finished. Figure 19-10 shows the analysis' COD of the *Multi-Tier Personal Data System*. In the figure, component *MTPDS_GUI* has two operations: *Calculate_AgeClick* and *Calculate_OverweightClick*; component *M_Subsystem_2* has two operations: *Calculate_Age* and *Calculate_Overweight*.

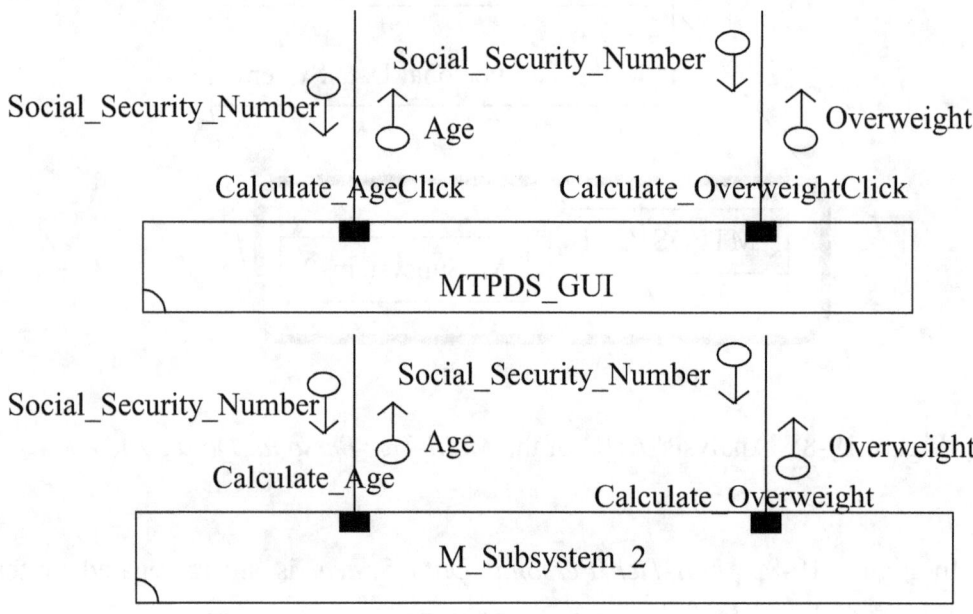

Figure 19-10 Analysis' COD of the *Multi-Tier Personal Data System*

The operation formula of *Calculate_AgeClick* is *Calculate_AgeClick(In Social_Security_Number; Out Age)*. The operation formula of *Calculate_OverweightClick* is *Calculate_OverweightClick(In Social_Security_Number; Out Overweight)*. The operation formula of *Calculate_Age* is *Calculate_Age(In Social_Security_Number; Out Age)*. The operation formula of *Calculate_Overweight* is *Calculate_Overweight(In Social_Security_Number; Out Overweight)*.

Figure 19-11 shows the primitive data type specification of the *Social_Security_Number* input parameter and the *Age*, *Overweight* output parameters.

Parameter	Data Type	Instances
Social_Security_Number	Text	424-87-3651, 512-24-3722
Age	Integer	28, 56
Overweight	Boolean	Yes, No

Figure 19-11 Primitive Data Type Specification

Analysis' CCD is the component connection diagram we obtain after the analysis phase is finished. Figure 19-12 shows the analysis' CCD of the *Multi-Tier Personal Data System*.

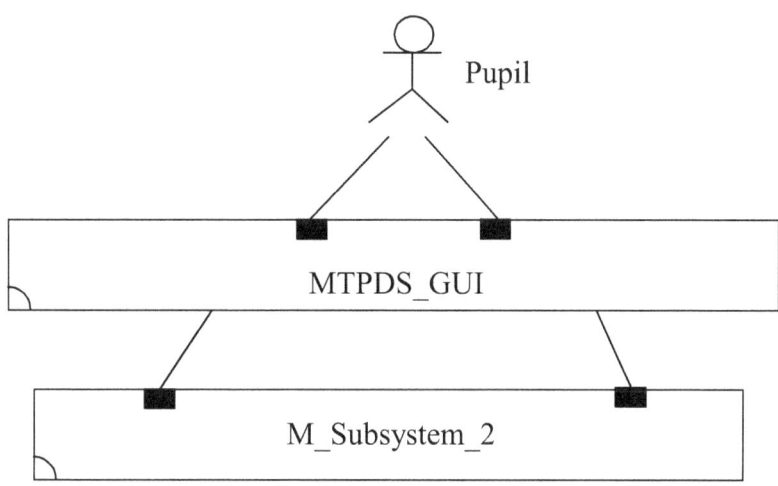

Figure 19-12   Analysis' CCD of the *Multi-Tier Personal Data System*

In Figure 19-12, actor *Pupil* has two connections with the *MTPDS_GUI* component; component *MTPDS_GUI* has two connections with the *M_Subsystem_2* component.

19-1-2 Analysis' Systems Behavior

The entire SBC analysis' systems behavior includes: a) *Analysis' SBCD* and b) *Analysis' IFD*.

Analysis' SBCD is the structure-behavior coalescence diagram we obtain after the analysis phase is finished. Figure 19-13 shows the analysis' SBCD of the *Multi-Tier Personal Data System* in which interactions among the *Pupil* actor and the *MTPDS_GUI*, *M_Subsystem_2* components shall draw forth the *AgeCalculation* and *OverweightCalculation* behaviors.

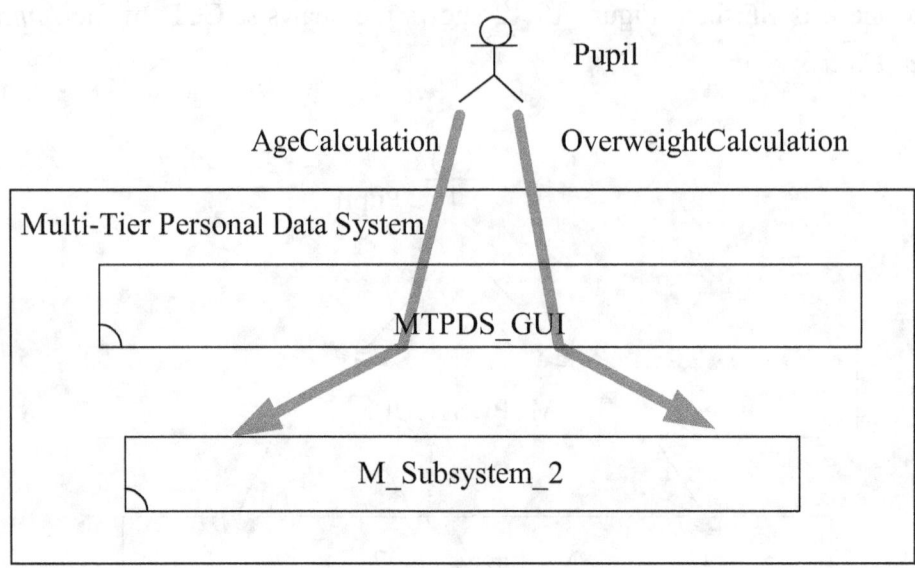

Figure 19-13   Analysis' SBCD of the *Multi-Tier Personal Data System*

The overall behavior of a software system is contained in its individual behaviors. After the analysis phase is finished, the overall analysis' behavior of the *Multi-Tier Personal Data System* includes the *AgeCalculation* and *OverweightCalculation* behaviors. In other words, the *AgeCalculation* and *OverweightCalculation* behaviors together provide the overall analysis' behavior of the *Multi-Tier Personal Data System* after the analysis phase is finished.

Be noticed that the *AgeCalculation* behavior and the *OverweightCalculation* behavior are mutually independent of each other. They shall be executed concurrently [Hoar85, Miln89, Miln99].

The major purpose of using the architectural approach, instead of separating the structure model from the behavior model, is to achieve a coalesced model. In Figure 19-13, we not only see its software structure, also see at the same time its software behavior in the analysis' SBCD of the *Multi-Tier Personal Data System*.

After the analysis phase is finished, the overall analysis' behavior of the *Multi-Tier Personal Data System* includes two behaviors: *AgeCalculation* and *OverweightCalculation*. Each of them is described by an individual IFD. Figure 19-14 shows the analysis' IFD of the *AgeCalculation* behavior. First, actor *Pupil* interacts with the *MTPDS_GUI* component through the *Calculate_AgeClick* operation call interaction, carrying the *Social_Security_Number* input parameter. Next, component *MTPDS_GUI* interacts with the *M_Subsystem_2* component through the

Calculate_Age operation call interaction, carrying the *Social_Security_Number* input parameter. Continuingly, component *MTPDS_GUI* interacts with the *M_Subsystem_2* component through the *Calculate_Age* operation return interaction, carrying the *Age* output parameter. Finally, actor *Pupil* interacts with the *MTPDS_GUI* component through the *Calculate_AgeClick* operation return interaction, carrying the *Age* output parameter.

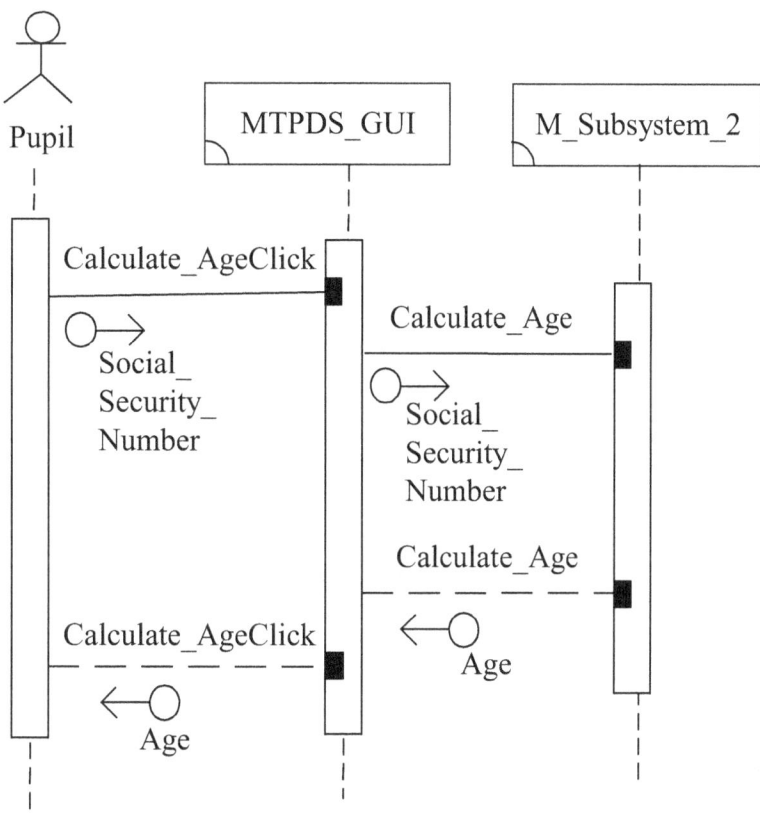

Figure 19-14   Analysis' IFD of the *AgeCalculation* Behavior

Figure 19-15 shows the analysis' IFD of the *OverweightCalculation* behavior. First, actor *Pupil* interacts with the *MTPDS_GUI* component through the *Calculate_OverweightClick* operation call interaction, carrying the *Social_Security_Number* input parameter. Next, component *MTPDS_GUI* interacts with the *M_Subsystem_2* component through the *Calculate_Overweight* operation call interaction, carrying the *Social_Security_Number* input parameter. Continuingly, component *MTPDS_GUI* interacts with the *M_Subsystem_2* component through the *Calculate_Overweight* operation return interaction, carrying the *Overweight* output parameter. Finally, actor *Pupil* interacts with the *MTPDS_GUI* component through the *Calculate_OverweightClick* operation return interaction, carrying the *Overweight* output parameter.

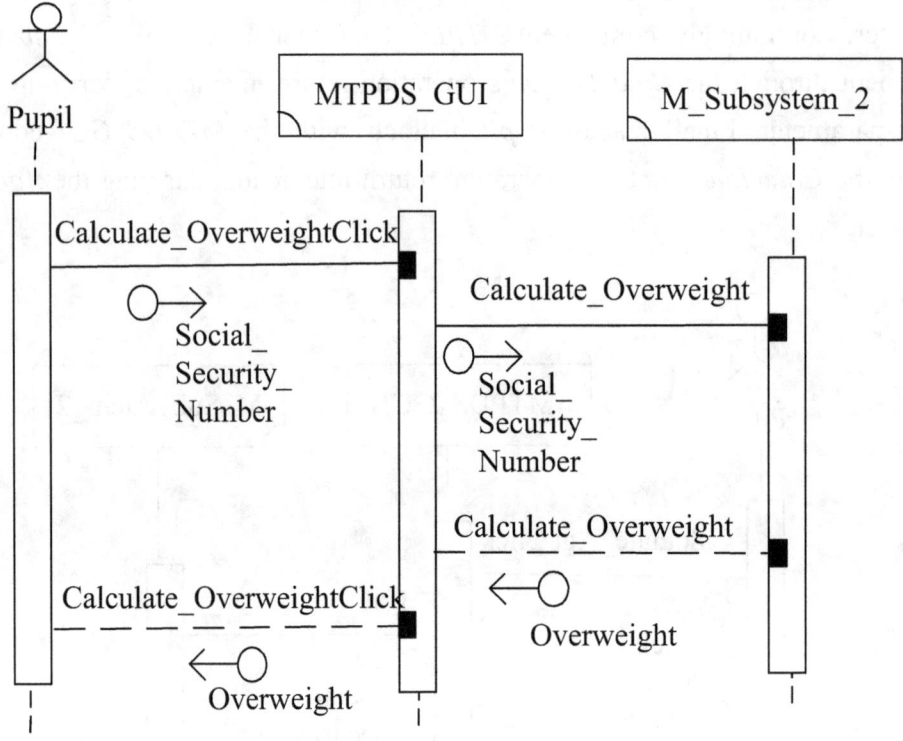

Figure 19-15   Analysis' IFD of the *OverweightCalculation* Behavior

19-2 Design View of the Multi-Tier Personal Data System

In the SBC software multi-level (hierarchical) view, an architect constructs the design's systems architecture for the designer to view. This design's systems architecture is called the design view as shown in Figure 19-16. Design view is one level down structural decomposition (with observation congruence verification) of the analysis view [Chao15a, Chao15b, Chao15c, Chao15d, Chao15e].

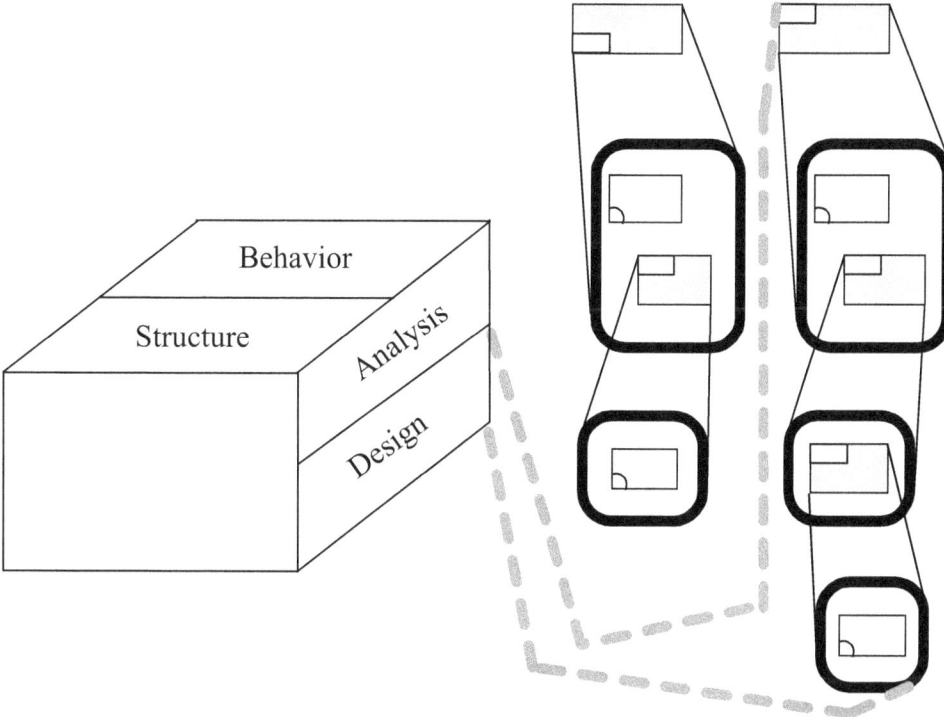

Figure 19-16   Design View

The design view consists of: a) design's systems structure and b) design's systems behavior.

19-2-1 Design's Systems Structure

The entire SBC design's systems structure includes: a) *Design's AHD*, b) *Design's FD*, c) *Design's COD* and d) *Design's CCD*.

We first draw the design's AHD of the *Multi-Tier Personal Data System*. As shown in Figure 19-17, *Multi-Tier Personal Data System* is composed of *MTPDS_GUI* and *M_Subsystem_2*; *M_Subsystem_2* is composed of *Age_Logic*, *Overweight_Logic* and *M_Subsystem_1*; *M_Subsystem_1* is composed of *Personal_Database*.

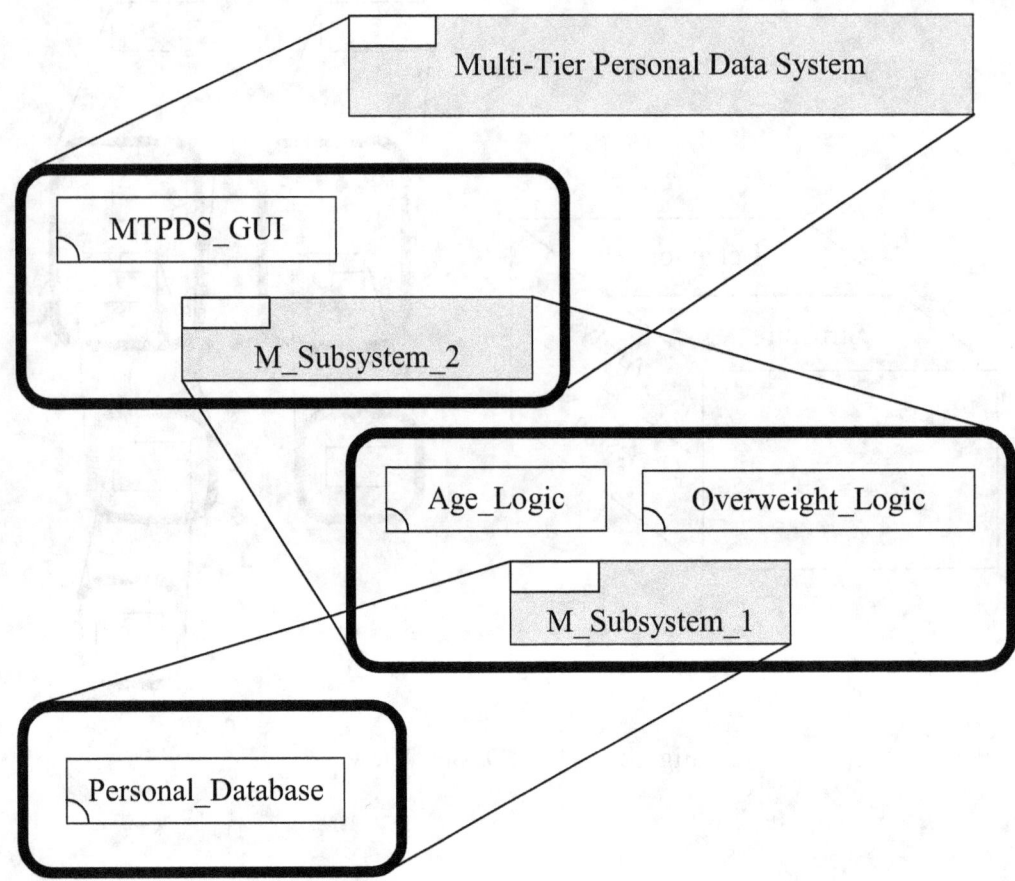

Figure 19-17   Design's AHD of the *Multi-Tier Personal Data System*

In Figure 19-17, *Multi-Tier Personal Data System*, *M_Subsystem_2* and *M_Subsystem_1* are aggregated systems while *MTPDS_GUI*, *Age_Logic*, *Overweight_Logic* and *Personal_Database* are non-aggregated systems.

Design's FD is the framework diagram we obtain after the design phase is finished. Figure 19-18 shows the design's FD of the *Multi-Tier Personal Data System*. In the figure, *Presentation_Layer* contains the *MTPDS_GUI* component; *Logic_Layer* contains the *Age_Logic* and *Overweight_Logic* components; *Data_Layer* contains the *Personal_Database* component.

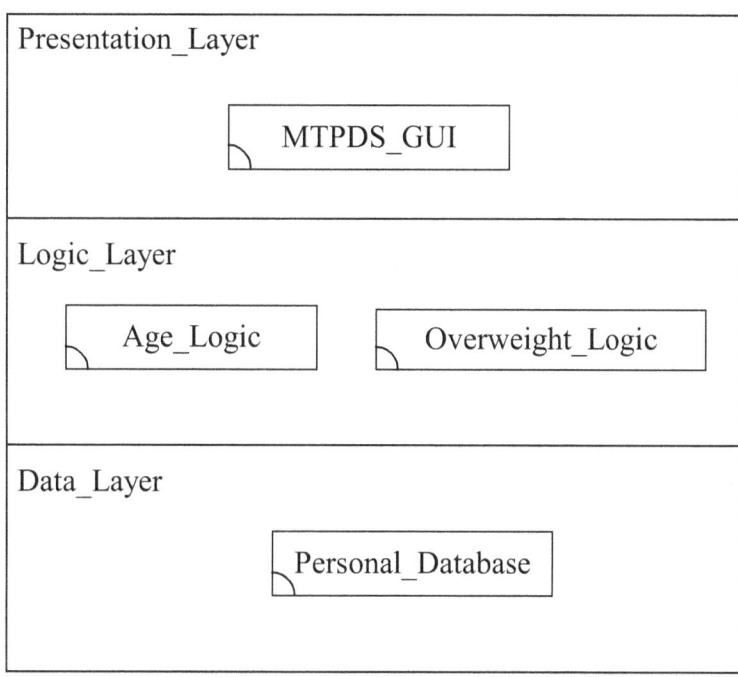

Figure 19-18   Design's FD of the *Multi-Tier Personal Data System*

Design's COD is the component operation diagram we obtain after the design phase is finished. Figure 19-19 shows the design's COD of the *Multi-Tier Personal Data System*. In the figure, component *MTPDS_GUI* has two operations: *Calculate_AgeClick* and *Calculate_OverweightClick*; component *Age_Logic* has one operation: *Calculate_Age*; component *Overweight_Logic* has one operation: *Calculate_Overweight*; component *Personal_Database* has two operations: *Sql_DateOfBirth_Select* and *Sql_SexHeightWeight_Select*.

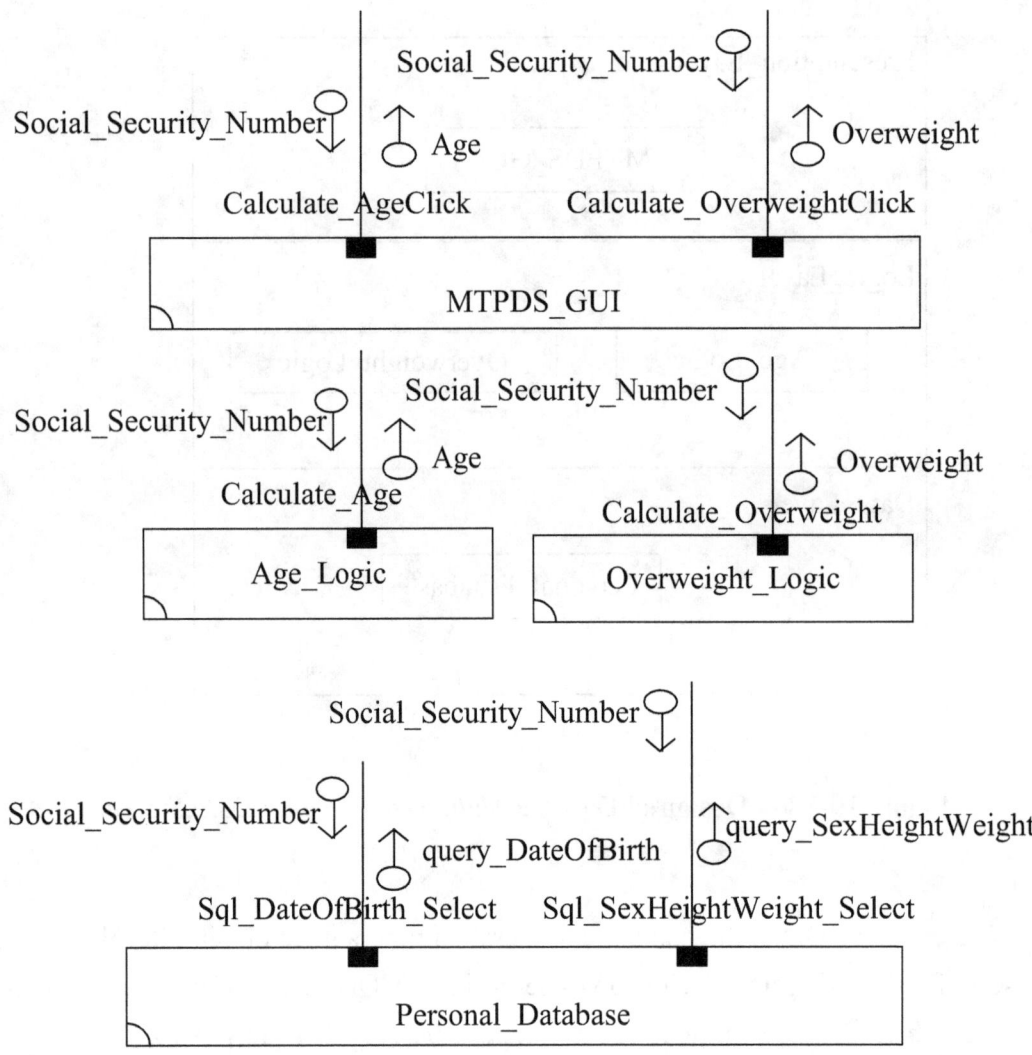

Figure 19-19   Design's COD of the *Multi-Tier Personal Data System*

The operation formula of *Calculate_AgeClick* is *Calculate_AgeClick(In Social_Security_Number; Out Age)*. The operation formula of *Calculate_OverweightClick* is *Calculate_OverweightClick(In Social_Security_Number; Out Overweight)*. The operation formula of *Calculate_Age* is *Calculate_Age(In Social_Security_Number; Out Age)*. The operation formula of *Calculate_Overweight* is *Calculate_Overweight(In Social_Security_Number; Out Overweight)*. The operation formula of *Sql_DateOfBirth_Select* is *Sql_DateOfBirth_Select(In Social_Security_Number; Out query_DateOfBirth)*. The operation formula of *Sql_SexHeightWeight_Select* is *Sql_SexHeightWeight_Select(In Social_Security_Number; Out query_SexHeightWeight)*.

Figure 19-20 shows the primitive data type specification of the *Social_Security_Number* input parameter and the *Age*, *Overweight* output parameters.

Parameter	Data Type	Instances
Social_Security_Number	Text	424-87-3651, 512-24-3722
Age	Integer	28, 56
Overweight	Boolean	Yes, No

Figure 19-20   Primitive Data Type Specification

Figure 19-21 shows the composite data type specification of the *query_DateOfBirth* output parameter occurring in the *Sql_DateOfBirth_Select(In Social_Security_Number; Out query_DateOfBirth)* operation formula.

Parameter	*query_DateOfBirth*

Data Type	TABLE of
	   Social_Security_Number : Text
	   Age : Integer
	 End TABLE;

Instances	424-87-3651 — 28 512-24-3722 — 56

Figure 19-21   Composite Data Type Specification

Figure 19-22 shows the composite data type specification of the *query_SexHeightWeight* output parameter occurring in the *Sql_SexHeightWeight_Select(In Social_Security_Number; Out query_SexHeightWeight)* operation formula.

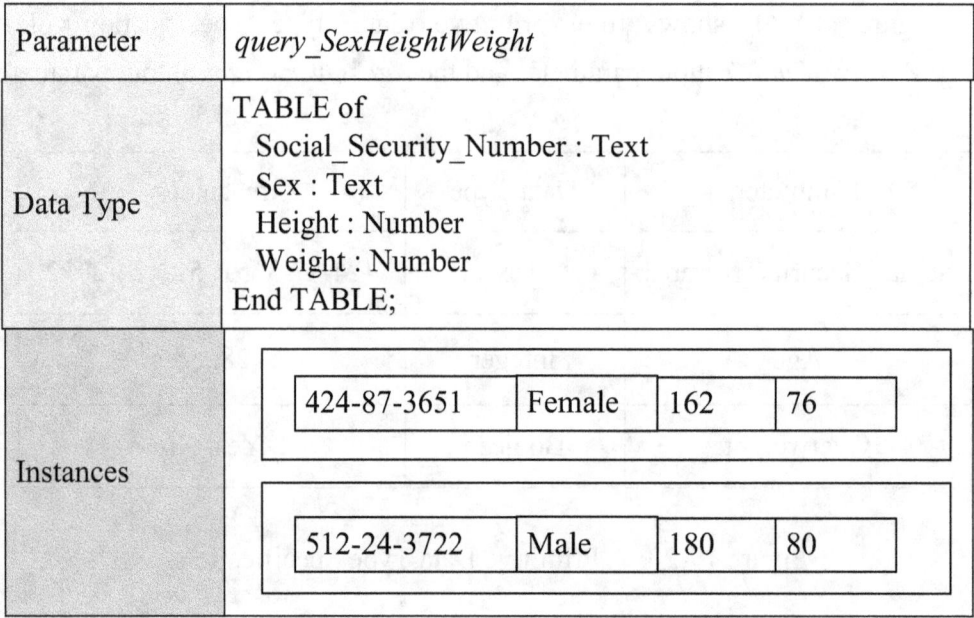

Figure 19-22   Composite Data Type Specification

Design's CCD is the component connection diagram we obtain after the design phase is finished. Figure 19-23 shows the design's CCD of the *Multi-Tier Personal Data System*.

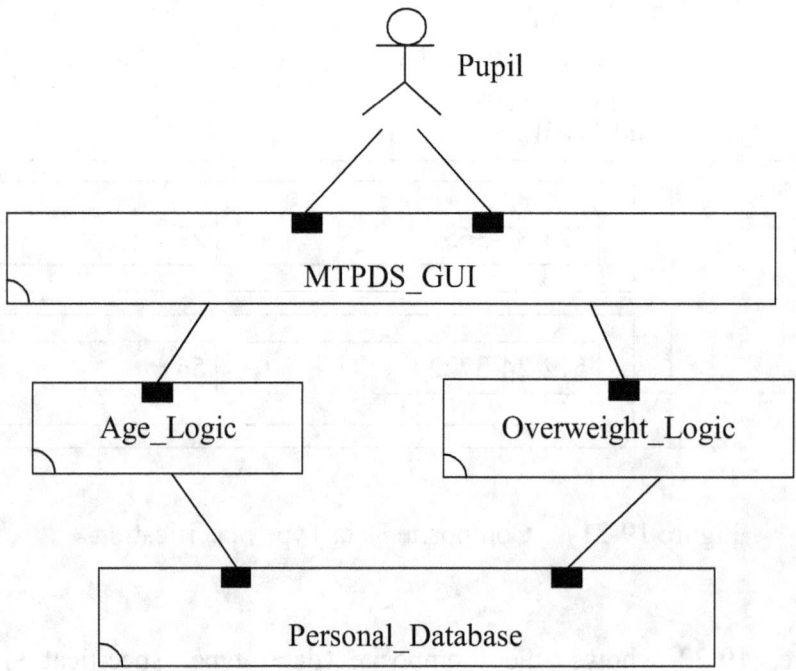

Figure 19-23   Design's CCD of the *Multi-Tier Personal Data System*

In Figure 19-23, actor *Pupil* has two connections with the *MTPDS_GUI* component; component *MTPDS_GUI* has one connection with each one of *Age_Logic* and *Overweight_Logic* component; component *Age_Logic* has a connection with the *Personal_Database* component; component *Overweight_Logic* has a connection with the *Personal_Database* component.

19-2-2 Design's Systems Behavior

The entire SBC design's systems behavior includes: a) *Design's SBCD* and b) *Design's IFD*.

Design's SBCD is the structure-behavior coalescence diagram we obtain after the design phase is finished. Figure 19-24 shows the design's SBCD of the *Multi-Tier Personal Data System* in which interactions among the *Pupil* actor and the *MTPDS_GUI*, *Age_Logic*, *Overweight_Logic*, *Personal_Database* components shall draw forth the *AgeCalculation* and *OverweightCalculation* behaviors.

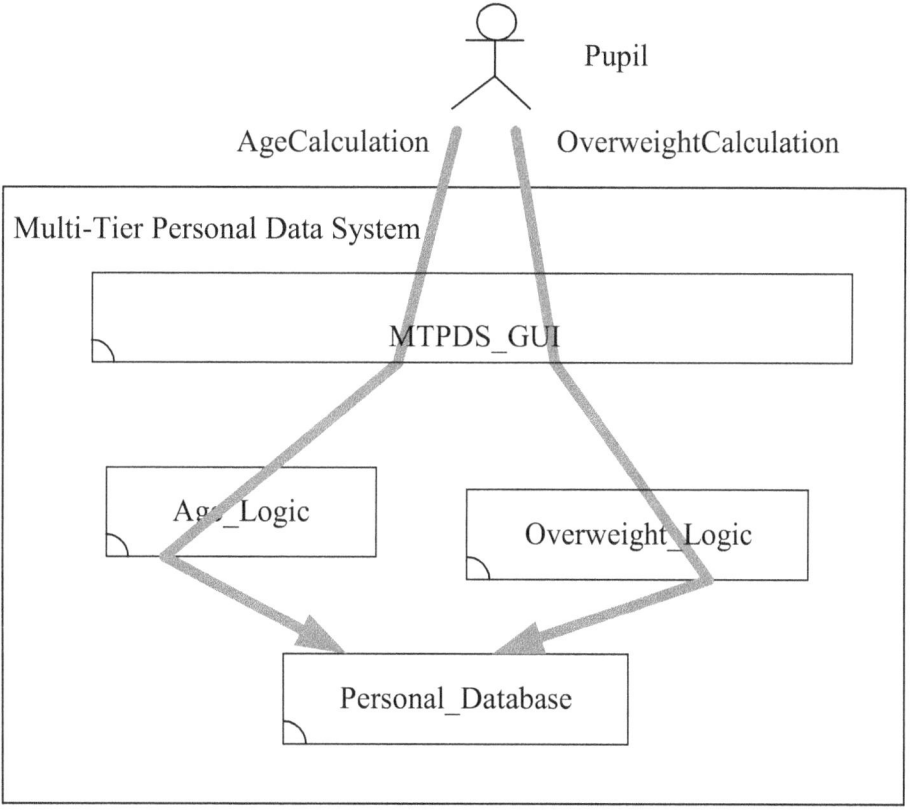

Figure 19-24   Design's SBCD of the *Multi-Tier Personal Data System*

The overall behavior of a software system is contained in its individual behaviors. After the design phase is finished, the overall design's behavior of the *Multi-Tier Personal Data System* includes the *AgeCalculation* and *OverweightCalculation* behaviors. In other words, the *AgeCalculation* and *OverweightCalculation* behaviors together provide the overall design's behavior of the *Multi-Tier Personal Data System* after the design phase is finished.

Be noticed that the *AgeCalculation* behavior and the *OverweightCalculation* behavior are mutually independent of each other. They shall be executed concurrently [Hoar85, Miln89, Miln99].

The major purpose of using the architectural approach, instead of separating the structure model from the behavior model, is to achieve a coalesced model. In Figure 19-24, we not only see its software structure, also see at the same time its software behavior in the design's SBCD of the *Multi-Tier Personal Data System*.

After the design phase is finished, the overall design's behavior of the *Multi-Tier Personal Data System* includes two behaviors: *AgeCalculation* and *OverweightCalculation*. Each of them is described by an individual IFD. Figure 19-25 shows the design's IFD of the *AgeCalculation* behavior. First, actor *Pupil* interacts with the *MTPDS_GUI* component through the *Calculate_AgeClick* operation call interaction, carrying the *Social_Security_Number* input parameter. Next, component *MTPDS_GUI* interacts with the *Age_Logic* component through the *Calculate_Age* operation call interaction, carrying the *Social_Security_Number* input parameter. Continuingly, component *Age_Logic* interacts with the *Personal_Database* component through the *Sql_DateOfBirth_Select* operation call interaction, carrying the *Social_Security_Number* input parameter and the *query_DateOfBirth* output parameter. Repeatedly, component *MTPDS_GUI* interacts with the *Age_Logic* component through the *Calculate_Age* operation return interaction, carrying the *Age* output parameter. Finally, actor *Pupil* interacts with the *MTPDS_GUI* component through the *Calculate_AgeClick* operation return interaction, carrying the *Age* output parameter.

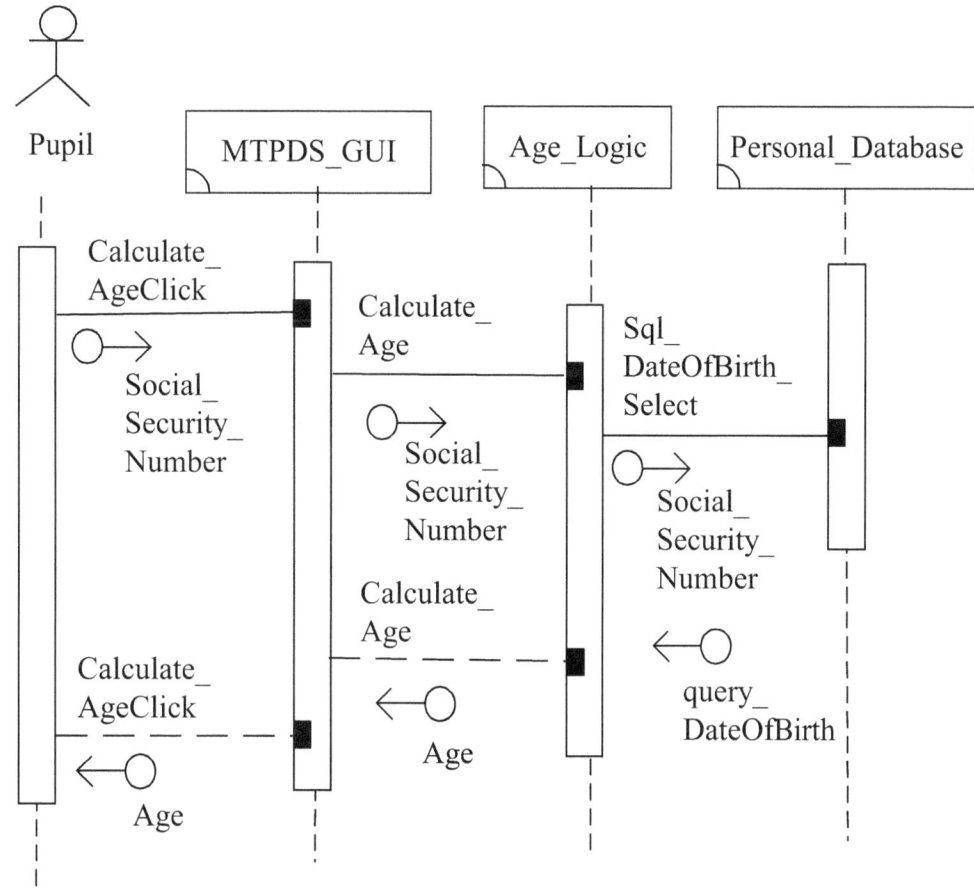

Figure 19-25   Design's IFD of the *AgeCalculation* Behavior

Figure 19-26 shows the design's IFD of the *OverweightCalculation* behavior. First, actor *Pupil* interacts with the *MTPDS_GUI* component through the *Calculate_OverweightClick* operation call interaction, carrying the *Social_Security_Number* input parameter. Next, component *MTPDS_GUI* interacts with the *Overweight_Logic* component through the *Calculate_Overweight* operation call interaction, carrying the *Social_Security_Number* input parameter. Continuingly, component *Overweight_Logic* interacts with the *Personal_Database* component through the *Sql_SexHeightWeight_Select* operation call interaction, carrying the *Social_Security_Number* input parameter and the *query_SexHeightWeight* output parameter. Repeatedly, component *MTPDS_GUI* interacts with the *Overweight_Logic* component through the *Calculate_Overweight* operation return interaction, carrying the *Overweight* output parameter. Finally, actor *Pupil* interacts with the *MTPDS_GUI* component through the *Calculate_OverweightClick* operation return interaction, carrying the *Overweight* output parameter.

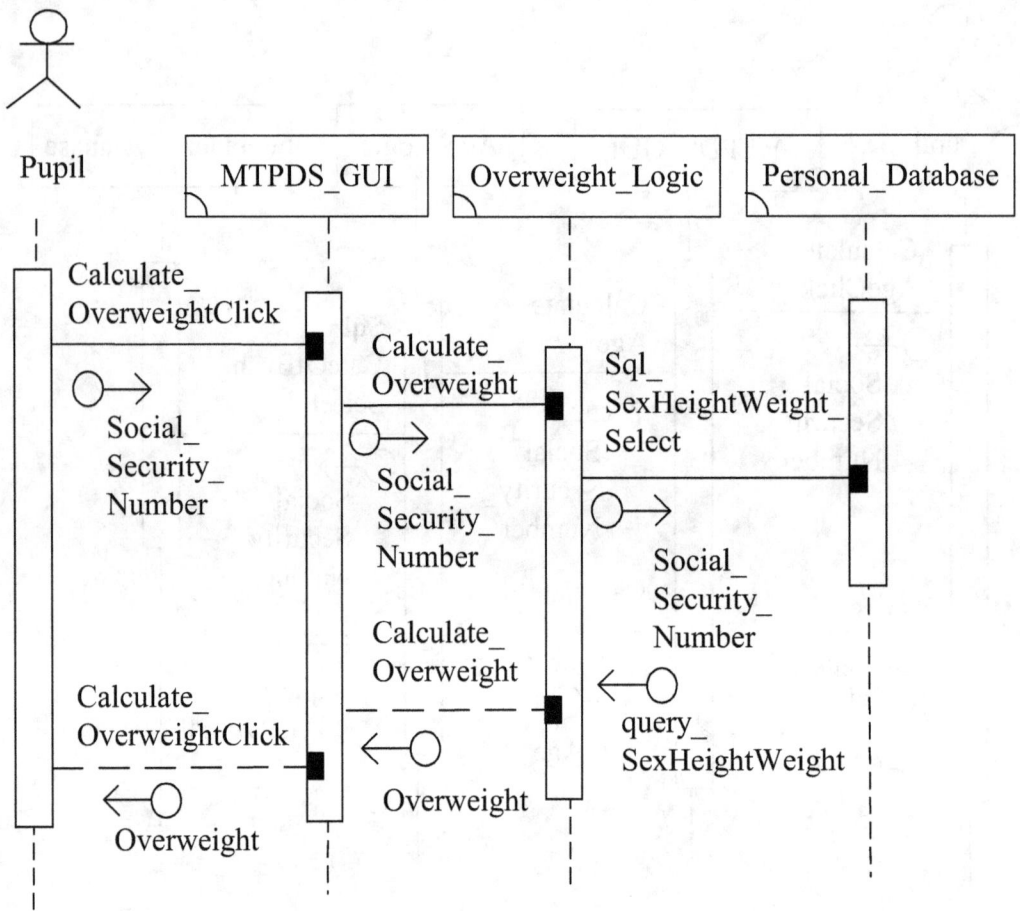

Figure 19-26   Design's IFD of the *OverweightCalculation* Behavior

Chapter 20: Stanford University -- Enterprise Architecture

This chapter examines the *Stanford University* which represents a case study of enterprise architecture [Bern05, Mino08, O'Rou03, Sche06, Sche08] for enterprise systems modeling and architecting. The curriculum of the *Stanford University* contains a set of courses: calculus, algebra, mechanics, quantum, database, networking, accounting and economics. A student may take any course offered by the *Stanford University*. The overall behavior of the *Stanford University* is majestically represented by eight behaviors: S*tudy_Calculus_Course*, S*tudy_Algebra_Course*, S*tudy_Mechanics_Course*, S*tudy_Quantum_Course*, S*tudy_Database_Course*, S*tudy_Networking_Course*, S*tudy_Economics_Course* and S*tudy_Accounting_Course* as shown in Figure 20-1.

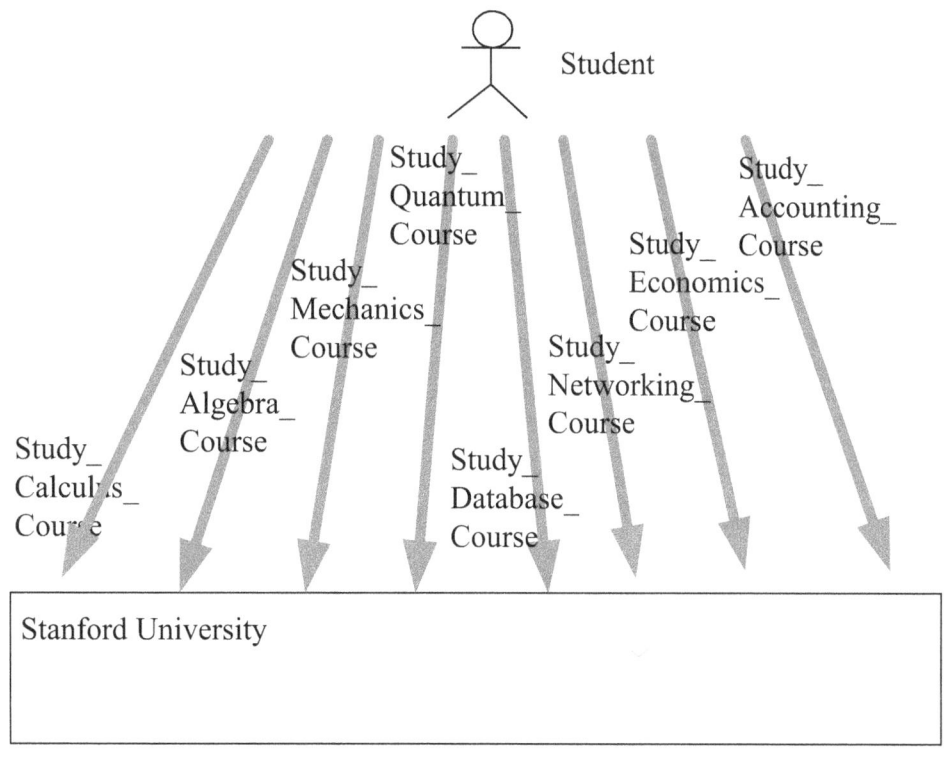

Figure 20-1   Eight Behaviors of the *Stanford University*

Using the SBC enterprise multi-level (hierarchical) view, an architect goes through: a) concept view, b) analysis view, c) design view and d) implementation view for the *Stanford University* enterprise systems architecture as shown in Figure 20-2.

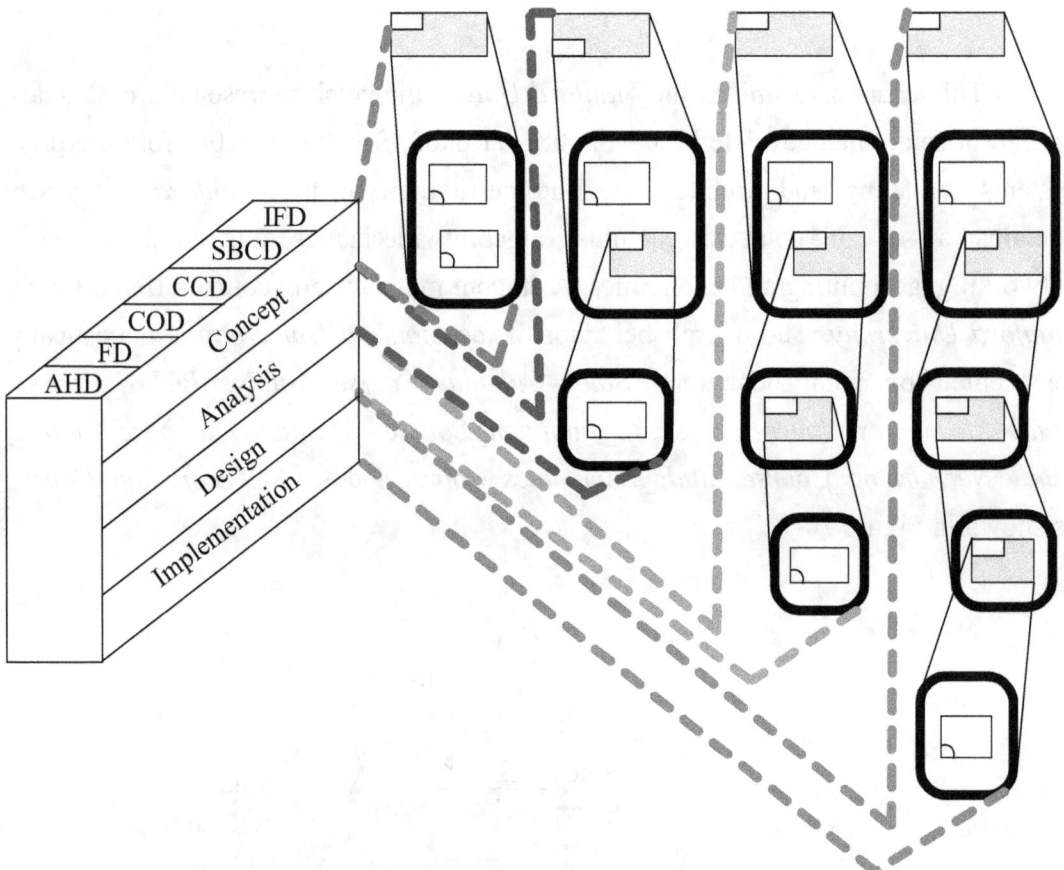

Figure 20-2   SBC Enterprise Multi-Level (Hierarchical) View

20-1 Concept View of the Stanford University

In the SBC enterprise multi-level (hierarchical) view, an architect constructs the concept's systems architecture for an administrator to view. This concept's systems architecture is called the concept view as shown in Figure 20-3.

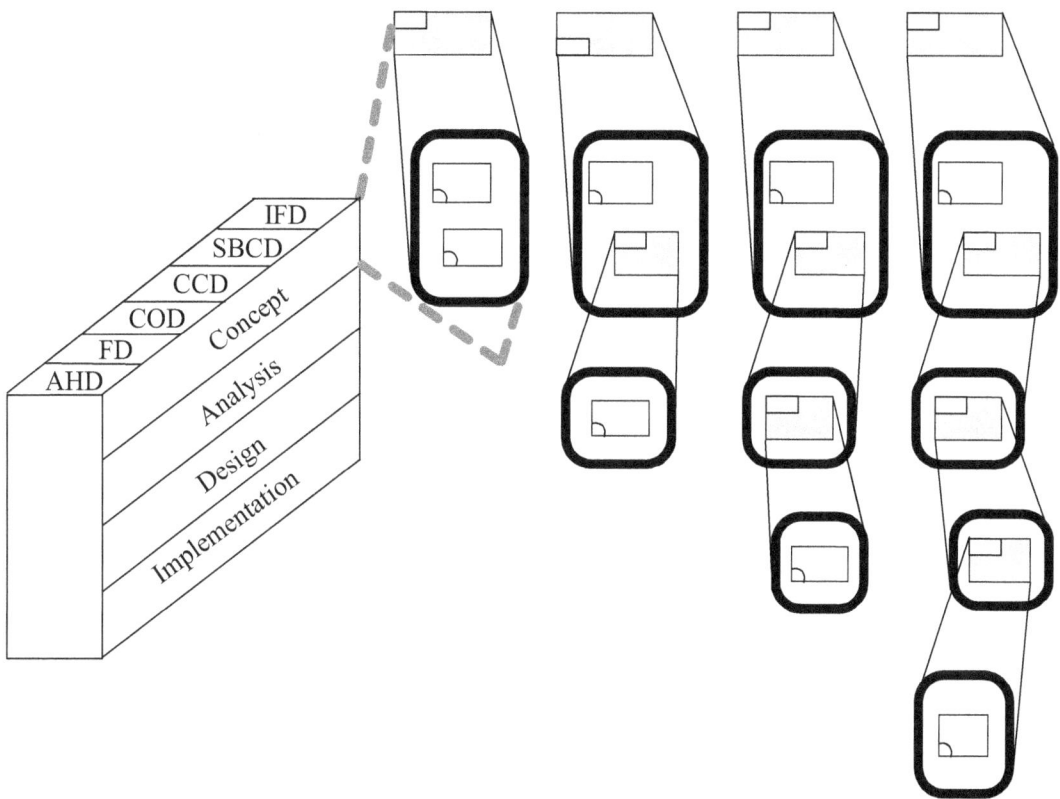

Figure 20-3   Concept View

The concept view consists of: a) concept's systems structure and b) concept's systems behavior.

20-1-1 Concept's Systems Structure

The entire SBC concept's systems structure includes: a) *Concept's AHD*, b) *Concept's FD*, c) *Concept's COD* and d) *Concept's CCD*.

We first draw the concept's AHD of the *Stanford University*. As shown in Figure 20-4, *Stanford University* is composed of *University_President*, *Science_College* and *Management_College*.

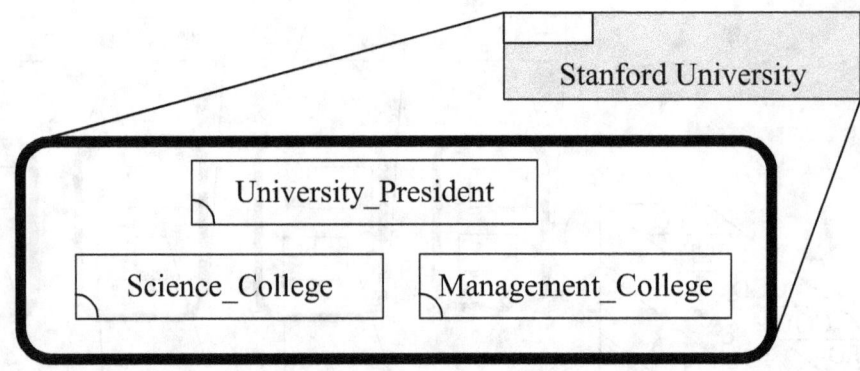

Figure 20-4   Concept's AHD of the *Stanford University*

In Figure 20-4, *Stanford University* is an aggregated system while *Science_College* and *Management_College* are non-aggregated systems.

Concept's FD is the framework diagram we obtain after the concept phase is finished. Figure 20-5 shows the concept's FD of the *Stanford University*. In the figure, *University_Layer* contains the *University_President*, *Science_College* and *Management_College* components.

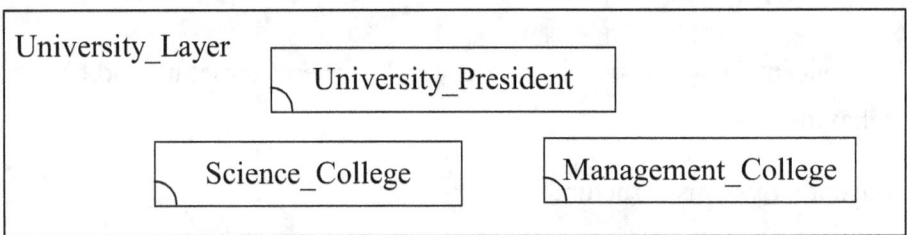

Figure 20-5   Cocnep's FD of the *Stanford University*

Concept's COD is the component operation diagram we obtain after the concept phase is finished. Figure 20-6 shows the concept's COD of the *Stanford University*. In the figure, component *University_President* has eight operations: *University_Teach_Calculus*, *University_Teach_Algebra*, *University_Teach_Mechanics*, *University_Teach_Quantum*, *University_Teach_Database*, *University_Teach_Networking*, *University_Teach_Economics* and *University_Teach_Accounting*; component *Science_College* has four operations: *College_Teach_Calculus*, *College_Teach_Algebra*, *College_Teach_Mechanics* and *College_Teach_Quantum*; component *Management_College* has four operations: *College_Teach_Database*,

College_Teach_Networking, *College_Teach_Economics* and *College_Teach_Accounting*.

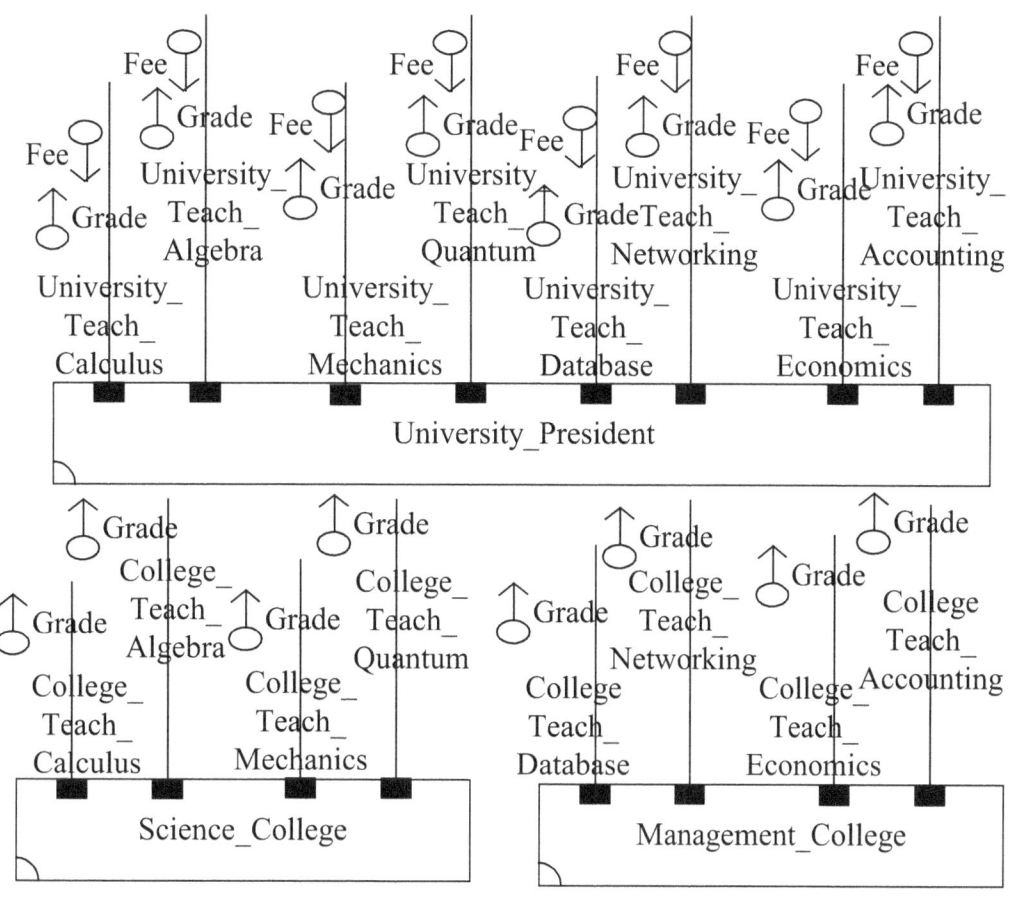

Figure 20-6   Concept's COD of the *Stanford University*

The operation formula of *University_Teach_Calculus* is *University_Teach_Calculus(In Fee; Out Grade)*. The operation formula of *University_Teach_Algebra* is *University_Teach_Algebra(In Fee; Out Grade)*. The operation formula of *University_Teach_Mechanics* is *University_Teach_Mechanics(In Fee; Out Grade)*. The operation formula of *University_Teach_Quantum* is *University_Teach_Quantum(In Fee; Out Grade)*. The operation formula of *University_Teach_Database* is *University_Teach_Database(In Fee; Out Grade)*. The operation formula of *University_Teach_Networking* is *University_Teach_Networking(In Fee; Out Grade)*. The operation formula of *University_Teach_Economics* is *University_Teach_Economics(In Fee; Out Grade)*. The operation formula of *University_Teach_Accounting* is *University_Teach_Accounting(In Fee; Out Grade)*.

The operation formula of *College_Teach_Calculus* is *College_Teach_Calculus(Out Grade)*. The operation formula of *College_Teach_Algebra* is *College_Teach_Algebra(Out Grade)*. The operation formula of *College_Teach_Mechanics* is *College_Teach_Mechanics(Out Grade)*. The operation formula of *College_Teach_Quantum* is *College_Teach_Quantum(Out Grade)*. The operation formula of *College_Teach_Database* is *College_Teach_Database(Out Grade)*. The operation formula of *College_Teach_Networking* is *College_Teach_Networking(Out Grade)*. The operation formula of *College_Teach_Economics* is *College_Teach_Economics(Out Grade)*. The operation formula of *College_Teach_Accounting* is *College_Teach_Accounting(Out Grade)*.

Figure 20-7 shows the primitive data type specification of the *Fee* input parameter and the *Grade* output parameter.

Parameter	Data Type	Instances
Fee	Intger	2000, 3500
Grade	Text	A+, B+, C-

Figure 20-7   Primitive Data Type Specification

Concept's CCD is the component connection diagram we obtain after the concept phase is finished. Figure 20-8 shows the concept's CCD of the *Stanford University*.

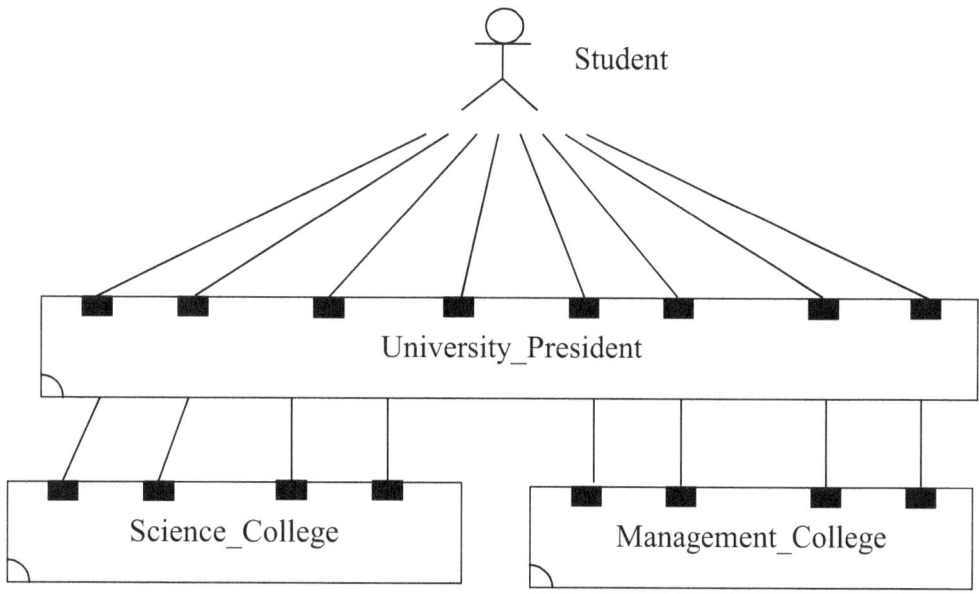

Figure 20-8   Concept's CCD of the *Stanford University*

In Figure 20-8, actor *Student* has eight connections with the *University_President* component; component *University_President* has four connections with each one of the *Science_College* and *Management_College* components.

20-1-2 Concept's Systems Behavior

The entire SBC concept's systems behavior includes: a) *Concept's SBCD* and b) *Concept's IFD*.

Concept's SBCD is the structure-behavior coalescence diagram we obtain after the concept phase is finished. Figure 20-9 shows the concept's SBCD of the *Stanford University* in which interactions among the *Student* actor and the *University_President*, *Science_College*, *Management_College* components shall draw forth the S*tudy_Calculus_Course*, S*tudy_Algebra_Course*, S*tudy_Mechanics_Course*, S*tudy_Quantum_Course*, S*tudy_Database_Course*, S*tudy_Networking_Course*, S*tudy_Economics_Course* and S*tudy_Accounting_Course* behaviors.

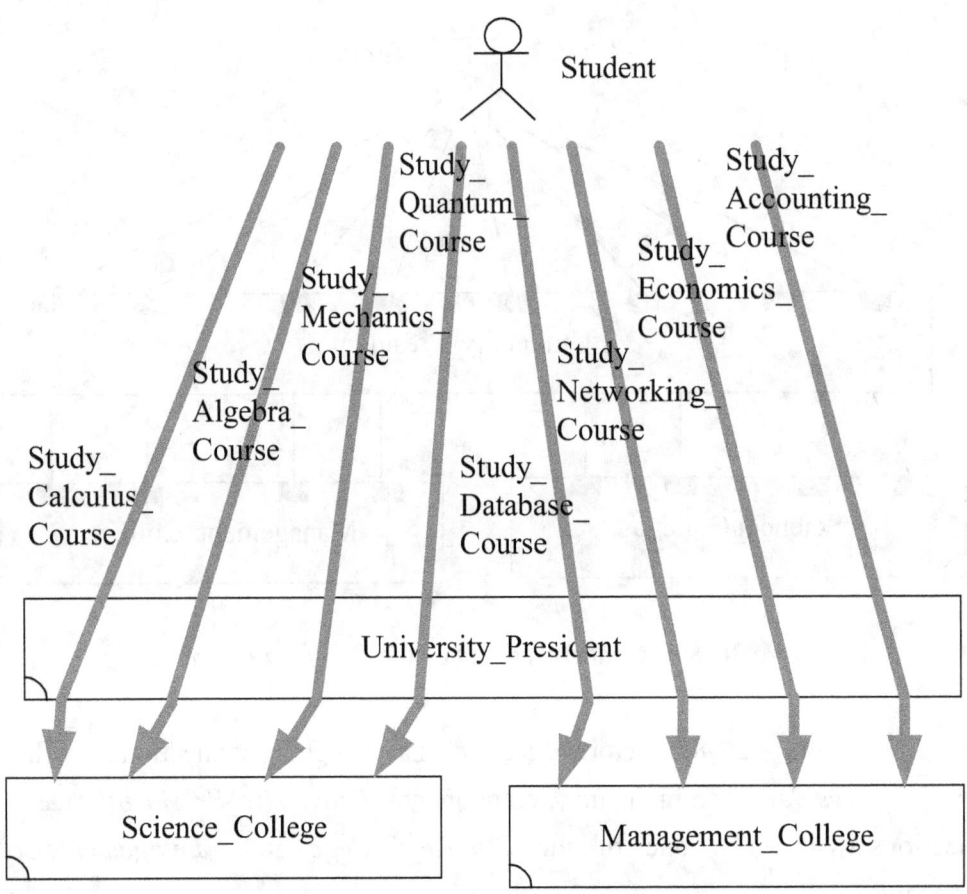

Figure 20-9   Concept's SBCD of the *Stanford University*

The overall behavior of an enterprise system is contained in its individual behaviors. After the concept phase is finished, the overall concept's behavior of the *Stanford University* includes the S*tudy_Calculus_Course*, S*tudy_Algebra_Course*, S*tudy_Mechanics_Course*, S*tudy_Quantum_Course*, S*tudy_Database_Course*, S*tudy_Networking_Course*, S*tudy_Economics_Course* and S*tudy_Accounting_Course* behaviors. In other words, the S*tudy_Calculus_Course*, S*tudy_Algebra_Course*, S*tudy_Mechanics_Course*, S*tudy_Quantum_Course*, S*tudy_Database_Course*, S*tudy_Networking_Course*, S*tudy_Economics_Course* and S*tudy_Accounting_Course* behaviors together provide the overall concept's behavior of the *Stanford University* after the concept phase is finished.

Be noticed that the S*tudy_Calculus_Course*, S*tudy_Algebra_Course*, S*tudy_Mechanics_Course*, S*tudy_Quantum_Course*, S*tudy_Database_Course*, S*tudy_Networking_Course*, S*tudy_Economics_Course* and S*tudy_Accounting_Course* behaviors are mutually independent of each other. They shall be executed concurrently [Hoar85, Miln89, Miln99].

The major purpose of using the architectural approach, instead of separating the structure model from the behavior model, is to achieve a coalesced model. In Figure 20-9, we not only see its organizational structure, also see at the same time its organizational behavior in the concept's SBCD of the *Stanford University*.

After the concept phase is finished, the overall concept's behavior of the *Stanford University* includes eight behaviors: S*tudy_Calculus_Course*, S*tudy_Algebra_Course*, S*tudy_Mechanics_Course*, S*tudy_Quantum_Course*, S*tudy_Database_Course*, S*tudy_Networking_Course*, S*tudy_Economics_Course* and S*tudy_Accounting_Course*. Each of them is described by an individual IFD. Figure 20-10 shows the concept's IFD of the S*tudy_Calculus_Course* behavior. First, actor *Student* interacts with the *University_President* component through the *University_Teach_Calculus* operation call interaction, carrying the *Fee* input parameter. Next, component *University_President* interacts with the *Science_College* component through the *College_Teach_Calculus* operation call interaction, carrying the *Grade* output parameter. Finally, actor *Student* interacts with the *University_President* component through the *University_Teach_Calculus* operation return interaction, carrying the *Grade* output parameter.

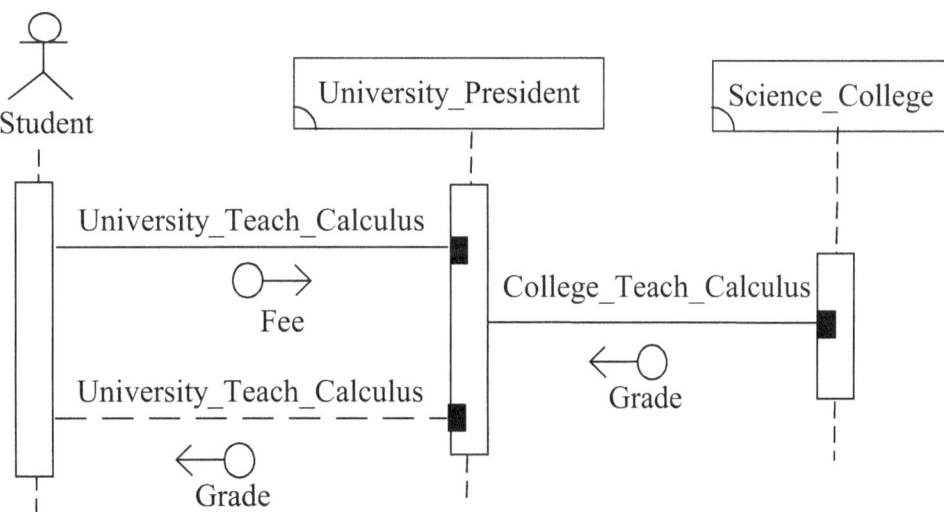

Figure 20-10   Concept's IFD of the *Study_Calculus_Course* Behavior

Figure 20-11 shows the concept's IFD of the *Study_Algebra_Course* behavior. First, actor *Student* interacts with the *University_President* component through the *University_Teach_Algebra* operation call interaction, carrying the *Fee* input parameter. Next, component *University_President* interacts with the *Science_College* component through the *College_Teach_Algebra* operation call interaction, carrying the *Grade* output parameter. Finally, actor *Student* interacts with the *University_President* component through the *University_Teach_Algebra* operation return interaction, carrying the *Grade* output parameter.

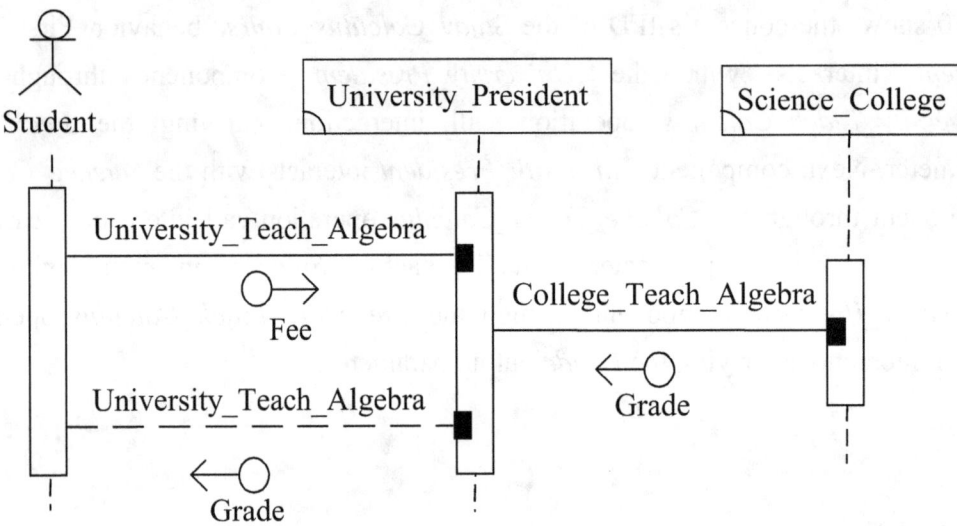

Figure 20-11   Concept's IFD of the *Study_Algebra_Course* Behavior

Figure 20-12 shows the concept's IFD of the *Study_Mechanics_Course* behavior. First, actor *Student* interacts with the *University_President* component through the *University_Teach_Mechanics* operation call interaction, carrying the *Fee* input parameter. Next, component *University_President* interacts with the *Science_College* component through the *College_Teach_Mechanics* operation call interaction, carrying the *Grade* output parameter. Finally, actor *Student* interacts with the *University_President* component through the *University_Teach_Mechanics* operation return interaction, carrying the *Grade* output parameter.

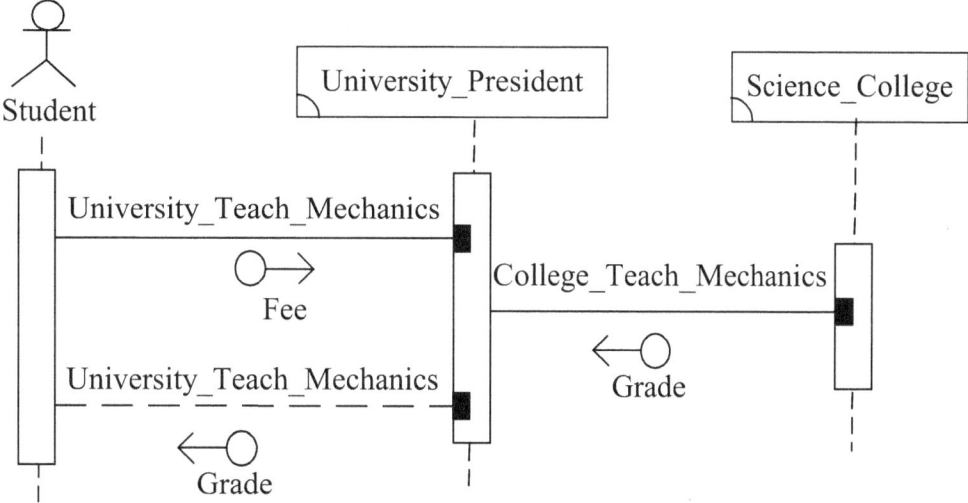

Figure 20-12　　Concept's IFD of the *Study_Mechanics_Course* Behavior

Figure 20-13 shows the concept's IFD of the S*tudy_Quantum_Course* behavior. First, actor *Student* interacts with the *University_President* component through the *University_Teach_Quantum* operation call interaction, carrying the *Fee* input parameter. Next, component *University_President* interacts with the *Science_College* component through the *College_Teach_Quantum* operation call interaction, carrying the *Grade* output parameter. Finally, actor *Student* interacts with the *University_President* component through the *University_Teach_Quantum* operation return interaction, carrying the *Grade* output parameter.

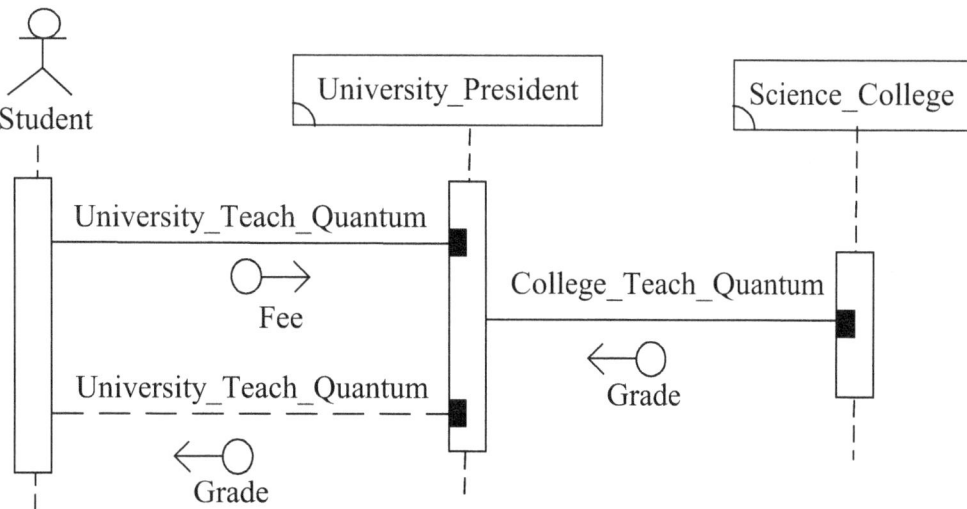

Figure 20-13　　Concept's IFD of the *Study_Quantum_Course* Behavior

Figure 20-14 shows the concept's IFD of the *Study_Database_Course* behavior. First, actor *Student* interacts with the *University_President* component through the *University_Teach_Database* operation call interaction, carrying the *Fee* input parameter. Next, component *University_President* interacts with the *Management_College* component through the *College_Teach_Database* operation call interaction, carrying the *Grade* output parameter. Finally, actor *Student* interacts with the *University_President* component through the *University_Teach_Database* operation return interaction, carrying the *Grade* output parameter.

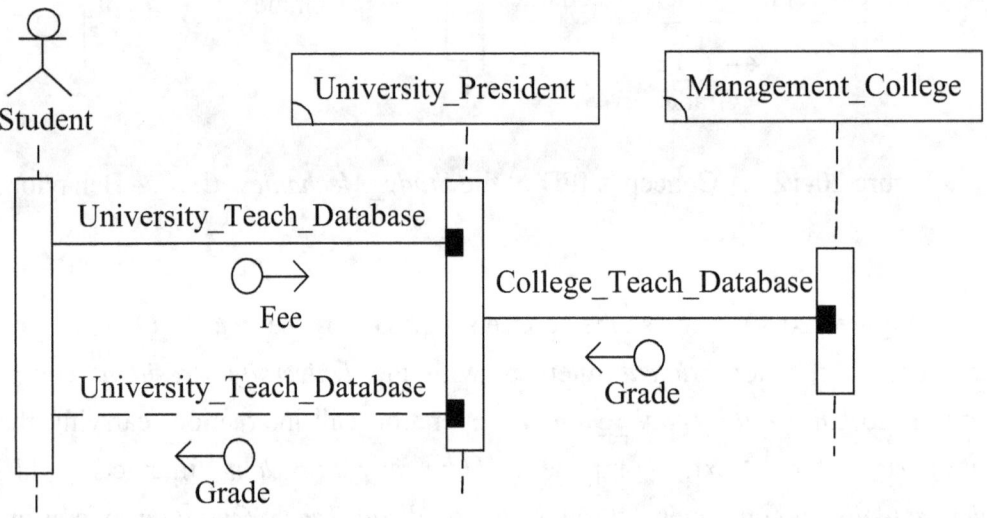

Figure 20-14   Concept's IFD of the *Study_Database_Course* Behavior

Figure 20-15 shows the concept's IFD of the *Study_Networking_Course* behavior. First, actor *Student* interacts with the *University_President* component through the *University_Teach_Networking* operation call interaction, carrying the *Fee* input parameter. Next, component *University_President* interacts with the *Management_College* component through the *College_Teach_Networking* operation call interaction, carrying the *Grade* output parameter. Finally, actor *Student* interacts with the *University_President* component through the *University_Teach_Networking* operation return interaction, carrying the *Grade* output parameter.

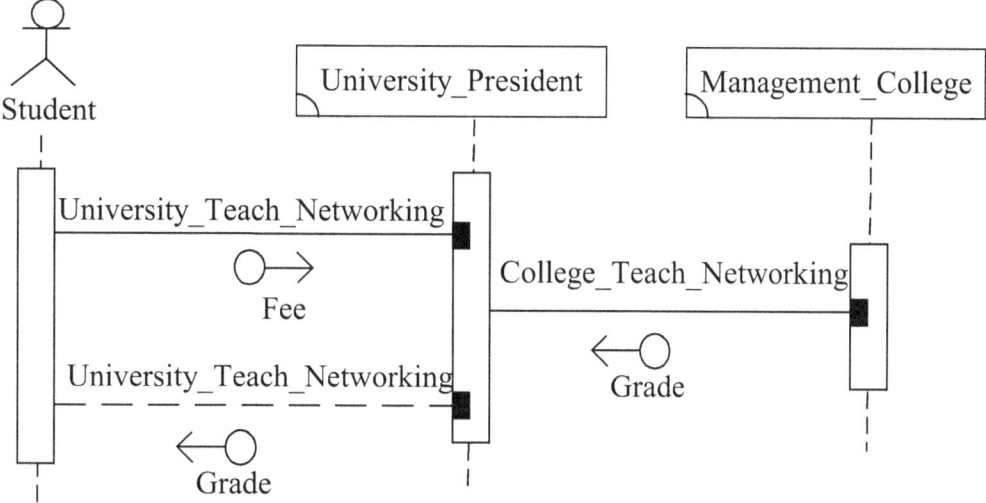

Figure 20-15   Concept's IFD of the *Study_Networking_Course* Behavior

Figure 20-16 shows the concept's IFD of the *Study_Economics_Course* behavior. First, actor *Student* interacts with the *University_President* component through the *University_Teach_Economics* operation call interaction, carrying the *Fee* input parameter. Next, component *University_President* interacts with the *Management_College* component through the *College_Teach_Economics* operation call interaction, carrying the *Grade* output parameter. Finally, actor *Student* interacts with the *University_President* component through the *University_Teach_Economics* operation return interaction, carrying the *Grade* output parameter.

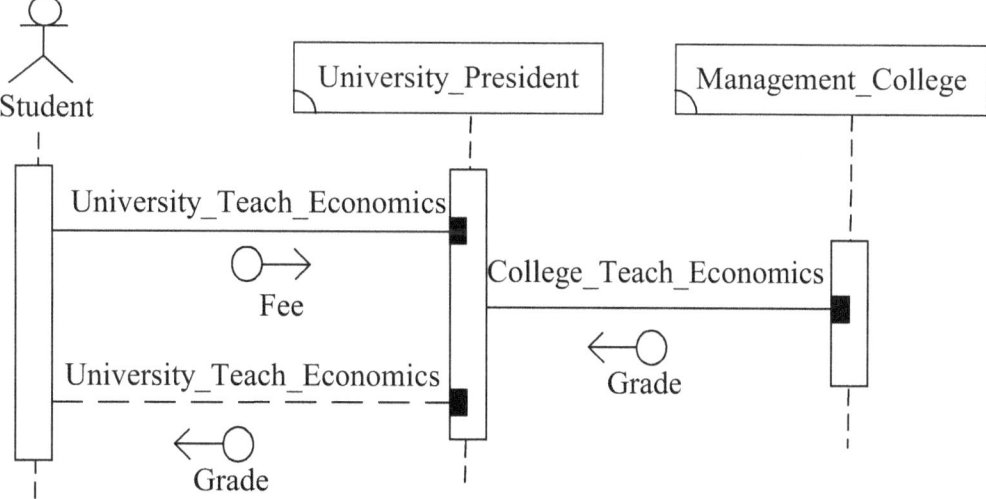

Figure 20-16   Concept's IFD of the *Study_Economics_Course* Behavior

Figure 20-17 shows the concept's IFD of the *Study_Accounting_Course* behavior. First, actor *Student* interacts with the *University_President* component through the *University_Teach_Accounting* operation call interaction, carrying the *Fee* input parameter. Next, component *University_President* interacts with the *Management_College* component through the *College_Teach_Accounting* operation call interaction, carrying the *Grade* output parameter. Finally, actor *Student* interacts with the *University_President* component through the *University_Teach_Accounting* operation return interaction, carrying the *Grade* output parameter.

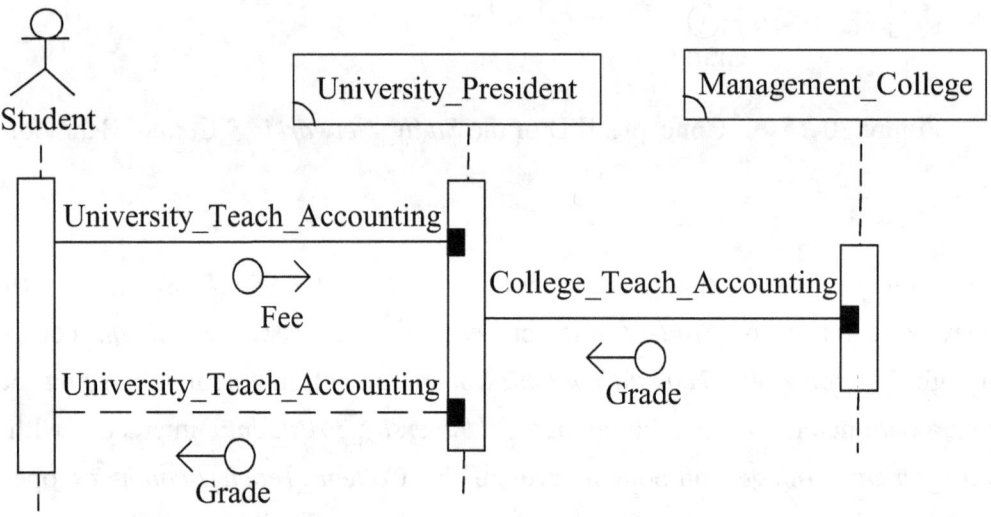

Figure 20-17   Concept's IFD of the *Study_Accounting_Course* Behavior

20-2 Analysis View of the Stanford University

In the SBC enterprise multi-level (hierarchical) view, an architect constructs the analysis' systems architecture for the analyzer to view. This analysis' systems architecture is called the analysis view as shown in Figure 20-18. Analysis view is one level down structural decomposition (with observation congruence verification) of the concept view [Chao15a, Chao15b, Chao15c, Chao15d, Chao15e].

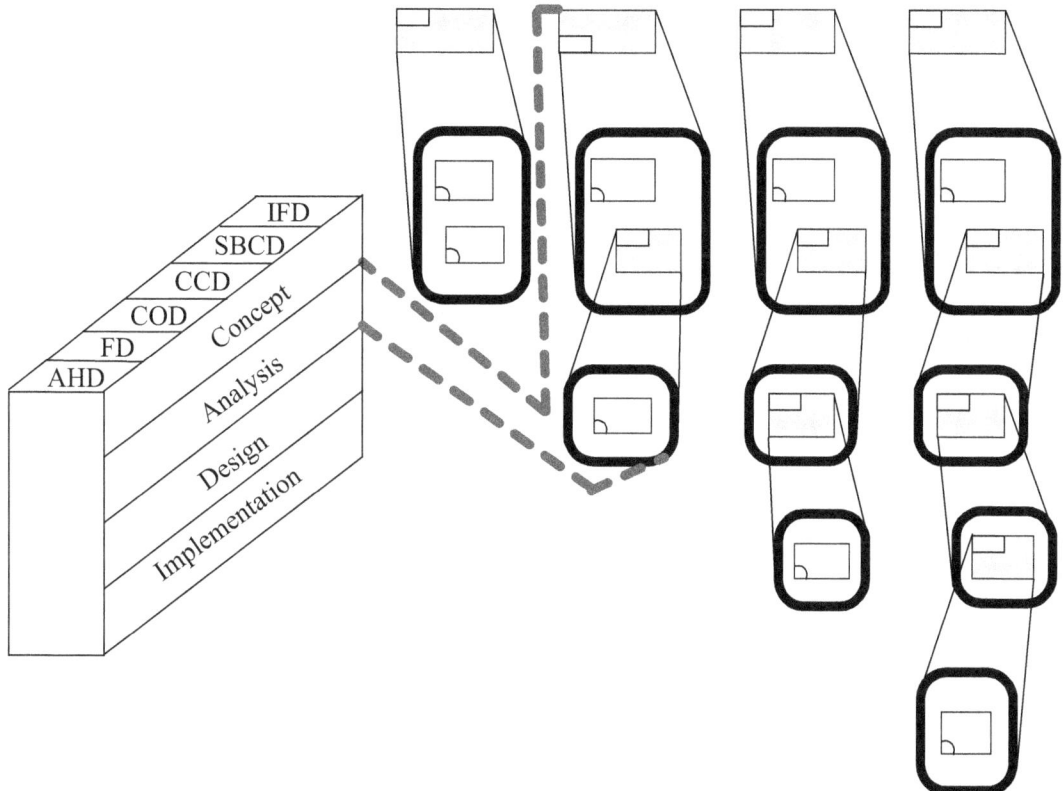

Figure 20-18   Analysis View

The analysis view consists of: a) analysis' systems structure and b) analysis' systems behavior.

20-2-1 Analysis' Systems Structure

The entire SBC analysis' systems structure includes: a) *Analysis' AHD*, b) *Analysis' FD*, c) *Analysis' COD* and d) *Analysis' CCD*.

We first draw the analysis' AHD of the *Stanford University*. As shown in Figure 20-19, *Stanford University* is composed of *University_President*, *Science_College* and *Management_College*; *Science_College* is composed of *Science_Dean*, *Mathematics_Department* and *Physics_Department*; *Management_College* is composed of *Management_Dean*, *MIS_Department* and *Finance_Department*.

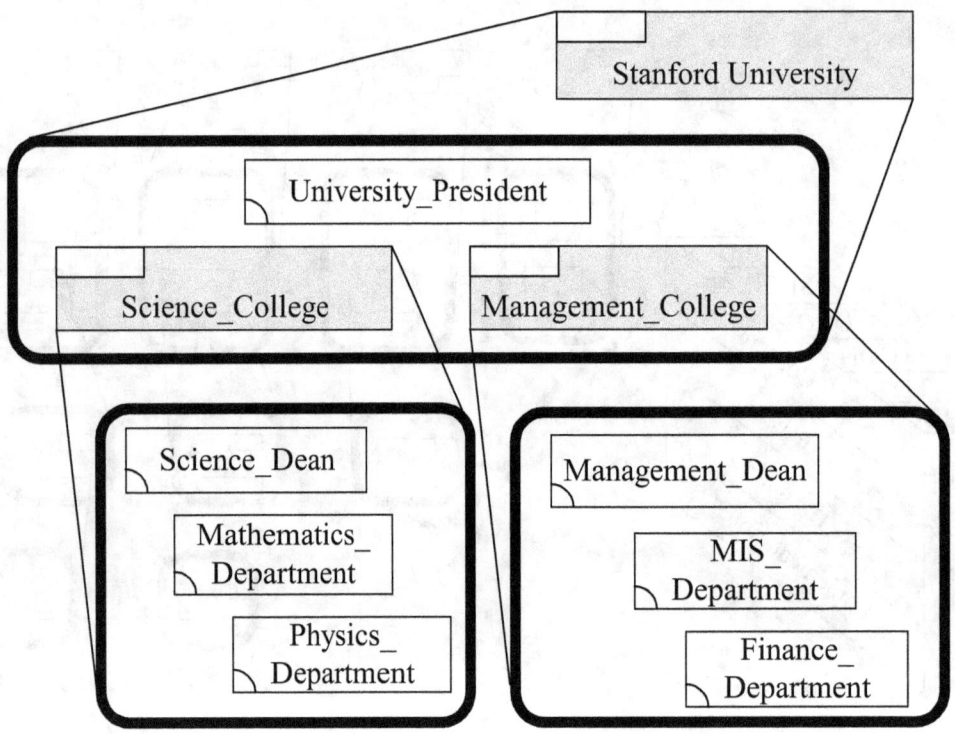

Figure 20-19　Analysis' AHD of the Stanford University

In Figure 20-19, *Stanford University*, *Science_College* and *Management_College* are aggregated systems while *University_President*, *Science_Dean*, *Mathematics_Department*, *Physics_Department*, *Management_Dean*, *MIS_Department* and *Finance_Department* are non-aggregated systems.

Analysis' FD is the framework diagram we obtain after the analysis phase is finished. Figure 20-20 shows the analysis' FD of the *Stanford University*. In the figure, *University_Layer* contains the *University_President* component; *College_Layer* contains the *Science_Dean*, *Management_Dean*, *Mathematics_Department*, *Physics_Department*, *MIS_Department* and *Finance_Department* components.

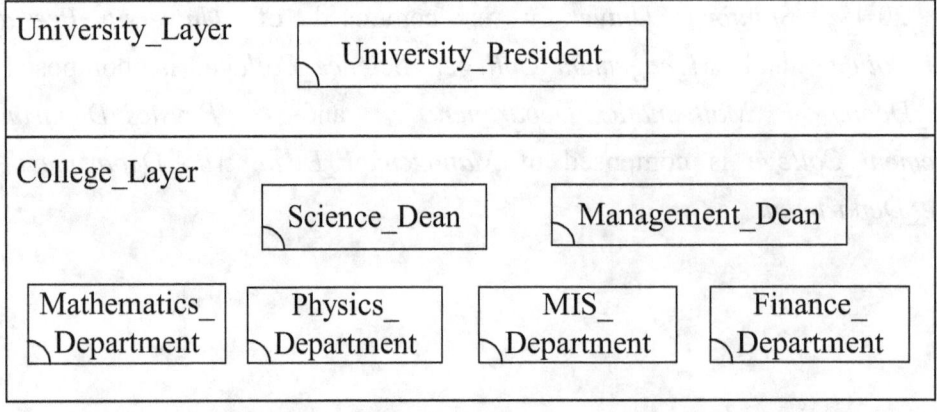

Figure 20-20　Analysis' FD of the *Stanford University*

Analysis' COD is the component operation diagram we obtain after the analysis phase is finished. Figure 20-21 shows the analysis' COD of the *Stanford University*. In the figure, component *University_President* has eight operations: *University_Teach_Calculus*, *University_Teach_Algebra*, *University_Teach_Mechanics*, *University_Teach_Quantum*, *University_Teach_Database*, *University_Teach_Networking*, *University_Teach_Economics* and *University_Teach_Accounting*; component *Science_Dean* has four operations: *College_Teach_Calculus*, *College_Teach_Algebra*, *College_Teach_Mechanics* and *College_Teach_Quantum*; component *Management_Dean* has four operations: *College_Teach_Database*, *College_Teach_Networking*, *College_Teach_Economics* and *College_Teach_Accounting*; *Mathematics_Department* has two operations: *Department_Teach_Calculus* and *Department_Teach_Algebra*; *Physics_Department* has two operations: *Department_Teach_Mechanics* and *Department_Teach_Quantum*; *MIS_Department* has two operations: *Department_Teach_Database* and *Department_Teach_Networking*; *Finance_Department* has two operations: *Department_Teach_Economics* and *Department_Teach_Accounting*.

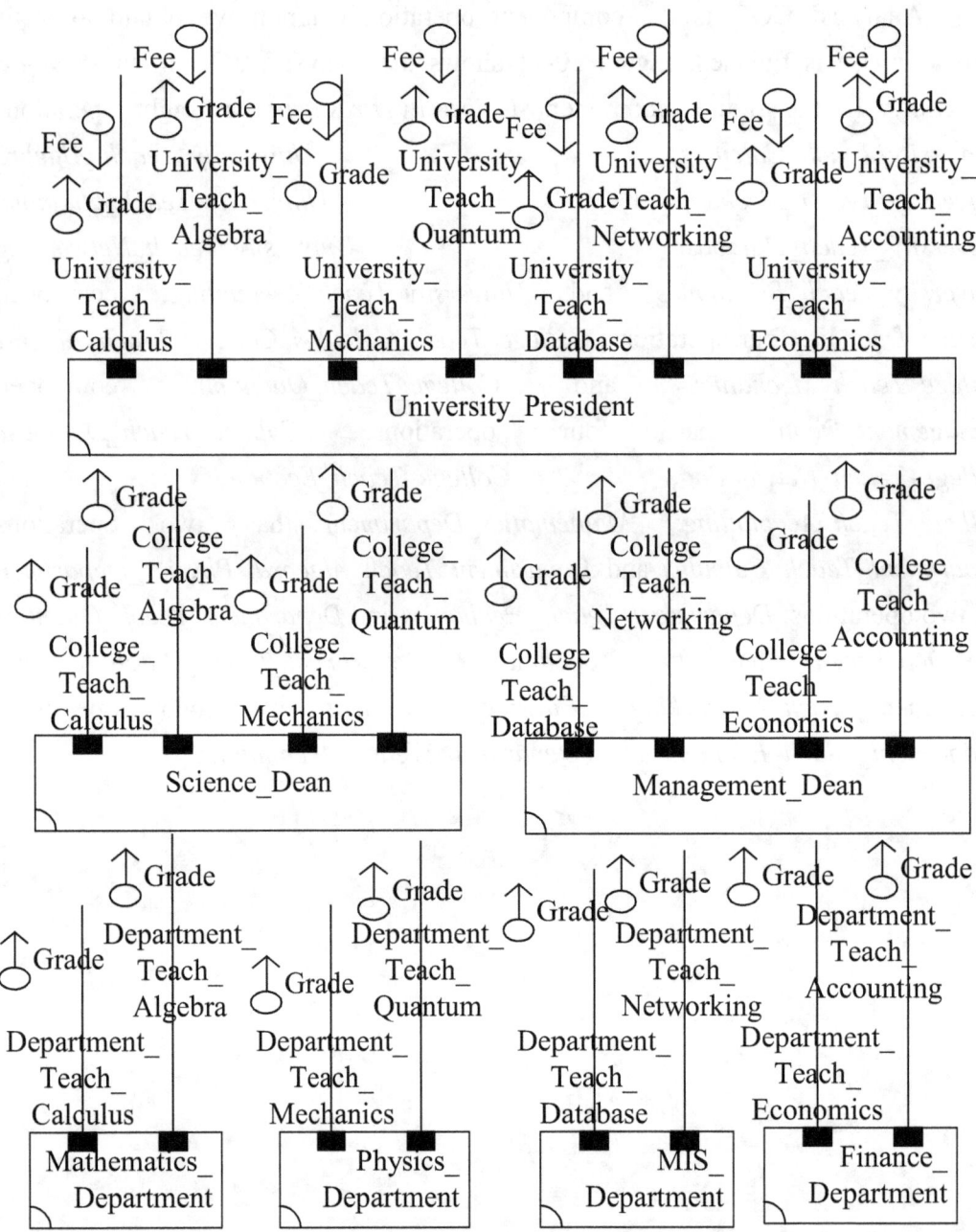

Figure 20-21 Analysis' COD of the *Stanford University*

The operation formula of *University_Teach_Calculus* is *University_Teach_Calculus(In Fee; Out Grade)*. The operation formula of *University_Teach_Algebra* is *University_Teach_Algebra(In Fee; Out Grade)*. The operation formula of *University_Teach_Mechanics* is *University_Teach_Mechanics(In Fee; Out Grade)*. The operation formula of *University_Teach_Quantum* is *University_Teach_Quantum(In Fee; Out Grade)*. The operation formula of *University_Teach_Database* is *University_Teach_Database(In*

Fee; Out Grade). The operation formula of *University_Teach_Networking* is *University_Teach_Networking(In Fee; Out Grade)*. The operation formula of *University_Teach_Economics* is *University_Teach_Economics(In Fee; Out Grade)*. The operation formula of *University_Teach_Accounting* is *University_Teach_Accounting(In Fee; Out Grade)*.

The operation formula of *College_Teach_Calculus* is *College_Teach_Calculus(Out Grade)*. The operation formula of *College_Teach_Algebra* is *College_Teach_Algebra(Out Grade)*. The operation formula of *College_Teach_Mechanics* is *College_Teach_Mechanics(Out Grade)*. The operation formula of *College_Teach_Quantum* is *College_Teach_Quantum(Out Grade)*. The operation formula of *College_Teach_Database* is *College_Teach_Database(Out Grade)*. The operation formula of *College_Teach_Networking* is *College_Teach_Networking(Out Grade)*. The operation formula of *College_Teach_Economics* is *College_Teach_Economics(Out Grade)*. The operation formula of *College_Teach_Accounting* is *College_Teach_Accounting(Out Grade)*.

The operation formula of *Department_Teach_Calculus* is *Department_Teach_Calculus(Out Grade)*. The operation formula of *Department_Teach_Algebra* is *Department_Teach_Algebra(Out Grade)*. The operation formula of *Department_Teach_Mechanics* is *Department_Teach_Mechanics(Out Grade)*. The operation formula of *Department_Teach_Quantum* is *Department_Teach_Quantum(Out Grade)*. The operation formula of *Department_Teach_Database* is *Department_Teach_Database(Out Grade)*. The operation formula of *Department_Teach_Networking* is *Department_Teach_Networking(Out Grade)*. The operation formula of *Department_Teach_Economics* is *Department_Teach_Economics(Out Grade)*. The operation formula of *Department_Teach_Accounting* is *Department_Teach_Accounting(Out Grade)*.

Figure 20-22 shows the primitive data type specification of the *Fee* input parameter and the *Grade* output parameter.

Parameter	Data Type	Instances
Fee	Intger	2000, 3500
Grade	Text	A+, B+, C-

Figure 20-22   Primitive Data Type Specification

Analysis' CCD is the component connection diagram we obtain after the analysis phase is finished. Figure 20-23 shows the analysis' CCD of the *Stanford University*.

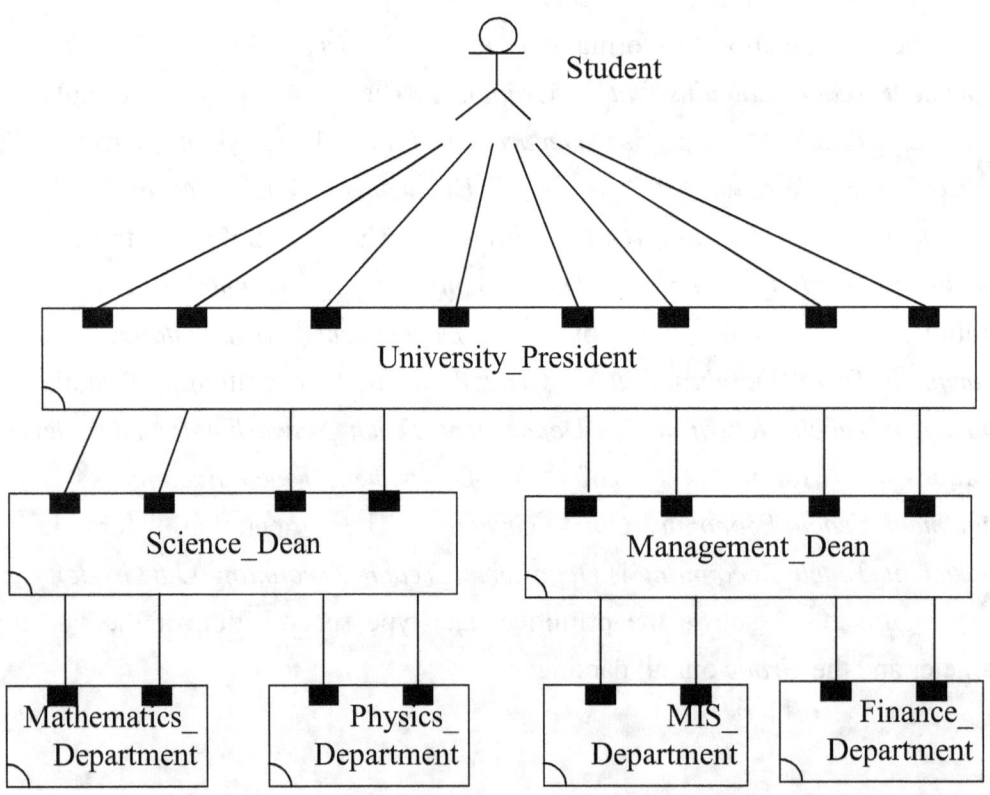

Figure 20-23   Analysis' CCD of the *Stanford University*

In Figure 20-23, actor *Student* has eight connections with the *University_President* component; component *University_President* has four connections with each one of the *Science_Dean* and *Management_Dean* components; component *Science_Dean* has two connections with each one of the *Mathematics_Department* and *Physics_Department* components; component *Management_Dean* has two connections with each one of the *MIS_Department* and *Finance_Department* components.

20-2-2 Analysis' Systems Behavior

The entire SBC analysis' systems behavior includes: a) *Analysis' SBCD* and b) *Analysis' IFD*.

Analysis' SBCD is the structure-behavior coalescence diagram we obtain after the analysis phase is finished. Figure 20-24 shows the analysis' SBCD of the *Stanford University* in which interactions among the *Student* actor and the *University_President, Science_Dean, Management_Dean, Mathematics_Department, Physics_Department, MIS_Department, Finance_Department* components shall draw forth the S*tudy_Calculus_Course*, S*tudy_Algebra_Course*, S*tudy_Mechanics_Course*, S*tudy_Quantum_Course*, S*tudy_Database_Course*, S*tudy_Networking_Course*, S*tudy_Economics_Course* and S*tudy_Accounting_Course* behaviors.

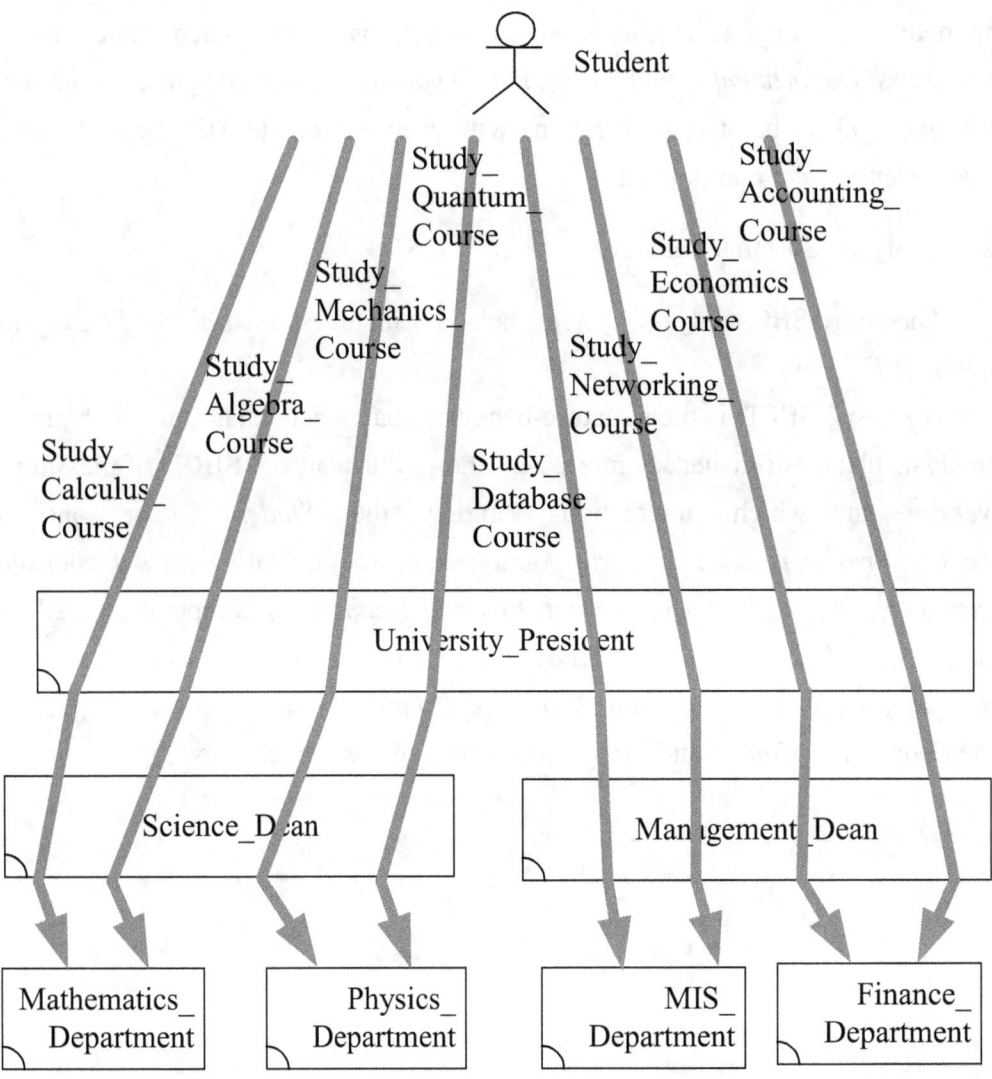

Figure 20-24   Analysis' SBCD of the *Stanford University*

The overall behavior of an enterprise system is contained in its individual behaviors. After the analysis phase is finished, the overall analysis' behavior of the *Stanford University* includes the S*tudy_Calculus_Course*, S*tudy_Algebra_Course*, S*tudy_Mechanics_Course*, S*tudy_Quantum_Course*, S*tudy_Database_Course*, S*tudy_Networking_Course*, S*tudy_Economics_Course* and S*tudy_Accounting_Course* behaviors. In other words, the S*tudy_Calculus_Course*, S*tudy_Algebra_Course*, S*tudy_Mechanics_Course*, S*tudy_Quantum_Course*, S*tudy_Database_Course*, S*tudy_Networking_Course*, S*tudy_Economics_Course* and S*tudy_Accounting_Course* behaviors together provide the overall analysis' behavior of the *Stanford University* after the analysis phase is finished.

Be noticed that the *Study_Calculus_Course*, *Study_Algebra_Course*, *Study_Mechanics_Course*, *Study_Quantum_Course*, *Study_Database_Course*, *Study_Networking_Course*, *Study_Economics_Course* and *Study_Accounting_Course* behaviors are mutually independent of each other. They shall be executed concurrently [Hoar85, Miln89, Miln99].

The major purpose of using the architectural approach, instead of separating the structure model from the behavior model, is to achieve a coalesced model. In Figure 20-24, we not only see its organizational structure, also see at the same time its organizational behavior in the analysis' SBCD of the *Stanford University*.

After the analysis phase is finished, the overall analysis' behavior of the *Stanford University* includes eight behaviors: *Study_Calculus_Course*, *Study_Algebra_Course*, *Study_Mechanics_Course*, *Study_Quantum_Course*, *Study_Database_Course*, *Study_Networking_Course*, *Study_Economics_Course* and *Study_Accounting_Course*. Each of them is described by an individual IFD. Figure 20-25 shows the analysis' IFD of the *Study_Calculus_Course* behavior. First, actor *Student* interacts with the *University_President* component through the *University_Teach_Calculus* operation call interaction, carrying the *Fee* input parameter. Next, component *University_President* interacts with the *Science_Dean* component through the *College_Teach_Calculus* operation call interaction. Continuingly, component *Science_Dean* interacts with the *Mathematics_Department* component through the *Department_Teach_Calculus* operation call interaction, carrying the *Grade* output parameter. Continuingly, component *University_President* interacts with the *Science_Dean* component through the *College_Teach_Calculus* operation return interaction, carrying the *Grade* output parameter. Finally, actor *Student* interacts with the *University_President* component through the *University_Teach_Calculus* operation return interaction, carrying the *Grade* output parameter.

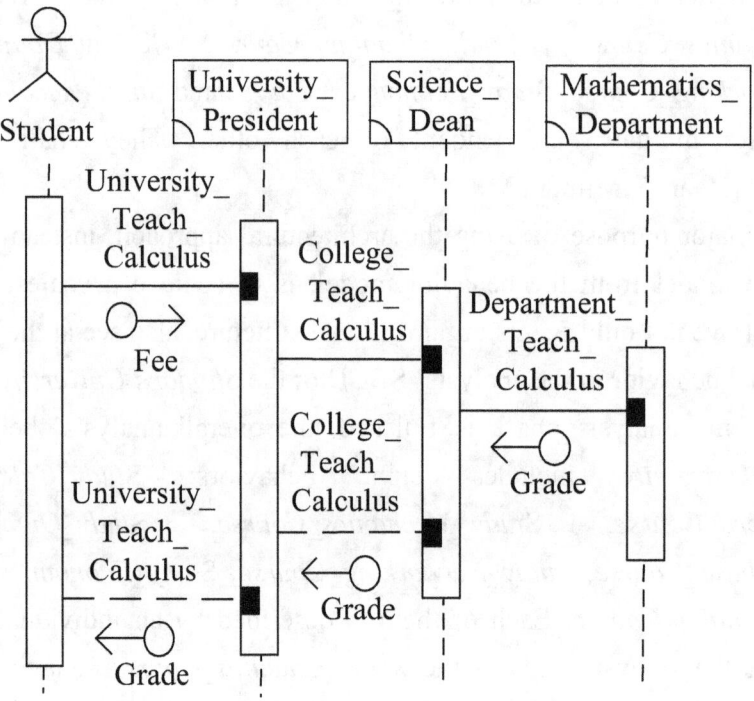

Figure 20-25   Analysis' IFD of the *Study_Calculus_Course* Behavior

Figure 20-26 shows the analysis' IFD of the *Study_Algebra_Course* behavior. First, actor *Student* interacts with the *University_President* component through the *University_Teach_Algebra* operation call interaction, carrying the *Fee* input parameter. Next, component *University_President* interacts with the *Science_Dean* component through the *College_Teach_Algebra* operation call interaction. Continuingly, component *Science_Dean* interacts with the *Mathematics_Department* component through the *Department_Teach_Algebra* operation call interaction, carrying the *Grade* output parameter. Continuingly, component *University_President* interacts with the *Science_Dean* component through the *College_Teach_Algebra* operation return interaction, carrying the *Grade* output parameter. Finally, actor *Student* interacts with the *University_President* component through the *University_Teach_Algebra* operation return interaction, carrying the *Grade* output parameter.

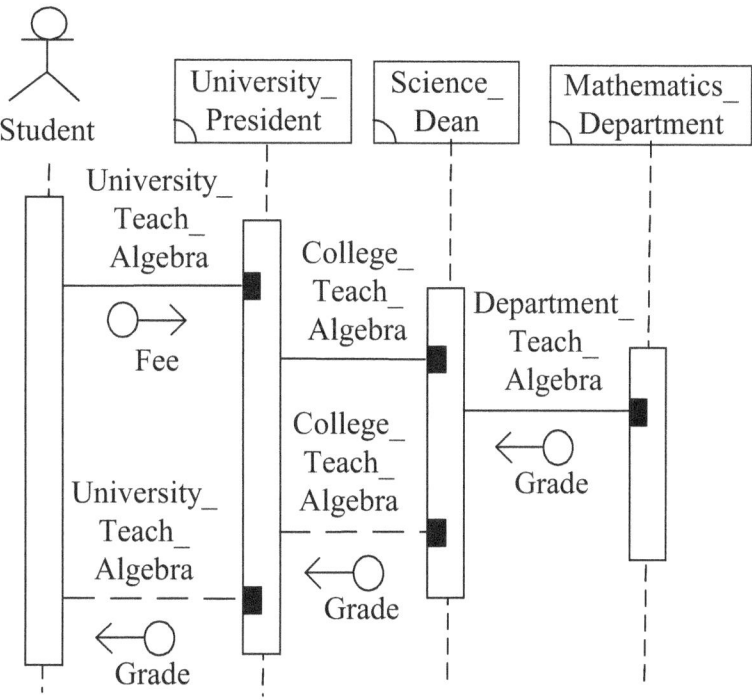

Figure 20-26   Analysis' IFD of the *Study_Algebra_Course* Behavior

Figure 20-27 shows the analysis' IFD of the *Study_Mechanics_Course* behavior. First, actor *Student* interacts with the *University_President* component through the *University_Teach_Mechanics* operation call interaction, carrying the *Fee* input parameter. Next, component *University_President* interacts with the *Science_Dean* component through the *College_Teach_Mechanics* operation call interaction. Continuingly, component *Science_Dean* interacts with the *Physics_Department* component through the *Department_Teach_Mechanics* operation call interaction, carrying the *Grade* output parameter. Continuingly, component *University_President* interacts with the *Science_Dean* component through the *College_Teach_Mechanics* operation return interaction, carrying the *Grade* output parameter. Finally, actor *Student* interacts with the *University_President* component through the *University_Teach_Mechanics* operation return interaction, carrying the *Grade* output parameter.

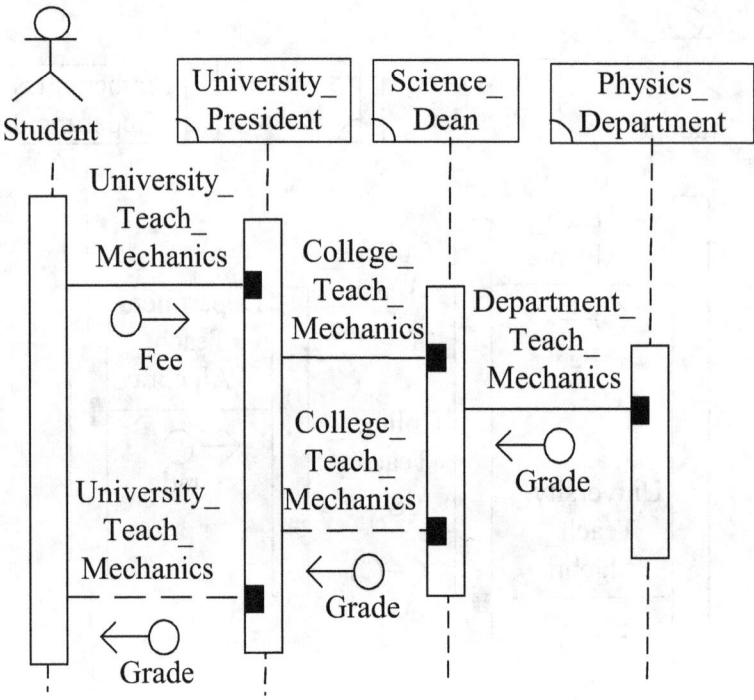

Figure 20-27   Analysis' IFD of the *Study_Mechanics_Course* Behavior

Figure 20-28 shows the analysis' IFD of the *Study_Quantum_Course* behavior. First, actor *Student* interacts with the *University_President* component through the *University_Teach_Quantum* operation call interaction, carrying the *Fee* input parameter. Next, component *University_President* interacts with the *Science_Dean* component through the *College_Teach_Quantum* operation call interaction. Continuingly, component *Science_Dean* interacts with the *Physics_Department* component through the *Department_Teach_Quantum* operation call interaction, carrying the *Grade* output parameter. Continuingly, component *University_President* interacts with the *Science_Dean* component through the *College_Teach_Quantum* operation return interaction, carrying the *Grade* output parameter. Finally, actor *Student* interacts with the *University_President* component through the *University_Teach_Quantum* operation return interaction, carrying the *Grade* output parameter.

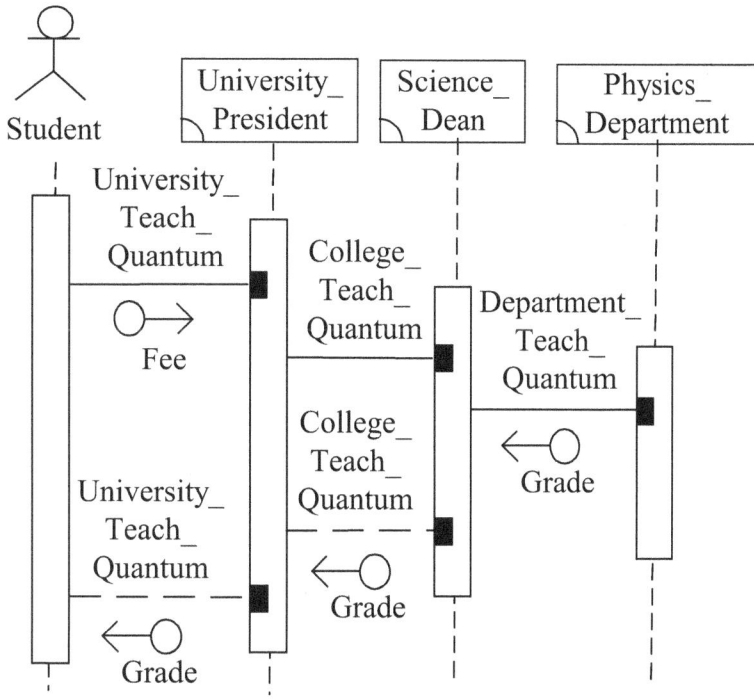

Figure 20-28   Analysis' IFD of the *Study_Quantum_Course* Behavior

Figure 20-29 shows the analysis' IFD of the S*tudy_Database_Course* behavior. First, actor *Student* interacts with the *University_President* component through the *University_Teach_Database* operation call interaction, carrying the *Fee* input parameter. Next, component *University_President* interacts with the *Management_Dean* component through the *College_Teach_Database* operation call interaction. Continuingly, component *Management_Dean* interacts with the *MIS_Department* component through the *Department_Teach_Database* operation call interaction, carrying the *Grade* output parameter. Continuingly, component *University_President* interacts with the *Management_Dean* component through the *College_Teach_Database* operation return interaction, carrying the *Grade* output parameter. Finally, actor *Student* interacts with the *University_President* component through the *University_Teach_Database* operation return interaction, carrying the *Grade* output parameter.

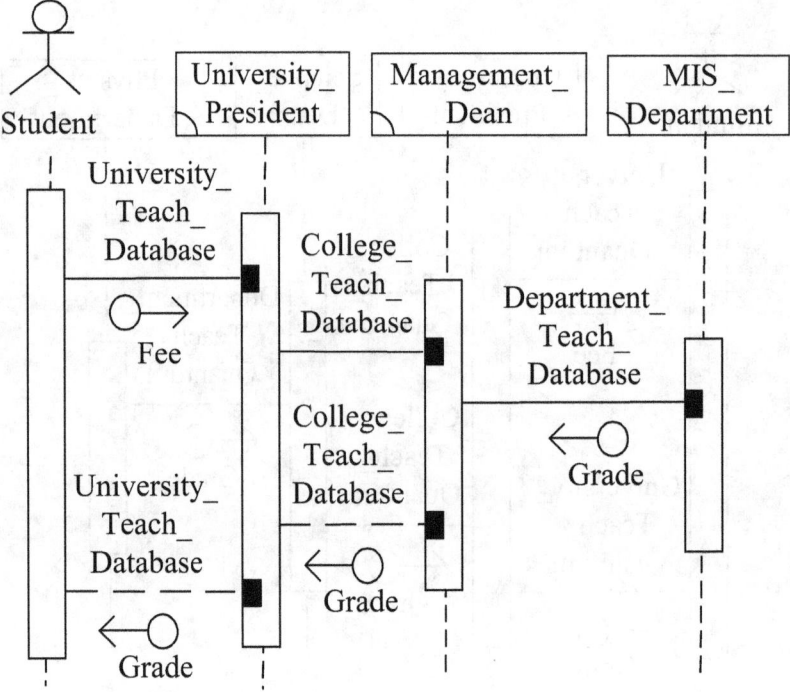

Figure 20-29　　Analysis' IFD of the *Study_Database_Course* Behavior

　　Figure 20-30 shows the analysis' IFD of the *Study_Networking_Course* behavior. First, actor *Student* interacts with the *University_President* component through the *University_Teach_Networking* operation call interaction, carrying the *Fee* input parameter. Next, component *University_President* interacts with the *Management_Dean* component through the *College_Teach_Networking* operation call interaction. Continuingly, component *Management_Dean* interacts with the *MIS_Department* component through the *Department_Teach_Networking* operation call interaction, carrying the *Grade* output parameter. Continuingly, component *University_President* interacts with the *Management_Dean* component through the *College_Teach_Networking* operation return interaction, carrying the *Grade* output parameter. Finally, actor *Student* interacts with the *University_President* component through the *University_Teach_Networking* operation return interaction, carrying the *Grade* output parameter.

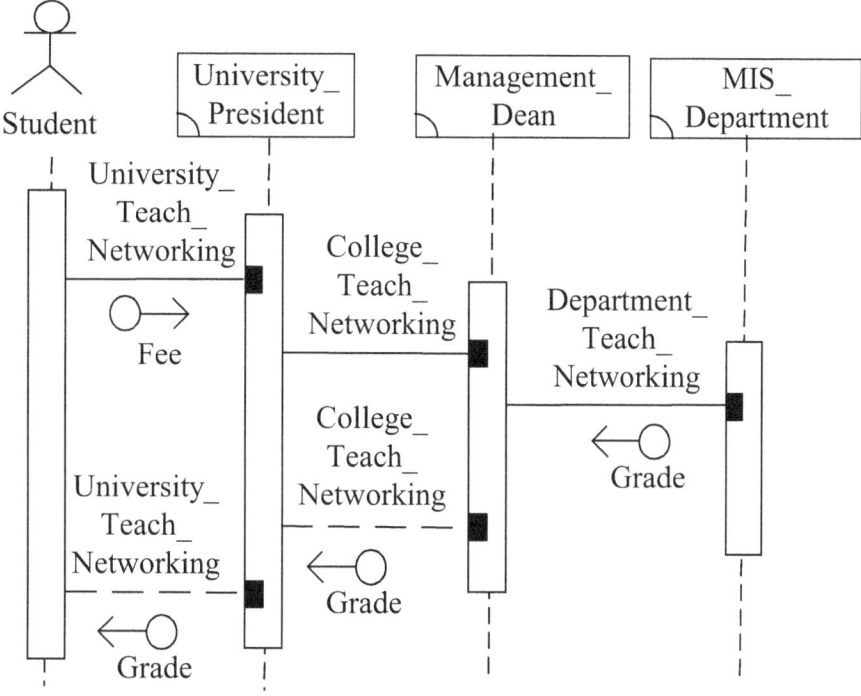

Figure 20-30   Analysis' IFD of the *Study_Networking_Course* Behavior

Figure 20-31 shows the analysis' IFD of the *Study_Economics_Course* behavior. First, actor *Student* interacts with the *University_President* component through the *University_Teach_Economics* operation call interaction, carrying the *Fee* input parameter. Next, component *University_President* interacts with the *Management_Dean* component through the *College_Teach_Economics* operation call interaction. Continuingly, component *Management_Dean* interacts with the *Finance_Department* component through the *Department_Teach_Economics* operation call interaction, carrying the *Grade* output parameter. Continuingly, component *University_President* interacts with the *Management_Dean* component through the *College_Teach_Economics* operation return interaction, carrying the *Grade* output parameter. Finally, actor *Student* interacts with the *University_President* component through the *University_Teach_Economics* operation return interaction, carrying the *Grade* output parameter.

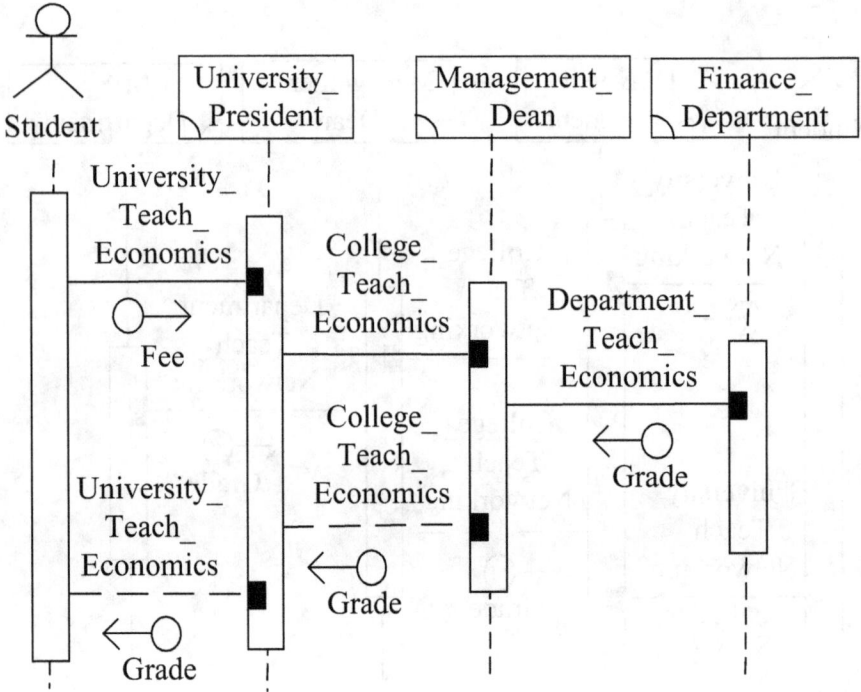

Figure 20-31   Analysis' IFD of the *Study_Economics_Course* Behavior

Figure 20-32 shows the analysis' IFD of the S*tudy_Accounting_Course* behavior. First, actor *Student* interacts with the *University_President* component through the *University_Teach_Accounting* operation call interaction, carrying the *Fee* input parameter. Next, component *University_President* interacts with the *Management_Dean* component through the *College_Teach_Accounting* operation call interaction. Continuingly, component *Management_Dean* interacts with the *Finance_Department* component through the *Department_Teach_Accounting* operation call interaction, carrying the *Grade* output parameter. Continuingly, component *University_President* interacts with the *Management_Dean* component through the *College_Teach_Accounting* operation return interaction, carrying the *Grade* output parameter. Finally, actor *Student* interacts with the *University_President* component through the *University_Teach_Accounting* operation return interaction, carrying the *Grade* output parameter.

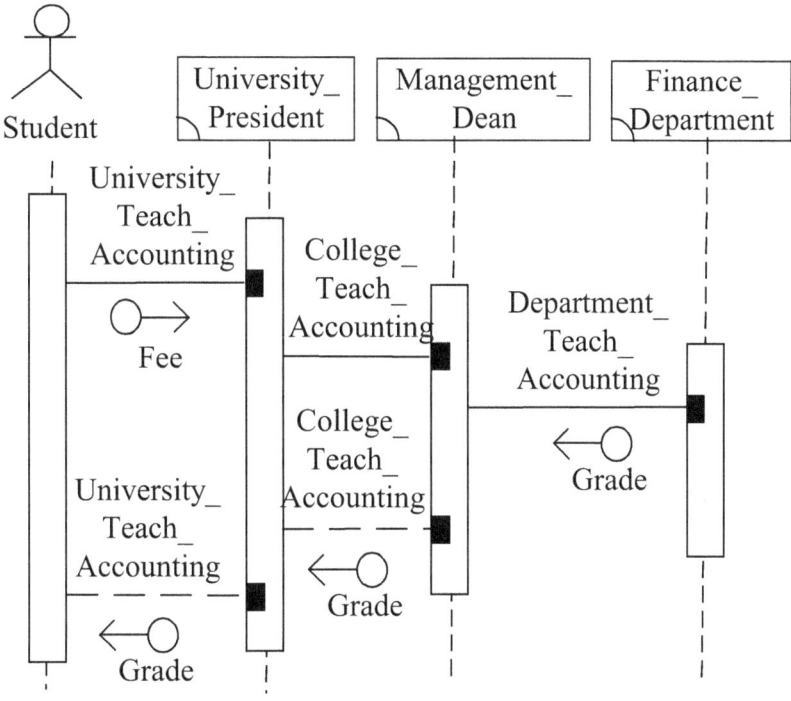

Figure 20-32   Analysis' IFD of the *Study_Accounting_Course* Behavior

20-3 Design View of the Stanford University

In the SBC enterprise multi-level (hierarchical) view, an architect constructs the design's systems architecture for the designer to view. This design's systems architecture is called the design view as shown in Figure 20-33. Design view is one level down structural decomposition (with observation congruence verification) of the analysis view [Chao15a, Chao15b, Chao15c, Chao15d, Chao15e].

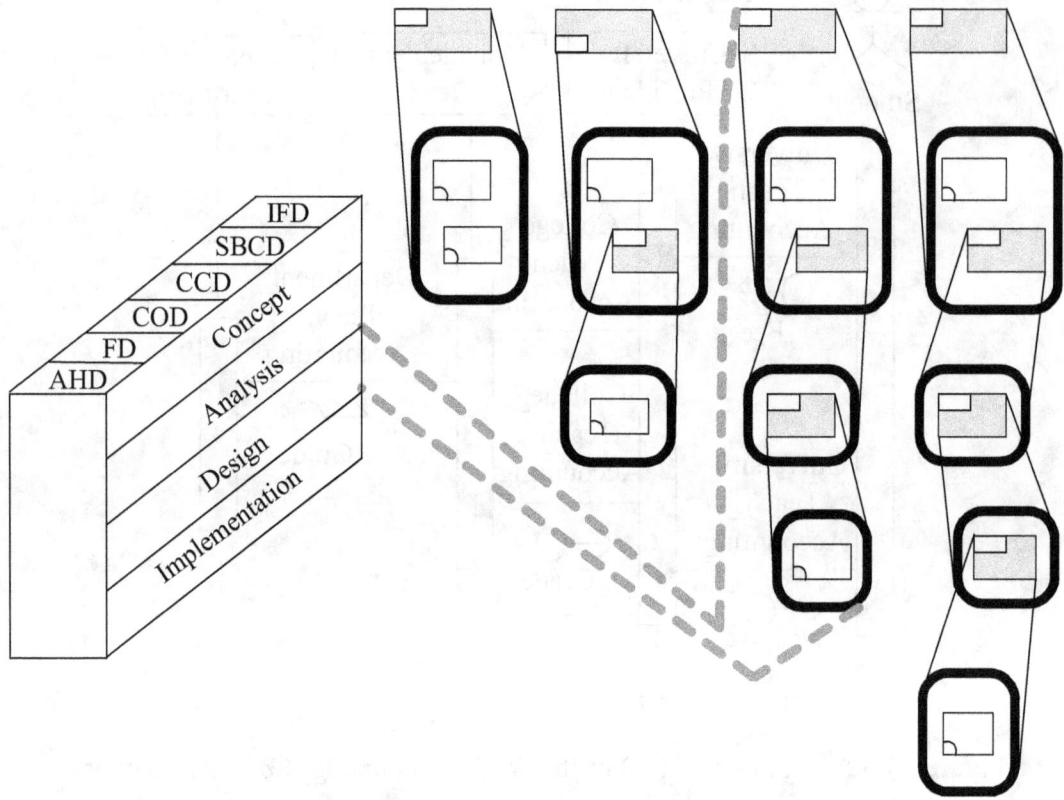

Figure 20-33   Design View

The design view consists of: a) design's systems structure and b) design's systems behavior.

20-3-1 Design's Systems Structure

The entire SBC design's systems structure includes: a) *Design's AHD*, b) *Design's FD*, c) *Design's COD* and d) *Design's CCD*.

We first draw the design's AHD of the *Stanford University*. As shown in Figure 20-34, *Stanford University* is composed of *University_President*, *Science_College* and *Management_College*; *Science_College* is composed of *Science_Dean*, *Mathematics_Department* and *Physics_Department*; *Management_College* is composed of *Management_Dean*, *MIS_Department* and *Finance_Department*; *Mathematics_Department* is composed of *Mathematics_Chairman* and *Mathematics_Lecturers*; *Physics_Department* is composed of *Physics_Chairman* and *Physics_Lecturers*; *MIS_Department* is composed of *MIS_Chairman* and *MIS_Lecturers*; *Finance_Department* is composed of *Finance_Chairman* and *Finance_Lecturers*.

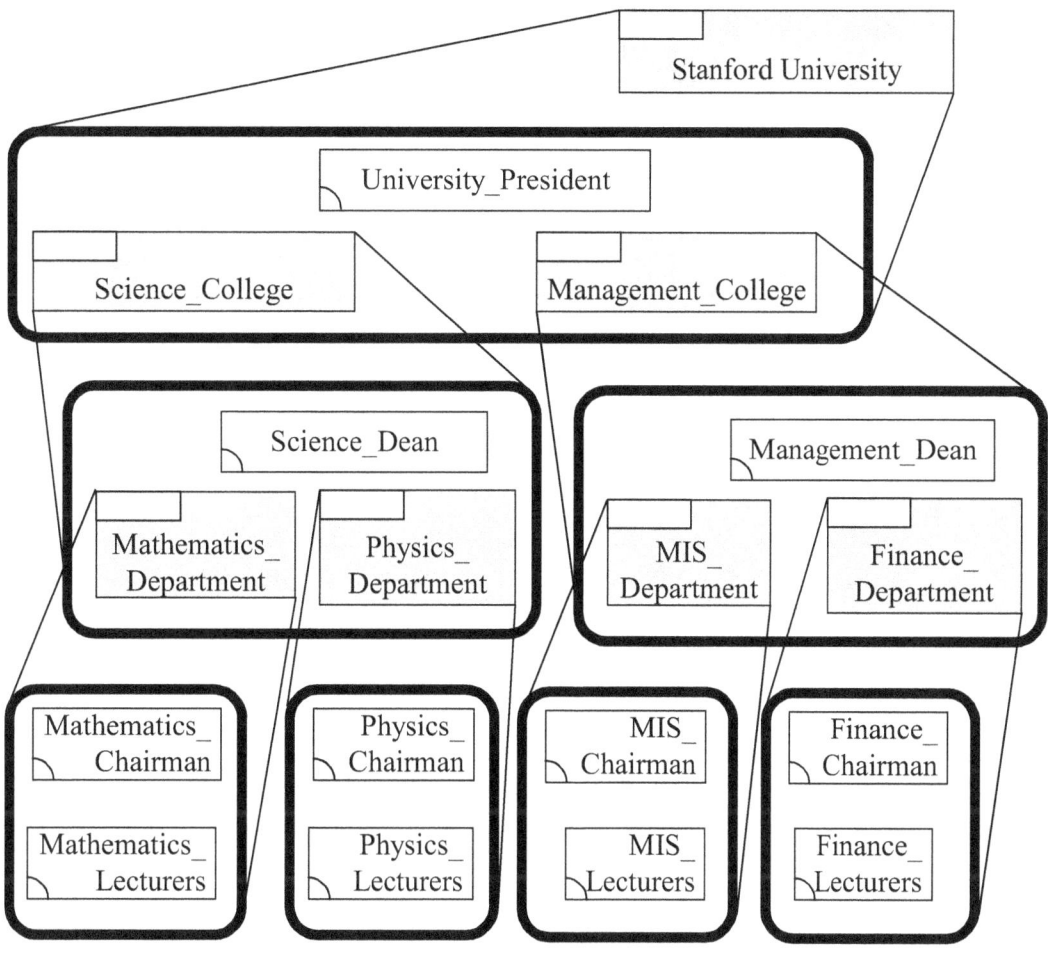

Figure 20-34   Design's AHD of the *Stanford University*

In Figure 20-34, *Stanford University*, *Science_College*, *Management_College*, *Mathematics_Department*, *Physics_Department*, *MIS_Department* and *Finance_Department* are aggregated systems while *University_President*, *Science_Dean*, *Management_Dean*, *Mathematics_Chairman*, *Physics_Chairman*, *MIS_Chairman*, *Finance_Chairman*, *Mathematics_Lecturers*, *Physics_Lecturers*, *MIS_Lecturers* and *Finance_Lecturers* are non-aggregated systems.

Analysis' FD is the framework diagram we obtain after the design phase is finished. Figure 20-35 shows the design's FD of the *Stanford University*. In the figure, *University_Layer* contains the *University_President* component; *College_Layer* contains the *Science_Dean* and *Management_Dean* components; *Department_Layer* contains the *Mathematics_Chairman*, *Physics_Chairman*, *MIS_Chairman* and *Finance_Chairman*, *Mathematics_Lecturers*, *Physics_Lecturers*, *MIS_Lecturers* and *Finance_Lecturers* components.

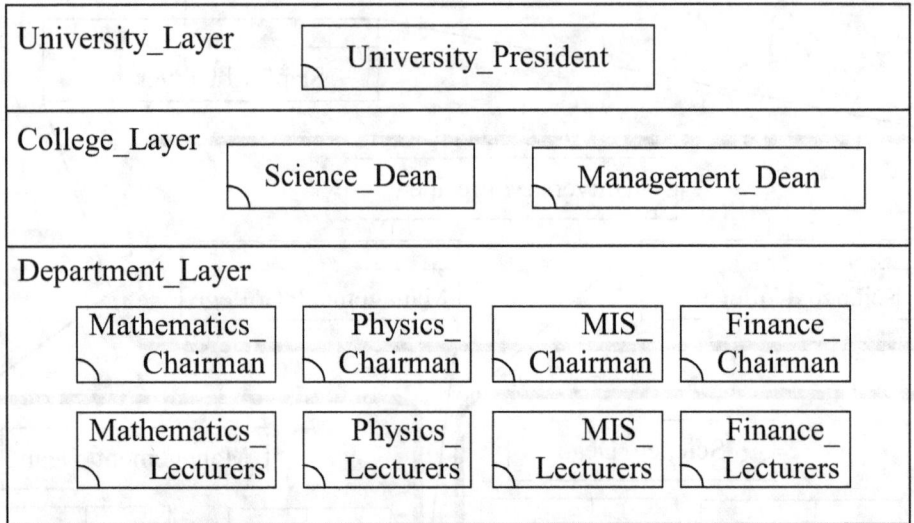

Figure 20-35   Design's FD of the *Stanford University*

Design's COD is the component operation diagram we obtain after the design phase is finished. Figure 20-36 shows the design's COD of the *Stanford University*. In the figure, component *University_President* has eight operations: *University_Teach_Calculus*, *University_Teach_Algebra*, *University_Teach_Mechanics*, *University_Teach_Quantum*, *University_Teach_Database*, *University_Teach_Networking*, *University_Teach_Economics* and *University_Teach_Accounting*; component *Science_Dean* has four operations: *College_Teach_Calculus*, *College_Teach_Algebra*, *College_Teach_Mechanics* and *College_Teach_Quantum*; component *Management_Dean* has four operations: *College_Teach_Database*, *College_Teach_Networking*, *College_Teach_Economics* and *College_Teach_Accounting*; *Mathematics_Chairman* has two operations: *Department_Teach_Calculus* and *Department_Teach_Algebra*; *Physics_Chairman* has two operations: *Department_Teach_Mechanics* and *Department_Teach_Quantum*; *MIS_Chairman* has two operations: *Department_Teach_Database* and *Department_Teach_Networking*; *Finance_Chairman* has two operations: *Department_Teach_Economics* and *Department_Teach_Accounting*; *Mathematics_Lecturers* has two operations: *Lecturers_Teach_Calculus* and *Lecturers_Teach_Algebra*; *Physics_Lecturers* has two operations: *Lecturers_Teach_Mechanics* and *Lecturers_Teach_Quantum*; *MIS_Lecturers* has two operations: *Lecturers_Teach_Database* and *Lecturers_Teach_Networking*; *Finance_Lecturers* has two operations: *Lecturers_Teach_Economics* and *Lecturers_Teach_Accounting*.

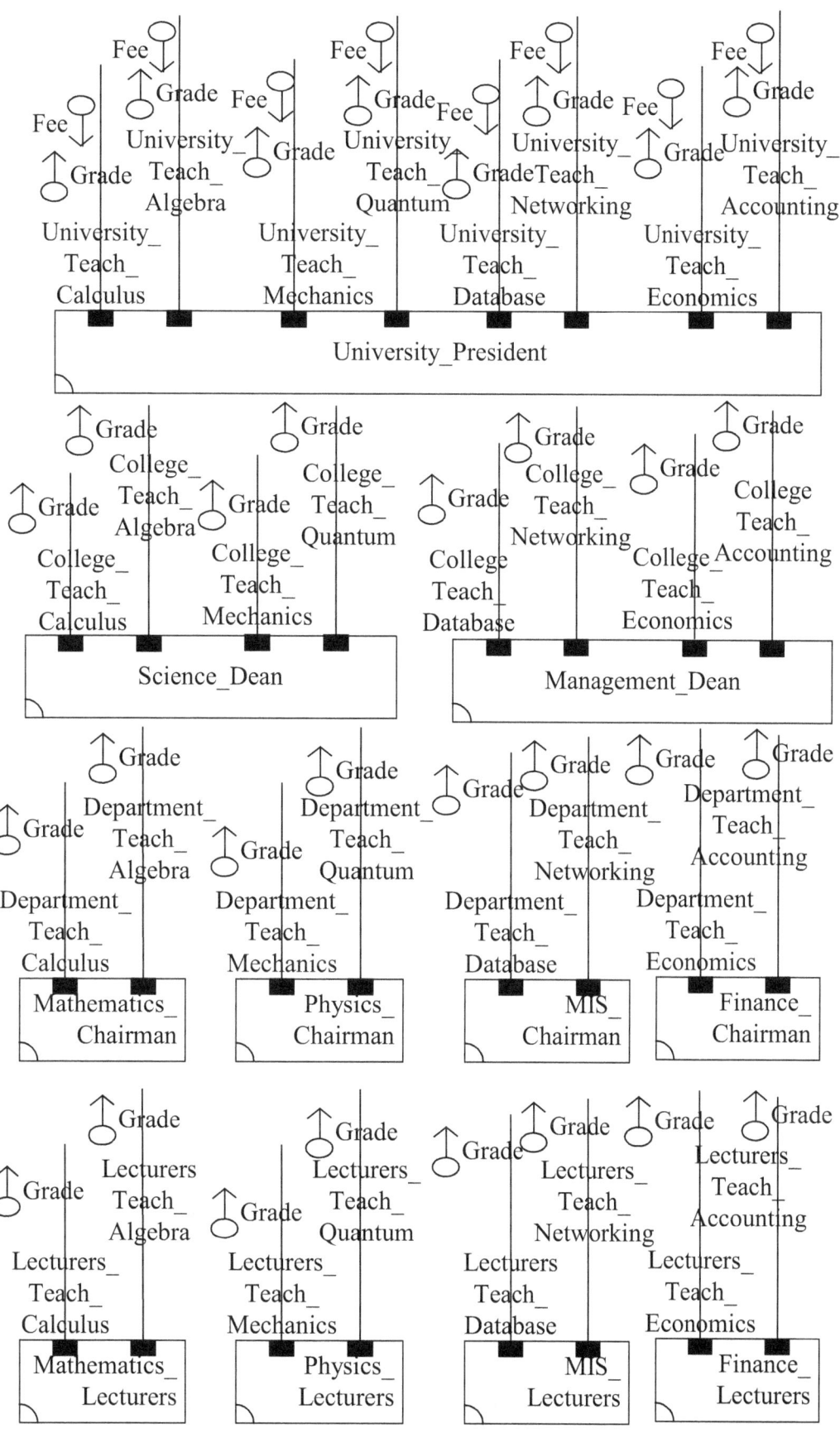

Figure 20-36   Design's COD of the *Stanford University*

The operation formula of *University_Teach_Calculus* is *University_Teach_Calculus(In Fee; Out Grade)*. The operation formula of *University_Teach_Algebra* is *University_Teach_Algebra(In Fee; Out Grade)*. The operation formula of *University_Teach_Mechanics* is *University_Teach_Mechanics(In Fee; Out Grade)*. The operation formula of *University_Teach_Quantum* is *University_Teach_Quantum(In Fee; Out Grade)*. The operation formula of *University_Teach_Database* is *University_Teach_Database(In Fee; Out Grade)*. The operation formula of *University_Teach_Networking* is *University_Teach_Networking(In Fee; Out Grade)*. The operation formula of *University_Teach_Economics* is *University_Teach_Economics(In Fee; Out Grade)*. The operation formula of *University_Teach_Accounting* is *University_Teach_Accounting(In Fee; Out Grade)*.

The operation formula of *College_Teach_Calculus* is *College_Teach_Calculus(Out Grade)*. The operation formula of *College_Teach_Algebra* is *College_Teach_Algebra(Out Grade)*. The operation formula of *College_Teach_Mechanics* is *College_Teach_Mechanics(Out Grade)*. The operation formula of *College_Teach_Quantum* is *College_Teach_Quantum(Out Grade)*. The operation formula of *College_Teach_Database* is *College_Teach_Database(Out Grade)*. The operation formula of *College_Teach_Networking* is *College_Teach_Networking(Out Grade)*. The operation formula of *College_Teach_Economics* is *College_Teach_Economics(Out Grade)*. The operation formula of *College_Teach_Accounting* is *College_Teach_Accounting(Out Grade)*.

The operation formula of *Department_Teach_Calculus* is *Department_Teach_Calculus(Out Grade)*. The operation formula of *Department_Teach_Algebra* is *Department_Teach_Algebra(Out Grade)*. The operation formula of *Department_Teach_Mechanics* is *Department_Teach_Mechanics(Out Grade)*. The operation formula of *Department_Teach_Quantum* is *Department_Teach_Quantum(Out Grade)*. The operation formula of *Department_Teach_Database* is *Department_Teach_Database(Out Grade)*. The operation formula of *Department_Teach_Networking* is *Department_Teach_Networking(Out Grade)*. The operation formula of *Department_Teach_Economics* is *Department_Teach_Economics(Out Grade)*. The operation formula of *Department_Teach_Accounting* is *Department_Teach_Accounting(Out Grade)*.

The operation formula of *Lecturers_Teach_Calculus* is *Lecturers_Teach_Calculus(Out Grade)*. The operation formula of *Lecturers_Teach_Algebra* is *Lecturers_Teach_Algebra(Out Grade)*. The operation formula of *Lecturers_Teach_Mechanics* is *Lecturers_Teach_Mechanics(Out Grade)*. The operation formula of *Lecturers_Teach_Quantum* is *Lecturers_Teach_Quantum(Out Grade)*. The operation formula of *Lecturers_Teach_Database* is *Lecturers_Teach_Database(Out Grade)*. The operation formula of *Lecturers_Teach_Networking* is *Lecturers_Teach_Networking(Out Grade)*. The operation formula of *Lecturers_Teach_Economics* is *Lecturers_Teach_Economics(Out Grade)*. The operation formula of *Lecturers_Teach_Accounting* is *Lecturers_Teach_Accounting(Out Grade)*.

Figure 20-37 shows the primitive data type specification of the *Fee* input parameter and the *Grade* output parameter.

Parameter	Data Type	Instances
Fee	Intger	2000, 3500
Grade	Text	A+, B+, C-

Figure 20-37   Primitive Data Type Specification

Design's CCD is the component connection diagram we obtain after the design phase is finished. Figure 20-38 shows the design's CCD of the *Stanford University*.

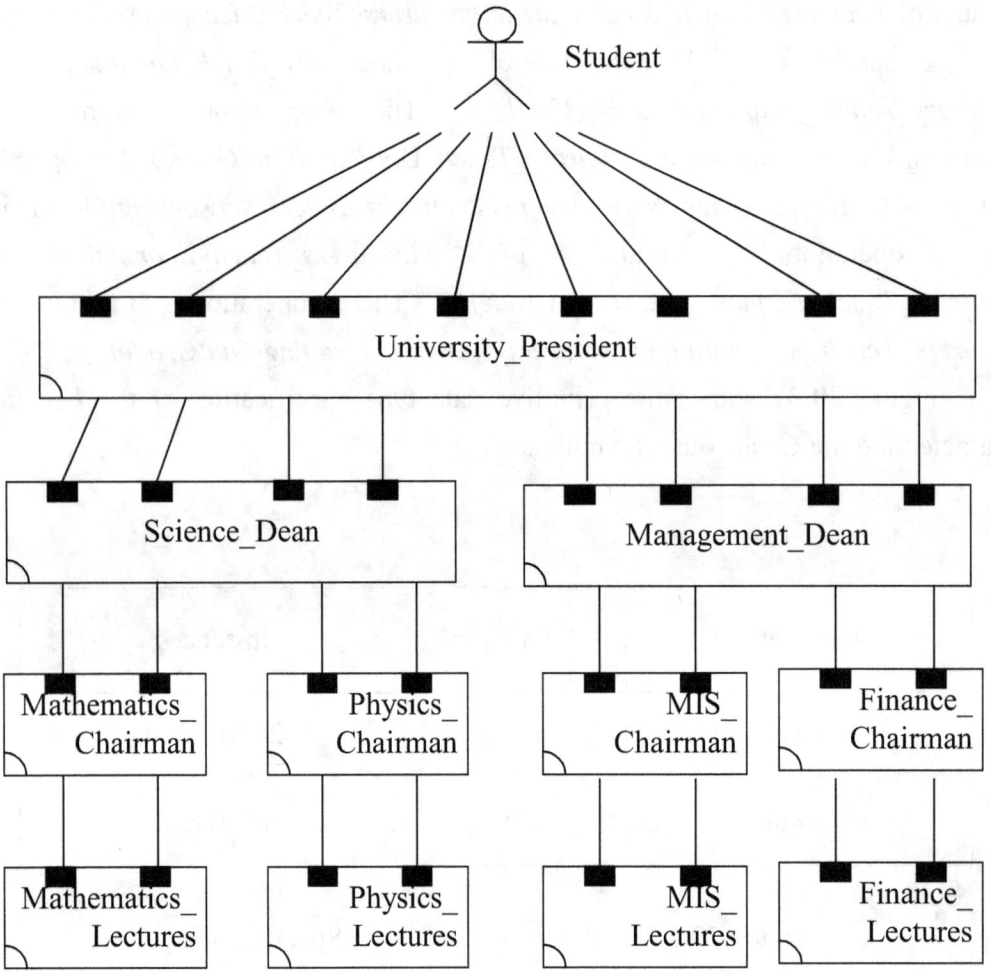

Figure 20-38   Design's CCD of the *Stanford University*

In Figure 20-38, actor *Student* has eight connections with the component *University_President*; component *University_President* has four connections with each of the components *Science_Dean* and *Management_Dean*; component *Science_Dean* has two connections with each of the components *Mathematics_Chairman* and *Physics_Chairman*; component *Management_Dean* has two connections with each of the components *MIS_Chairman* and *Finance_Chairman*; component *Mathematics_Chairman* has two connections with the component *Mathematics_Lecturers*; component *Physics_Chairman* has two connections with the component *Physics_Lecturers*; component *MIS_Chairman* has two connections with the component *MIS_Lecturers*; component *Finance_Chairman* has two connections with the component *Finance_Lecturers*.

20-3-2 Design's Systems Behavior

The entire SBC design's systems behavior includes: a) *Design's SBCD* and b) *Design's IFD*.

Design's SBCD is the structure-behavior coalescence diagram we obtain after the design phase is finished. Figure 20-39 shows the design's SBCD of the *Stanford University* in which interactions among the *Student* actor and the *University_President*, *Science_Dean*, *Management_Dean*, *Mathematics_Chairman*, *Physics_Chairman*, *MIS_Chairman*, *Finance_Chairman*, *Mathematics_Lecturers*, *Physics_Lecturers*, *MIS_Lecturers*, *Finance_Lecturers* components shall draw forth the S*tudy_Calculus_Course*, S*tudy_Algebra_Course*, S*tudy_Mechanics_Course*, S*tudy_Quantum_Course*, S*tudy_Database_Course*, S*tudy_Networking_Course*, S*tudy_Economics_Course* and S*tudy_Accounting_Course* behaviors.

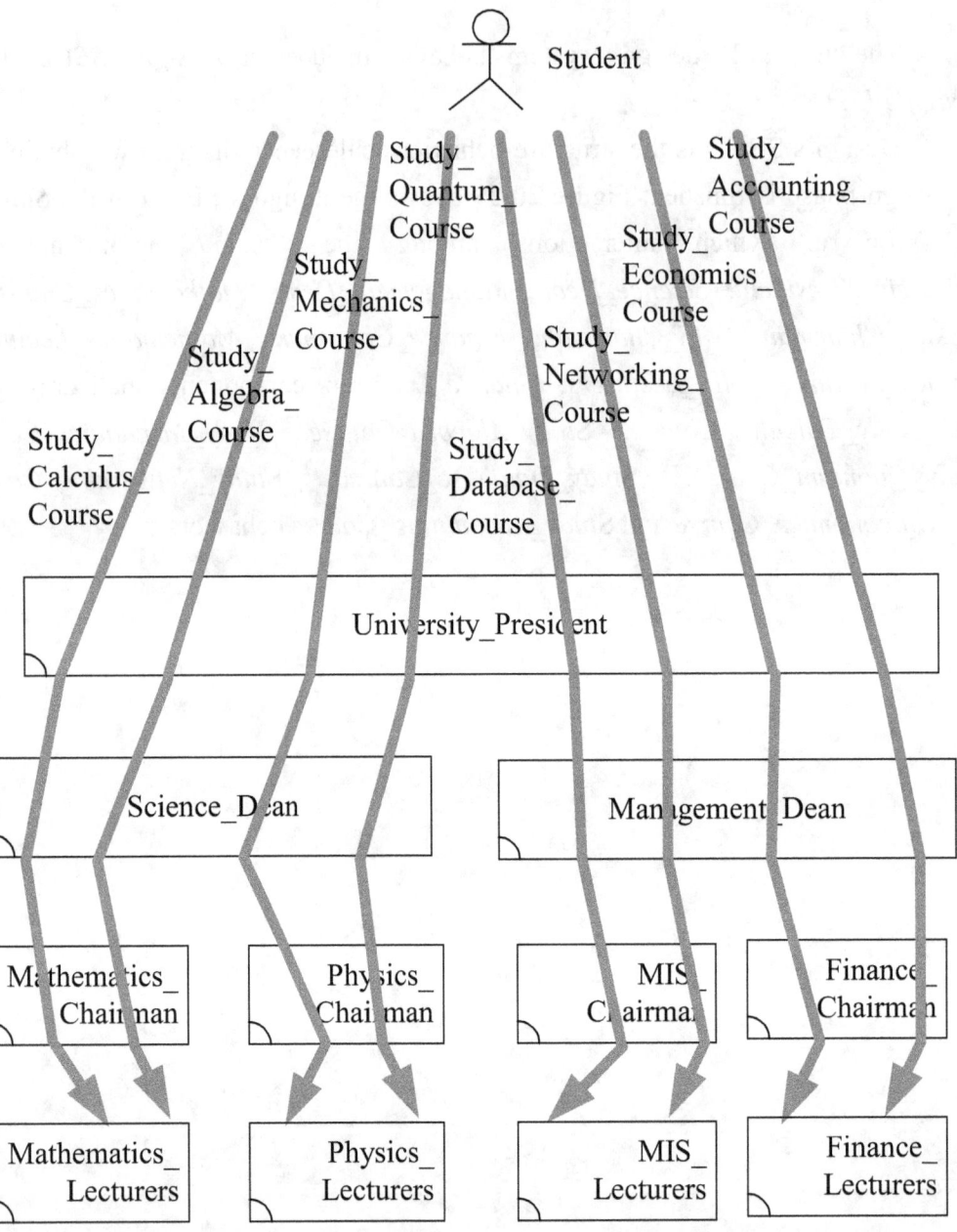

Figure 20-39   Design's SBCD of the *Stanford University*

The overall behavior of an enterprise system is contained in its individual behaviors. After the design phase is finished, the overall design's behavior of the *Stanford University* includes the S*tudy_Calculus_Course*, S*tudy_Algebra_Course*, S*tudy_Mechanics_Course*, S*tudy_Quantum_Course*, S*tudy_Database_Course*, S*tudy_Networking_Course*, S*tudy_Economics_Course* and S*tudy_Accounting_Course* behaviors. In other words, the S*tudy_Calculus_Course*, S*tudy_Algebra_Course*, S*tudy_Mechanics_Course*, S*tudy_Quantum_Course*, S*tudy_Database_Course*, S*tudy_Networking_Course*, S*tudy_Economics_Course* and S*tudy_Accounting_Course* behaviors together provide the overall design's behavior of the *Stanford University* after the design phase is finished.

Be noticed that the S*tudy_Calculus_Course*, S*tudy_Algebra_Course*, S*tudy_Mechanics_Course*, S*tudy_Quantum_Course*, S*tudy_Database_Course*, S*tudy_Networking_Course*, S*tudy_Economics_Course* and S*tudy_Accounting_Course* behaviors are mutually independent of each other. They shall be executed concurrently [Hoar85, Miln89, Miln99].

The major purpose of using the architectural approach, instead of separating the structure model from the behavior model, is to achieve a coalesced model. In Figure 20-39, we not only see its organizational structure, also see at the same time its organizational behavior in the design's SBCD of the *Stanford University*.

After the design phase is finished, the overall design's behavior of the *Stanford University* includes eight behaviors: S*tudy_Calculus_Course*, S*tudy_Algebra_Course*, S*tudy_Mechanics_Course*, S*tudy_Quantum_Course*, S*tudy_Database_Course*, S*tudy_Networking_Course*, S*tudy_Economics_Course* and S*tudy_Accounting_Course*. Each of them is described by an individual IFD. Figure 20-40 shows the design's IFD of the S*tudy_Calculus_Course* behavior. First, actor *Student* interacts with the *University_President* component through the *University_Teach_Calculus* operation call interaction, carrying the *Fee* input parameter. Next, component *University_President* interacts with the *Science_Dean* component through the *College_Teach_Calculus* operation call interaction. Continuingly, component *Science_Dean* interacts with the *Mathematics_Chairman* component through the *Department_Teach_Calculus* operation call interaction. Again, component *Mathematics_Chairman* interacts with the *Mathematics_Lecturers* component through the *Lecturers_Teach_Calculus* operation call interaction, carrying the *Grade* output parameter. Continuingly, component *Science_Dean* interacts with the *Mathematics_Chairman* component through the *Department_Teach_Calculus* operation return interaction, carrying the *Grade* output parameter. Continuingly, component *University_President* interacts with the *Science_Dean* component through the *College_Teach_Calculus* operation return interaction, carrying the *Grade* output parameter. Finally, actor *Student* interacts with the *University_President* component through the *University_Teach_Calculus* operation return interaction, carrying the *Grade* output parameter.

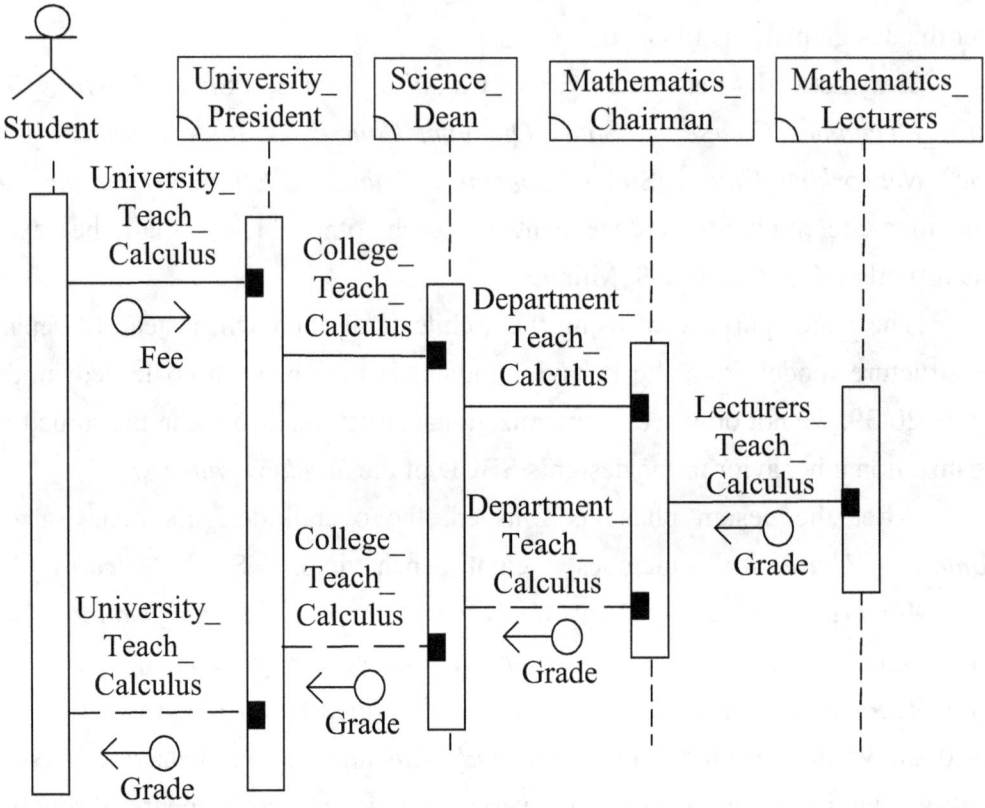

Figure 20-40   Design's IFD of the *Study_Calculus_Course* Behavior

Figure 20-41 shows the design's IFD of the S*tudy_Algebra_Course* behavior. First, actor *Student* interacts with the *University_President* component through the *University_Teach_Algebra* operation call interaction, carrying the *Fee* input parameter. Next, component *University_President* interacts with the *Science_Dean* component through the *College_Teach_Algebra* operation call interaction. Continuingly, component *Science_Dean* interacts with the *Mathematics_Chairman* component through the *Department_Teach_Algebra* operation call interaction. Again, component *Mathematics_Chairman* interacts with the *Mathematics_Lecturers* component through the *Lecturers_Teach_Algebra* operation call interaction, carrying the *Grade* output parameter. Continuingly, component *Science_Dean* interacts with the *Mathematics_Chairman* component through the *Department_Teach_Algebra* operation return interaction, carrying the *Grade* output parameter. Continuingly, component *University_President* interacts with the *Science_Dean* component through the *College_Teach_Algebra* operation return interaction, carrying the *Grade* output parameter. Finally, actor *Student* interacts with the *University_President* component through the *University_Teach_Algebra* operation return interaction, carrying the *Grade* output parameter.

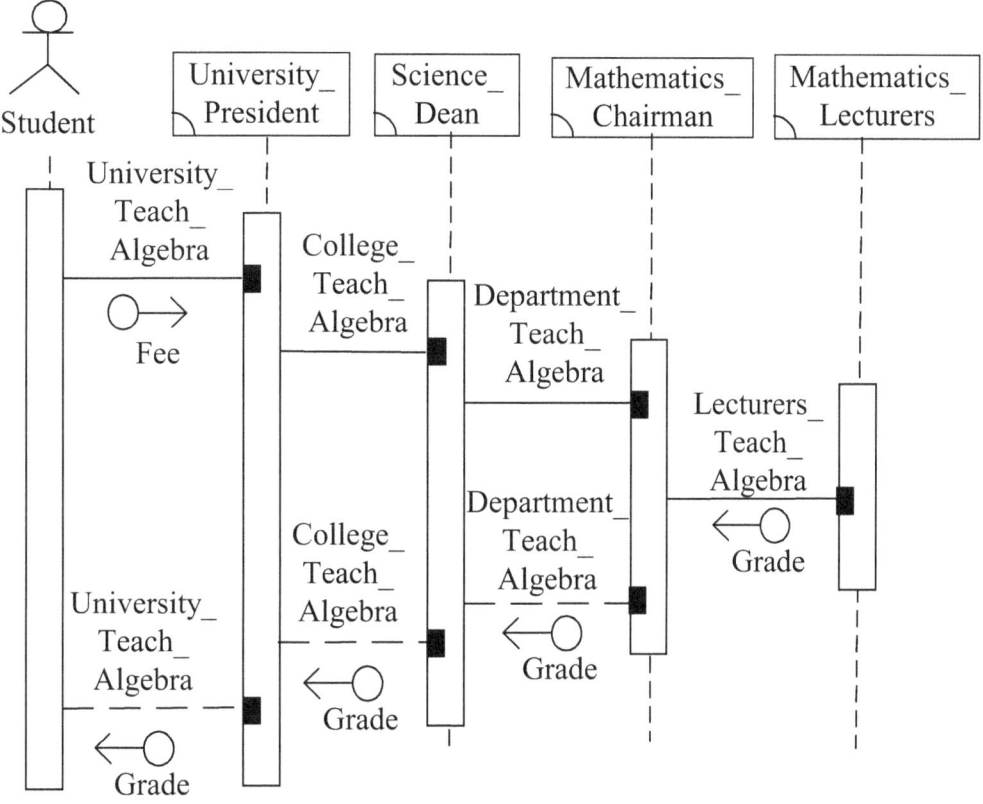

Figure 20-41   Design's IFD of the *Study_Algebra_Course* Behavior

Figure 20-42 shows the design's IFD of the S*tudy_Mechanics_Course* behavior. First, actor *Student* interacts with the *University_President* component through the *University_Teach_Mechanics* operation call interaction, carrying the *Fee* input parameter. Next, component *University_President* interacts with the *Science_Dean* component through the *College_Teach_Mechanics* operation call interaction. Continuingly, component *Science_Dean* interacts with the *Physics_Chairman* component through the *Department_Teach_Mechanics* operation call interaction. Again, component *Physics_Chairman* interacts with the *Physics_Lecturers* component through the *Lecturers_Teach_Mechanics* operation call interaction, carrying the *Grade* output parameter. Continuingly, component *Science_Dean* interacts with the *Physics_Chairman* component through the *Department_Teach_Mechanics* operation return interaction, carrying the *Grade* output parameter. Continuingly, component *University_President* interacts with the *Science_Dean* component through the *College_Teach_Mechanics* operation return interaction, carrying the *Grade* output parameter. Finally, actor *Student* interacts with the *University_President* component through the *University_Teach_Mechanics* operation return interaction, carrying the *Grade* output parameter.

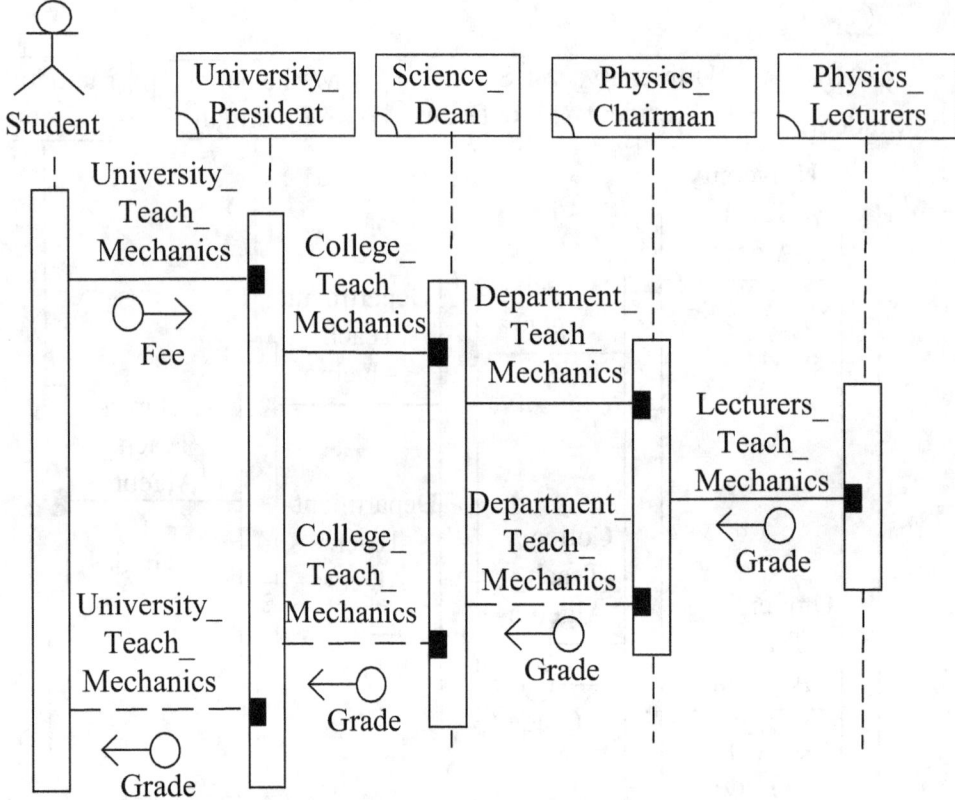

Figure 20-42   Design's IFD of the *Study_Mechanics_Course* Behavior

Figure 20-43 shows the design's IFD of the *Study_Quantum_Course* behavior. First, actor *Student* interacts with the *University_President* component through the *University_Teach_Quantum* operation call interaction, carrying the *Fee* input parameter. Next, component *University_President* interacts with the *Science_Dean* component through the *College_Teach_Quantum* operation call interaction. Continuingly, component *Science_Dean* interacts with the *Physics_Chairman* component through the *Department_Teach_Quantum* operation call interaction. Again, component *Physics_Chairman* interacts with the *Physics_Lecturers* component through the *Lecturers_Teach_Quantum* operation call interaction, carrying the *Grade* output parameter. Continuingly, component *Science_Dean* interacts with the *Physics_Chairman* component through the *Department_Teach_Quantum* operation return interaction, carrying the *Grade* output parameter. Continuingly, component *University_President* interacts with the *Science_Dean* component through the *College_Teach_Quantum* operation return interaction, carrying the *Grade* output parameter. Finally, actor *Student* interacts with the *University_President* component through the *University_Teach_Quantum* operation return interaction, carrying the *Grade* output parameter.

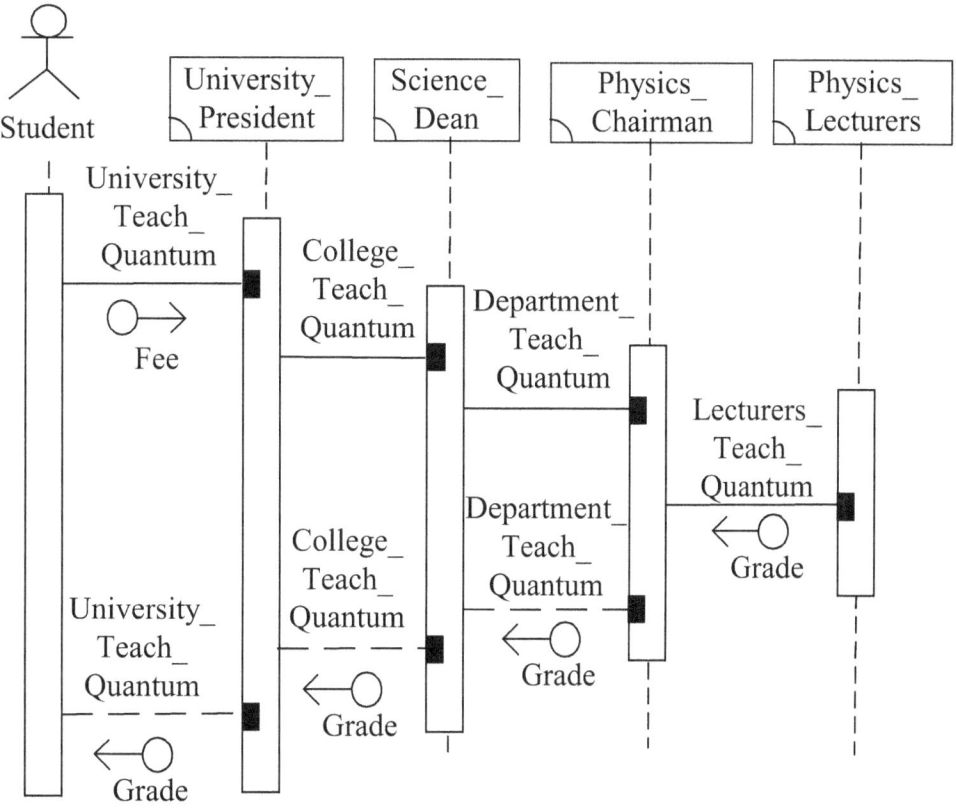

Figure 20-43   Design's IFD of the *Study_Quantum_Course* Behavior

Figure 20-44 shows the design's IFD of the S*tudy_Database_Course* behavior. First, actor *Student* interacts with the *University_President* component through the *University_Teach_Database* operation call interaction, carrying the *Fee* input parameter. Next, component *University_President* interacts with the *Management_Dean* component through the *College_Teach_Database* operation call interaction. Continuingly, component *Management_Dean* interacts with the *MIS_Chairman* component through the *Department_Teach_Database* operation call interaction. Again, component *MIS_Chairman* interacts with the *MIS_Lecturers* component through the *Lecturers_Teach_Database* operation call interaction, carrying the *Grade* output parameter. Continuingly, component *Management_Dean* interacts with the *MIS_Chairman* component through the *Department_Teach_Database* operation return interaction, carrying the *Grade* output parameter. Continuingly, component *University_President* interacts with the *Management_Dean* component through the *College_Teach_Database* operation return interaction, carrying the *Grade* output parameter. Finally, actor *Student* interacts with the *University_President* component through the *University_Teach_Database* operation return interaction, carrying the *Grade* output parameter.

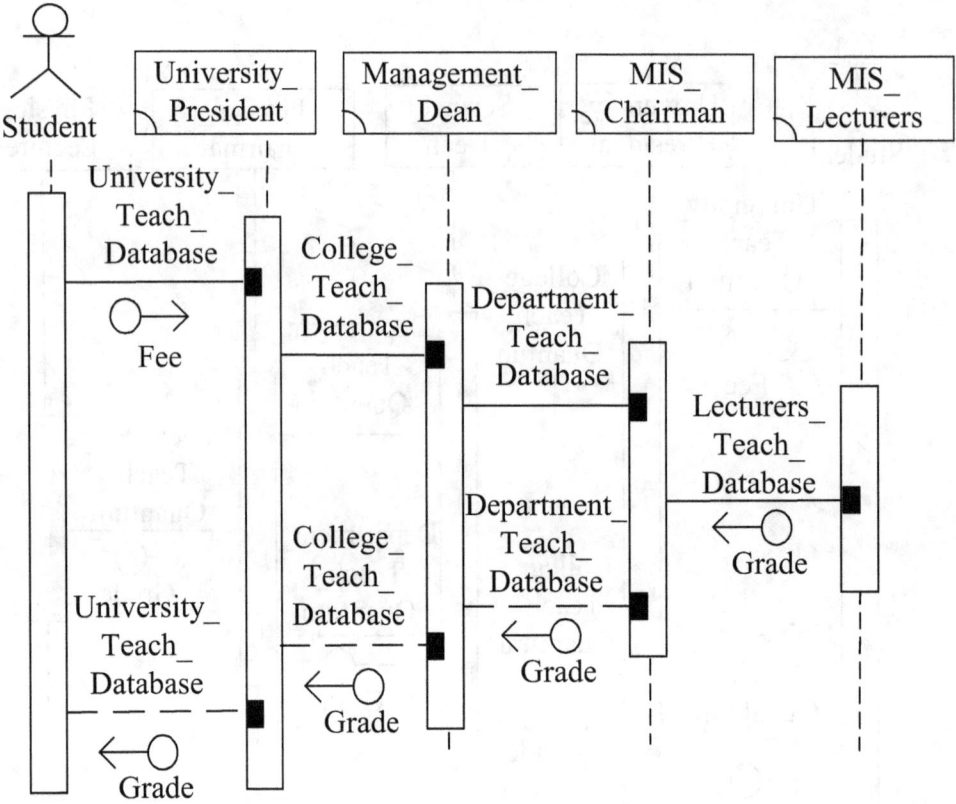

Figure 20-44   Design's IFD of the *Study_Database_Course* Behavior

Figure 20-45 shows the design's IFD of the *Study_Networking_Course* behavior. First, actor *Student* interacts with the *University_President* component through the *University_Teach_Networking* operation call interaction, carrying the *Fee* input parameter. Next, component *University_President* interacts with the *Management_Dean* component through the *College_Teach_Networking* operation call interaction. Continuingly, component *Management_Dean* interacts with the *MIS_Chairman* component through the *Department_Teach_Networking* operation call interaction. Again, component *MIS_Chairman* interacts with the *MIS_Lecturers* component through the *Lecturers_Teach_Networking* operation call interaction, carrying the *Grade* output parameter. Continuingly, component *Management_Dean* interacts with the *MIS_Chairman* component through the *Department_Teach_Networking* operation return interaction, carrying the *Grade* output parameter. Continuingly, component *University_President* interacts with the *Management_Dean* component through the *College_Teach_Networking* operation return interaction, carrying the *Grade* output parameter. Finally, actor *Student* interacts with the *University_President* component through the *University_Teach_Networking* operation return interaction, carrying the *Grade* output parameter.

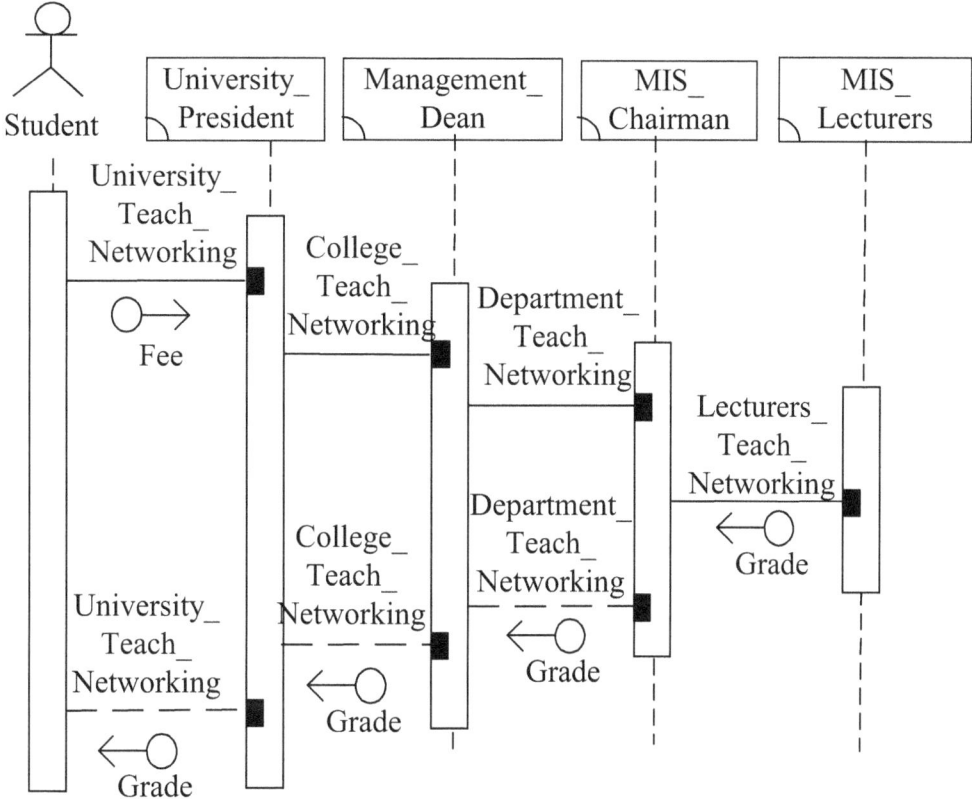

Figure 20-45 Design's IFD of the *Study_Networking_Course* Behavior

Figure 20-46 shows the design's IFD of the *Study_Economics_Course* behavior. First, actor *Student* interacts with the *University_President* component through the *University_Teach_Economics* operation call interaction, carrying the *Fee* input parameter. Next, component *University_President* interacts with the *Management_Dean* component through the *College_Teach_Economics* operation call interaction. Continuingly, component *Management_Dean* interacts with the *Finance_Chairman* component through the *Department_Teach_Economics* operation call interaction. Again, component *Finance_Chairman* interacts with the *Finance_Lecturers* component through the *Lecturers_Teach_Economics* operation call interaction, carrying the *Grade* output parameter. Continuingly, component *Management_Dean* interacts with the *Finance_Chairman* component through the *Department_Teach_Economics* operation return interaction, carrying the *Grade* output parameter. Continuingly, component *University_President* interacts with the *Management_Dean* component through the *College_Teach_Economics* operation return interaction, carrying the *Grade* output parameter. Finally, actor *Student* interacts with the *University_President* component through the *University_Teach_Economics* operation return interaction, carrying the *Grade* output parameter.

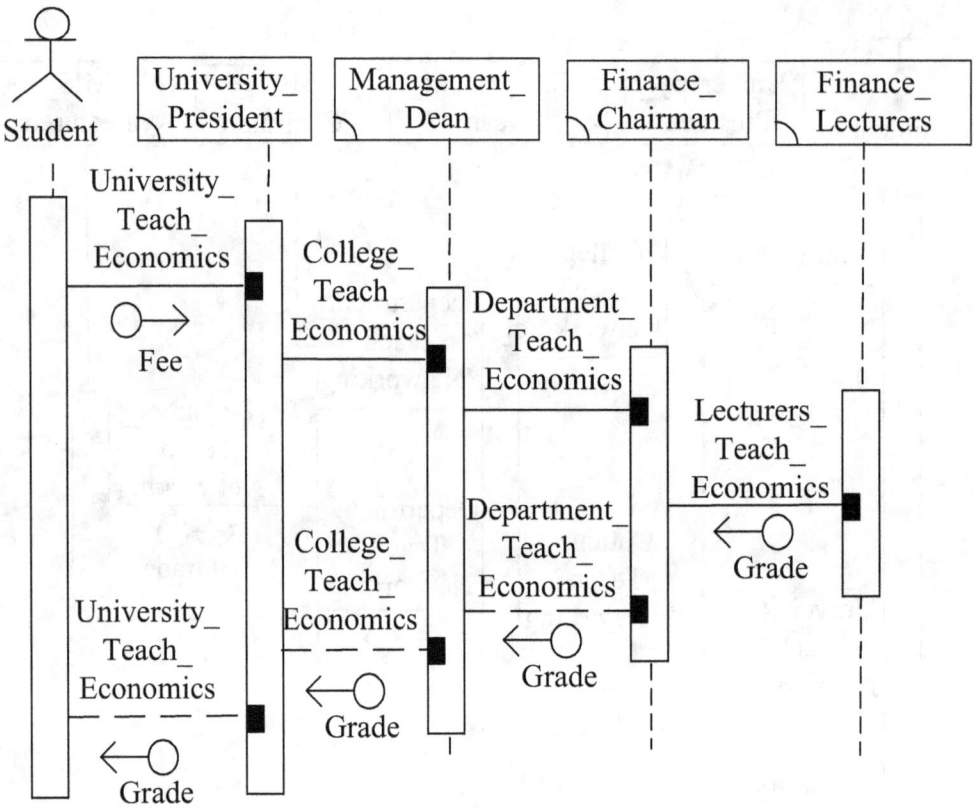

Figure 20-46   Design's IFD of the *Study_Economics_Course* Behavior

Figure 20-47 shows the design's IFD of the *Study_Accounting_Course* behavior. First, actor *Student* interacts with the *University_President* component through the *University_Teach_Accounting* operation call interaction, carrying the *Fee* input parameter. Next, component *University_President* interacts with the *Management_Dean* component through the *College_Teach_Accounting* operation call interaction. Continuingly, component *Management_Dean* interacts with the *Finance_Chairman* component through the *Department_Teach_Accounting* operation call interaction. Again, component *Finance_Chairman* interacts with the *Finance_Lecturers* component through the *Lecturers_Teach_Accounting* operation call interaction, carrying the *Grade* output parameter. Continuingly, component *Management_Dean* interacts with the *Finance_Chairman* component through the *Department_Teach_Accounting* operation return interaction, carrying the *Grade* output parameter. Continuingly, component *University_President* interacts with the *Management_Dean* component through the *College_Teach_Accounting* operation return interaction, carrying the *Grade* output parameter. Finally, actor *Student* interacts with the *University_President* component through the *University_Teach_Accounting* operation return interaction, carrying the *Grade* output parameter.

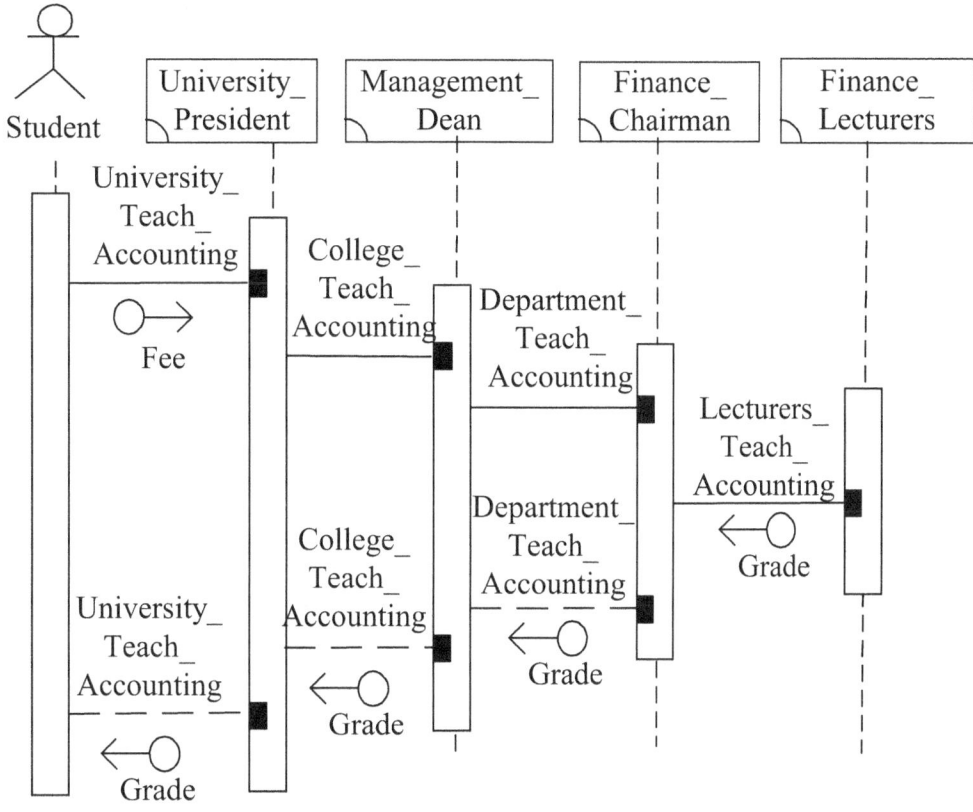

Figure 20-47   Design's IFD of the *Study_Accounting_Course* Behavior

20-4 Implementation View of the Stanford University

In the SBC enterprise multi-level (hierarchical) view, an architect constructs the implementation's systems architecture for the implementer to view. This implementation's systems architecture is called the implementation view as shown in Figure 20-48. Implementation view is one level down structural decomposition (with observation congruence verification) of the design view [Chao15a, Chao15b, Chao15c, Chao15d, Chao15e].

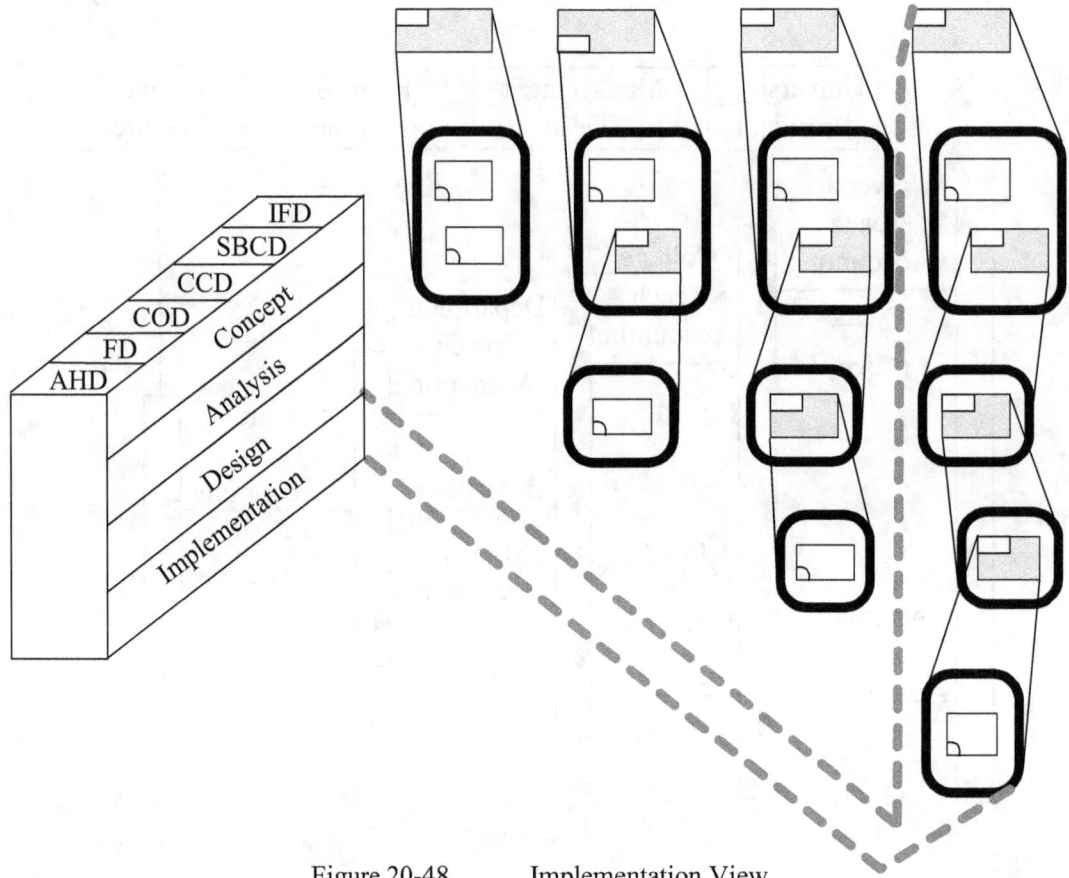

Figure 20-48   Implementation View

The implementation view consists of: a) implementation's systems structure and b) implementation's systems behavior.

20-4-1 Implementation's Systems Structure

The entire SBC implementation's systems structure includes: a) *Implementation's AHD*, b) *Implementation's FD*, c) *Implementation's COD* and d) *Implementation's CCD*.

We first draw the implementation's AHD of the *Stanford University*. As shown in Figure 20-49, *Stanford University* is composed of *University_President*, *Science_College* and *Management_College*; *Science_College* is composed of *Science_Dean*, *Mathematics_Department* and *Physics_Department*; *Management_College* is composed of *Management_Dean*, *MIS_Department* and *Finance_Department*; *Mathematics_Department* is composed of *Mathematics_Chairman* and *Mathematics_Lecturers*; *Physics_Department* is composed of *Physics_Chairman* and *Physics_Lecturers*; *MIS_Department* is composed of *MIS_Chairman* and *MIS_Lecturers*; *Finance_Department* is composed of *Finance_Chairman* and *Finance_Lecturers*; *Mathematics_Lecturers* is composed of *Calculus_Lecturer* and *Algebra_Lecturer*; *Physics_Lecturers* is composed of *Mechanics_Lecturer* and *Quantum_Lecturer*; *MIS_Lecturers* is composed of *Database_Lecturer* and *Networking_Lecturer*; *Finance_Lecturers* is composed of *Economics_Lecturer* and *Accounting_Lecturer*.

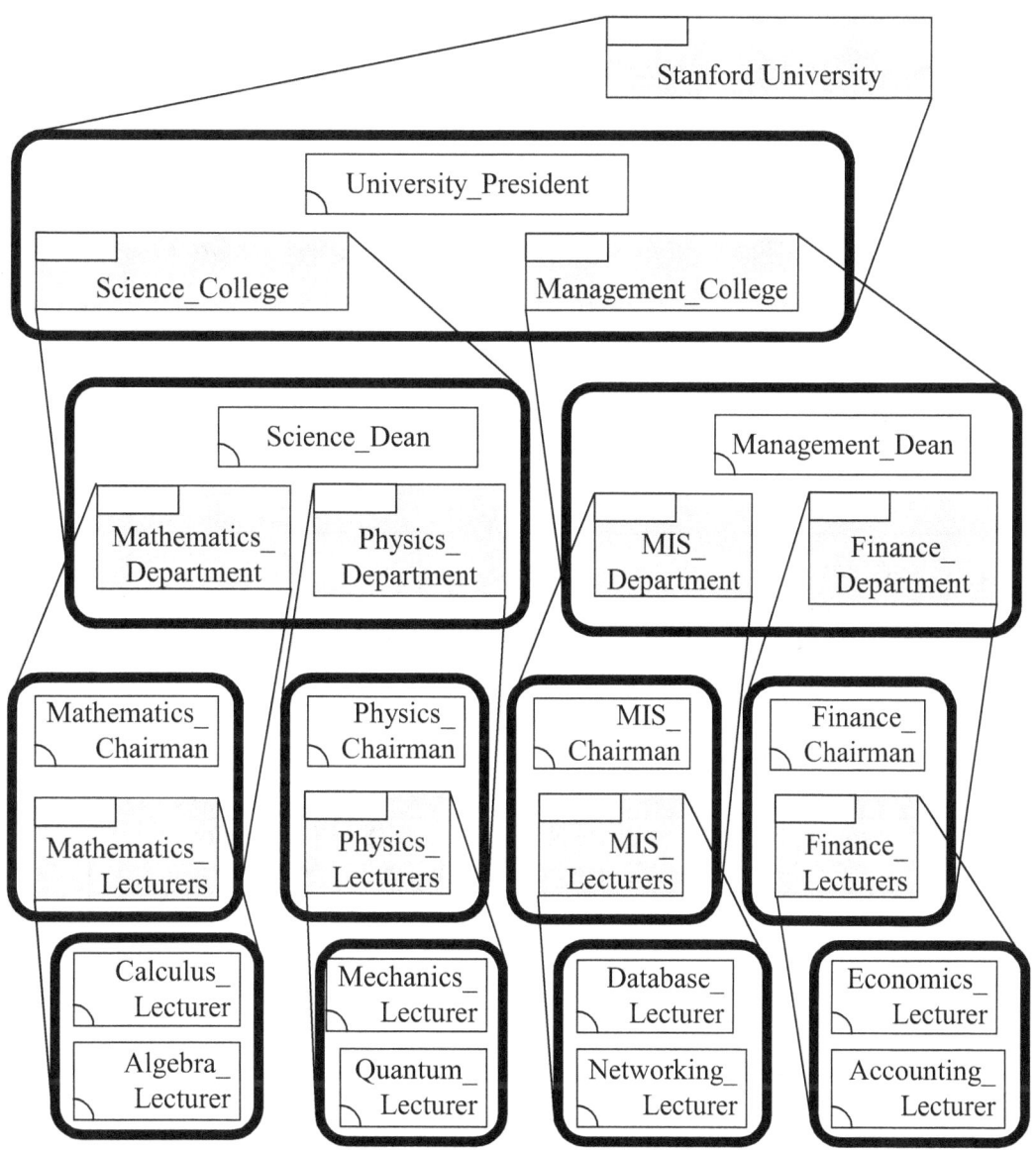

Figure 20-49　Implementation's AHD of the *Stanford University*

In Figure 20-49, *Stanford University, Science_College, Management_College, Mathematics_Department, Physics_Department, MIS_Department, Finance_Department, Mathematics_Lecturers, Physics_Lecturers, MIS_Lecturers* and *Finance_Lecturers* are aggregated systems while *University_President, Science_Dean, Management_Dean, Mathematics_Chairman, Physics_Chairman,*

MIS_Chairman, *Finance_Chairman*, *Calculus_Lecturer*, *Algebra_Lecturer*, *Mechanics_Lecturer*, *Quantum_Lecturer*, *Database_Lecturer*, *Networking_Lecturer*, *Economics_Lecturer* and *Accounting_Lecturer* are non-aggregated systems.

Implementation's FD is the framework diagram we obtain after the implementation phase is finished. Figure 20-50 shows the implementation's FD of the *Stanford University*. In the figure, *University_Layer* contains the *University_President* component; *College_Layer* contains the *Science_Dean* and *Management_Dean* components; *Department_Layer* contains the *Mathematics_Chairman*, *Physics_Chairman*, *MIS_Chairman* and *Finance_Chairman* components; *Lecturer_Layer* contains the *Calculus_Lecturer*, *Algebra_Lecturer*, *Mechanics_Lecturer*, *Quantum_Lecturer*, *Database_Lecturer*, *Networking_Lecturer*, *Economics_Lecturer* and *Accounting_Lecturer* components.

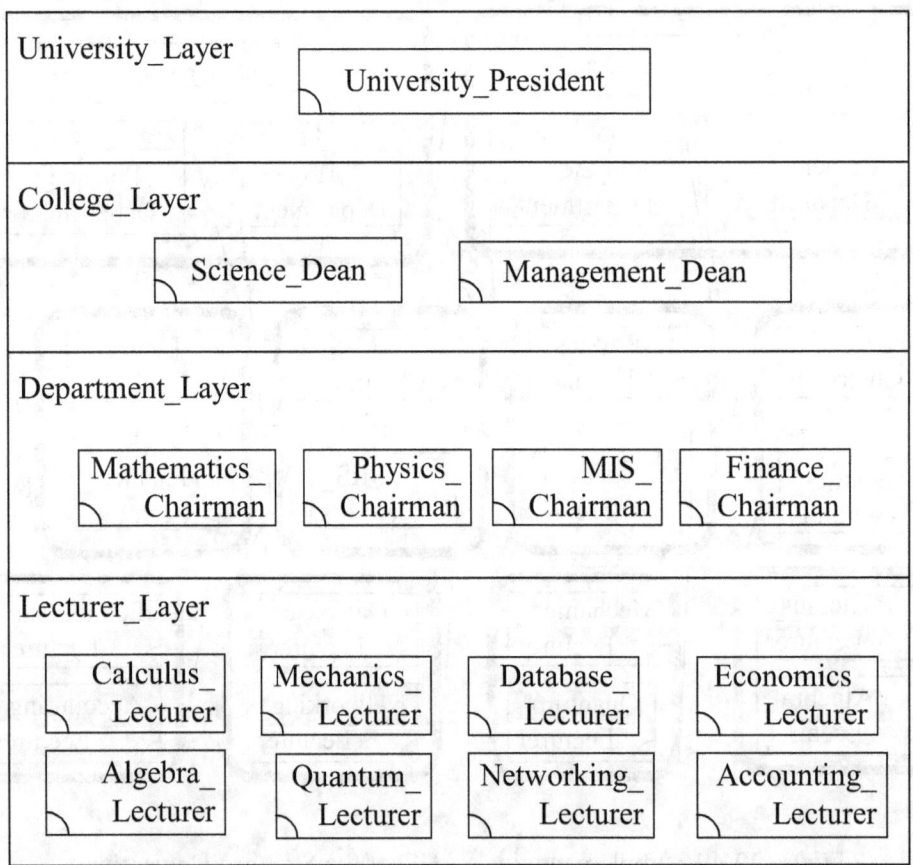

Figure 20-50 Implementation's FD of the *Stanford University*

Implementation's COD is the component operation diagram we obtain after the implementation phase is finished. Figure 20-51 shows the implementation's COD of the *Stanford University*. In the figure, component *University_President* has eight operations: *University_Teach_Calculus*, *University_Teach_Algebra*, *University_Teach_Mechanics*, *University_Teach_Quantum*, *University_Teach_Database*, *University_Teach_Networking*, *University_Teach_Economics* and *University_Teach_Accounting*; component *Science_Dean* has four operations: *College_Teach_Calculus*, *College_Teach_Algebra*, *College_Teach_Mechanics* and *College_Teach_Quantum*; component *Management_Dean* has four operations: *College_Teach_Database*, *College_Teach_Networking*, *College_Teach_Economics* and *College_Teach_Accounting*; *Mathematics_Chairman* has two operations: *Department_Teach_Calculus* and *Department_Teach_Algebra*; *Physics_Chairman* has two operations: *Department_Teach_Mechanics* and *Department_Teach_Quantum*; *MIS_Chairman* has two operations: *Department_Teach_Database* and *Department_Teach_Networking*; *Finance_Chairman* has two operations: *Department_Teach_Economics* and *Department_Teach_Accounting*; component *Calculus_Lecturer* has one operation: *Lecturers_Teach_Calculus*; component *Algebra_Lecturer* has one operation: *Lecturers_Teach_Algebra*; component *Mechanics_Lecturer* has one operation: *Lecturers_Teach_Mechanics*; component *Quantum_Lecturer* has one operation: *Lecturers_Teach_Quantum*; component *Database_Lecturer* has one operation: *Lecturers_Teach_Database*; component *Networking_Lecturer* has one operation: *Lecturers_Teach_Networking*; component *Economics_Lecturer* has one operation: *Lecturers_Teach_Economics*; component *Accounting_Lecturer* has one operation: *Lecturers_Teach_Accounting*.

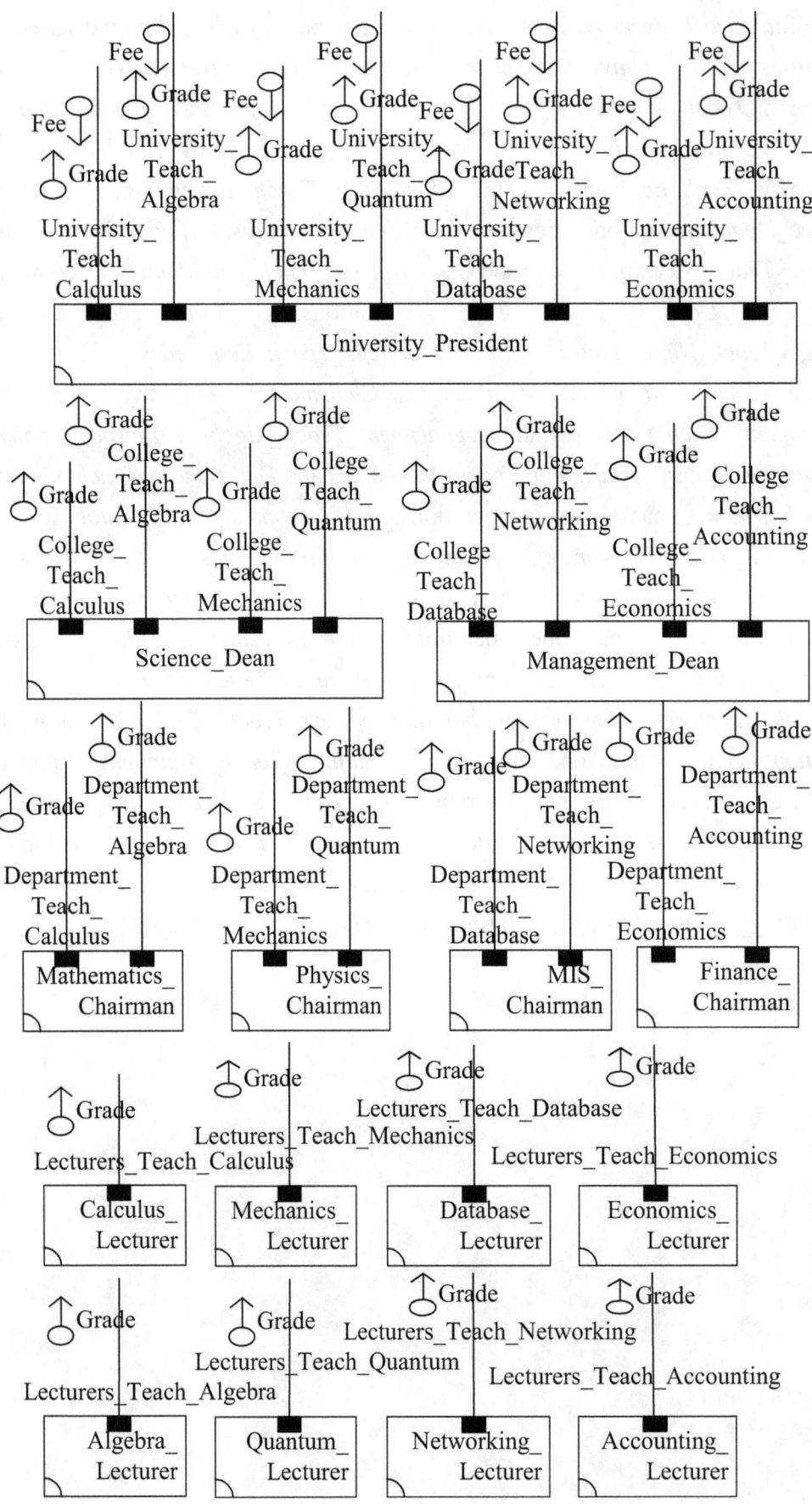

Figure 20-51 Implementation's COD of the *Stanford University*

The operation formula of *University_Teach_Calculus* is *University_Teach_Calculus(In Fee; Out Grade)*. The operation formula of *University_Teach_Algebra* is *University_Teach_Algebra(In Fee; Out Grade)*. The operation formula of *University_Teach_Mechanics* is *University_Teach_Mechanics(In Fee; Out Grade)*. The operation formula of *University_Teach_Quantum* is *University_Teach_Quantum(In Fee; Out Grade)*. The operation formula of *University_Teach_Database* is *University_Teach_Database(In Fee; Out Grade)*. The operation formula of *University_Teach_Networking* is *University_Teach_Networking(In Fee; Out Grade)*. The operation formula of *University_Teach_Economics* is *University_Teach_Economics(In Fee; Out Grade)*. The operation formula of *University_Teach_Accounting* is *University_Teach_Accounting(In Fee; Out Grade)*.

The operation formula of *College_Teach_Calculus* is *College_Teach_Calculus(Out Grade)*. The operation formula of *College_Teach_Algebra* is *College_Teach_Algebra(Out Grade)*. The operation formula of *College_Teach_Mechanics* is *College_Teach_Mechanics(Out Grade)*. The operation formula of *College_Teach_Quantum* is *College_Teach_Quantum(Out Grade)*. The operation formula of *College_Teach_Database* is *College_Teach_Database(Out Grade)*. The operation formula of *College_Teach_Networking* is *College_Teach_Networking(Out Grade)*. The operation formula of *College_Teach_Economics* is *College_Teach_Economics(Out Grade)*. The operation formula of *College_Teach_Accounting* is *College_Teach_Accounting(Out Grade)*.

The operation formula of *Department_Teach_Calculus* is *Department_Teach_Calculus(Out Grade)*. The operation formula of *Department_Teach_Algebra* is *Department_Teach_Algebra(Out Grade)*. The operation formula of *Department_Teach_Mechanics* is *Department_Teach_Mechanics(Out Grade)*. The operation formula of *Department_Teach_Quantum* is *Department_Teach_Quantum(Out Grade)*. The operation formula of *Department_Teach_Database* is *Department_Teach_Database(Out Grade)*. The operation formula of *Department_Teach_Networking* is *Department_Teach_Networking(Out Grade)*. The operation formula of *Department_Teach_Economics* is *Department_Teach_Economics(Out Grade)*. The operation formula of *Department_Teach_Accounting* is *Department_Teach_Accounting(Out Grade)*.

The operation formula of *Lecturers_Teach_Calculus* is *Lecturers_Teach_Calculus(Out Grade)*. The operation formula of *Lecturers_Teach_Algebra* is *Lecturers_Teach_Algebra(Out Grade)*. The operation formula of *Lecturers_Teach_Mechanics* is *Lecturers_Teach_Mechanics(Out Grade)*. The operation formula of *Lecturers_Teach_Quantum* is *Lecturers_Teach_Quantum(Out Grade)*. The operation formula of *Lecturers_Teach_Database* is *Lecturers_Teach_Database(Out Grade)*. The operation formula of *Lecturers_Teach_Networking* is *Lecturers_Teach_Networking(Out Grade)*. The operation formula of *Lecturers_Teach_Economics* is *Lecturers_Teach_Economics(Out Grade)*. The operation formula of *Lecturers_Teach_Accounting* is *Lecturers_Teach_Accounting(Out Grade)*.

Figure 20-52 shows the primitive data type specification of the *Fee* input parameter and the *Grade* output parameter.

Parameter	Data Type	Instances
Fee	Intger	2000, 3500
Grade	Text	A+, B+, C-

Figure 20-52   Primitive Data Type Specification

Implementation's CCD is the component connection diagram we obtain after the implementation phase is finished. Figure 20-53 shows the implementation's CCD of the *Stanford University*.

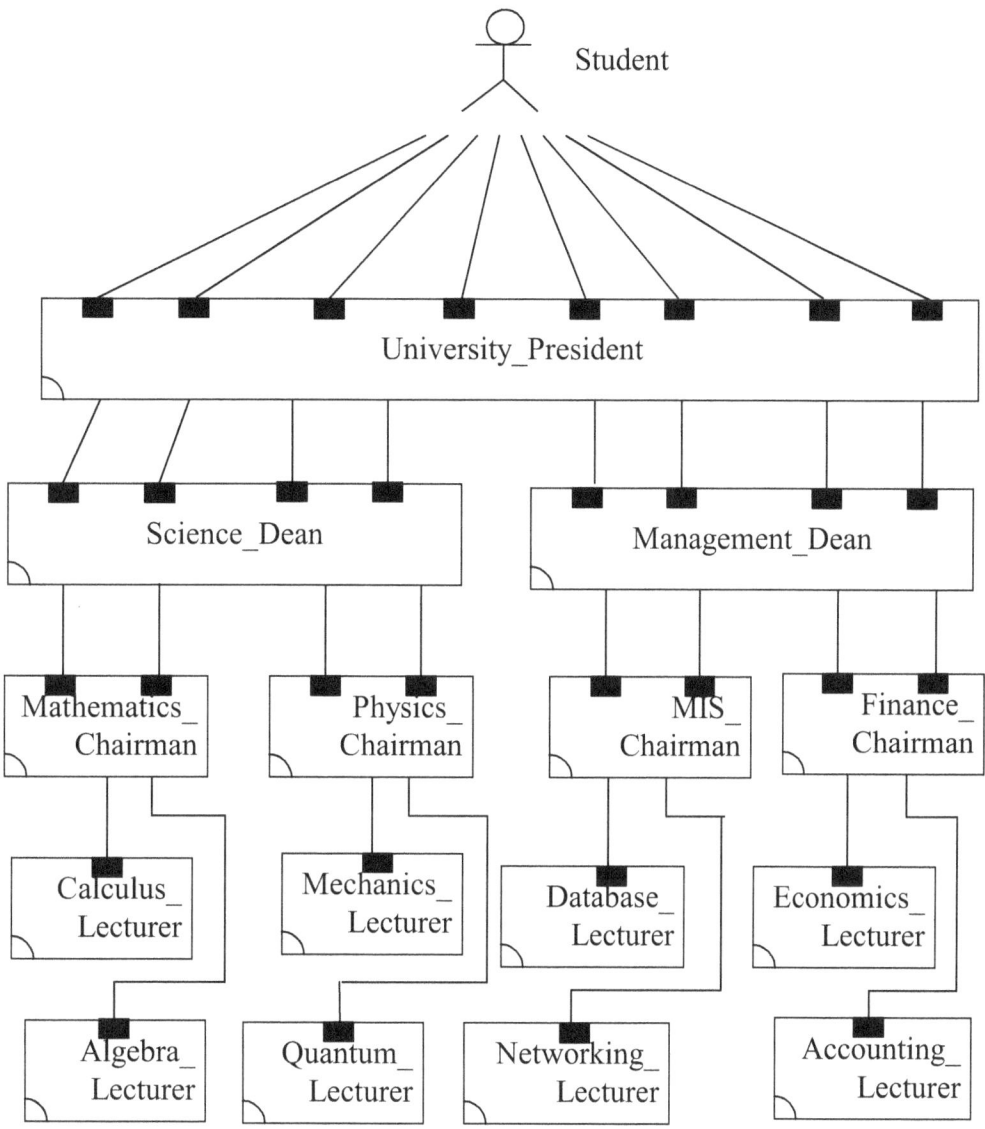

Figure 20-53   Implementation's CCD of the *Stanford University*

In Figure 20-53, actor *Student* has eight connections with the *University_President* component; component *University_President* has four connections with each one of the *Science_Dean* and *Management_Dean* components; component *Science_Dean* has two connections with each one of the *Mathematics_Chairman* and *Physics_Chairman* components; component *Management_Dean* has two connections with each one of the *MIS_Chairman* and *Finance_Chairman* components; component *Mathematics_Chairman* has a connection with each one of the *Calculus_Lecturer* and *Algebra_Lecturer* components; component *Physics_Chairman* has a connection with each one of the *Mechanics_Lecturer* and *Quantum_Lecturer* components; component *MIS_Chairman* has a connection with each one of the *Database_Lecturer* and *Networking_Lecturer* components; component *Finance_Chairman* has a connection with each one of the *Economics_Lecturer* and *Accounting_Lecturer* components.

20-4-2 Implementation's Systems Behavior

The entire SBC implementation's systems behavior includes: a) *Implementation's SBCD* and b) *Implementation's IFD*.

Implementation's SBCD is the structure-behavior coalescence diagram we obtain after the implementation phase is finished. Figure 20-54 shows the implementation's SBCD of the *Stanford University* in which interactions among the *Student* actor and the *University_President, Science_Dean, Management_Dean, Mathematics_Chairman, Physics_Chairman, MIS_Chairman, Finance_Chairman, Calculus_Lecturer, Algebra_Lecturer, Mechanics_Lecturer, Quantum_Lecturer, Database_Lecturer, Networking_Lecturer, Economics_Lecturer, Accounting_Lecturer* components shall draw forth the S*tudy_Calculus_Course,* S*tudy_Algebra_Course,* S*tudy_Mechanics_Course,* S*tudy_Quantum_Course,* S*tudy_Database_Course,* S*tudy_Networking_Course,* S*tudy_Economics_Course* and S*tudy_Accounting_Course* behaviors.

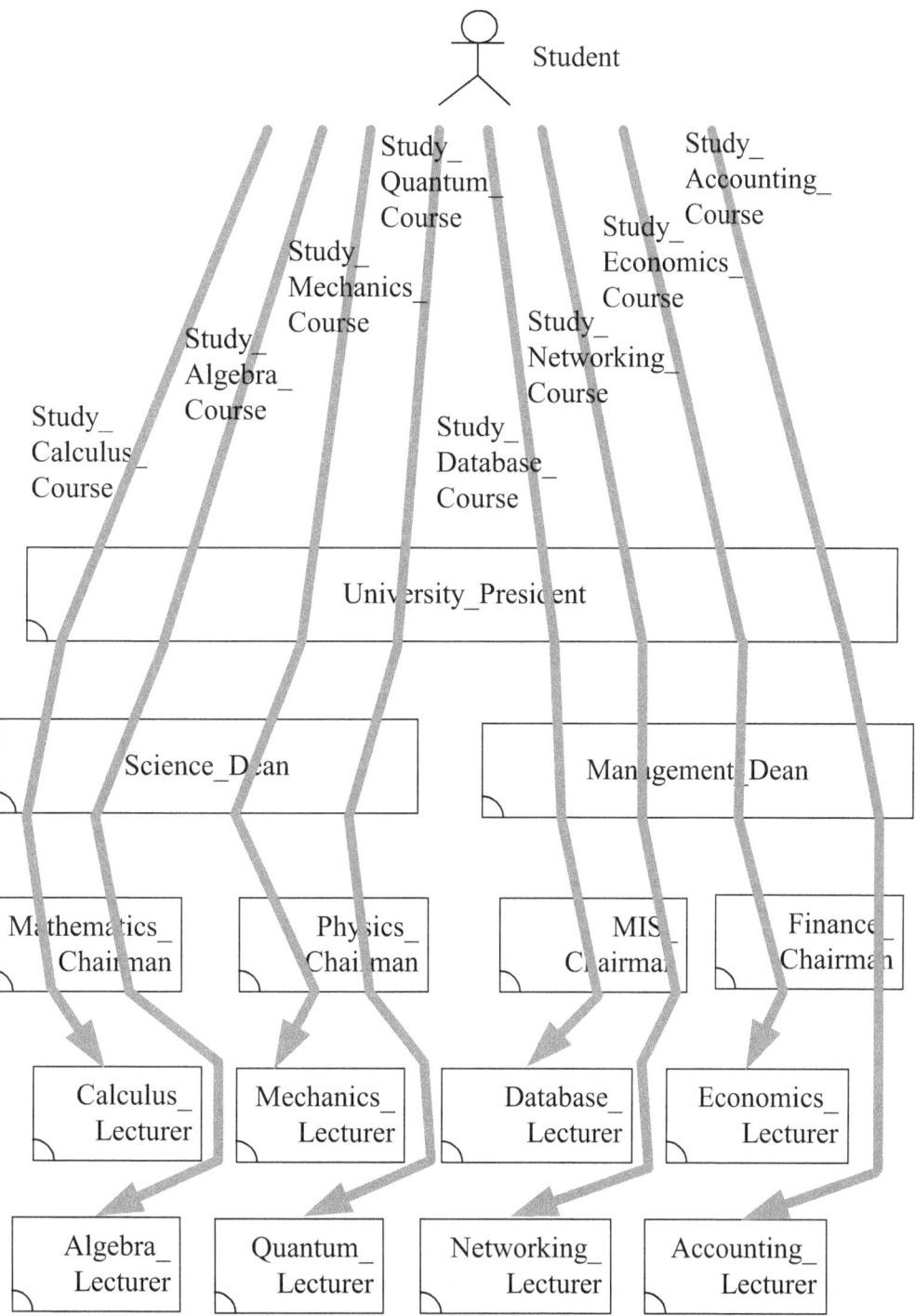

Figure 20-54 Implementation's SBCD of the *Stanford University*

The overall behavior of an enterprise system is contained in its individual behaviors. After the implementation phase is finished, the overall implementation's behavior of the *Stanford University* includes the S*tudy_Calculus_Course*, S*tudy_Algebra_Course*, S*tudy_Mechanics_Course*, S*tudy_Quantum_Course*,

S*tudy_Database_Course*, S*tudy_Networking_Course*, S*tudy_Economics_Course* and S*tudy_Accounting_Course* behaviors. In other words, the S*tudy_Calculus_Course*, S*tudy_Algebra_Course*, *Study_Mechanics_Course*, *Study_Quantum_Course*, S*tudy_Database_Course*, S*tudy_Networking_Course*, S*tudy_Economics_Course* and S*tudy_Accounting_Course* behaviors together provide the overall implementation's behavior of the *Stanford University* after the implementation phase is finished.

Be noticed that the S*tudy_Calculus_Course*, S*tudy_Algebra_Course*, S*tudy_Mechanics_Course*, S*tudy_Quantum_Course*, S*tudy_Database_Course*, S*tudy_Networking_Course*, S*tudy_Economics_Course* and S*tudy_Accounting_Course* behaviors are mutually independent of each other. They shall be executed concurrently [Hoar85, Miln89, Miln99].

The major purpose of using the architectural approach, instead of separating the structure model from the behavior model, is to achieve a coalesced model. In Figure 20-54, we not only see its organizational structure, also see at the same time its organizational behavior in the implementation's SBCD of the *Stanford University*.

After the implementation phase is finished, the overall implementation's behavior of the *Stanford University* includes eight behaviors: S*tudy_Calculus_Course*, S*tudy_Algebra_Course*, *Study_Mechanics_Course*, *Study_Quantum_Course*, S*tudy_Database_Course*, S*tudy_Networking_Course*, *Study_Economics_Course* and S*tudy_Accounting_Course*. Each of them is described by an individual IFD. Figure 20-55 shows the implementation's IFD of the S*tudy_Calculus_Course* behavior. First, actor *Student* interacts with the *University_President* component through the *University_Teach_Calculus* operation call interaction, carrying the *Fee* input parameter. Next, component *University_President* interacts with the *Science_Dean* component through the *College_Teach_Calculus* operation call interaction. Continuingly, component *Science_Dean* interacts with the *Mathematics_Chairman* component through the *Department_Teach_Calculus* operation call interaction. Again, component *Mathematics_Chairman* interacts with the *Calculus_Lecturer* component through the *Lecturers_Teach_Calculus* operation call interaction, carrying the *Grade* output parameter. Continuingly, component *Science_Dean* interacts with the *Mathematics_Chairman* component through the *Department_Teach_Calculus* operation return interaction, carrying the *Grade* output parameter. Continuingly, component *University_President* interacts with the *Science_Dean* component through the *College_Teach_Calculus* operation return interaction, carrying the *Grade* output parameter. Finally, actor *Student* interacts with the *University_President* component through the *University_Teach_Calculus* operation return interaction, carrying the *Grade* output parameter.

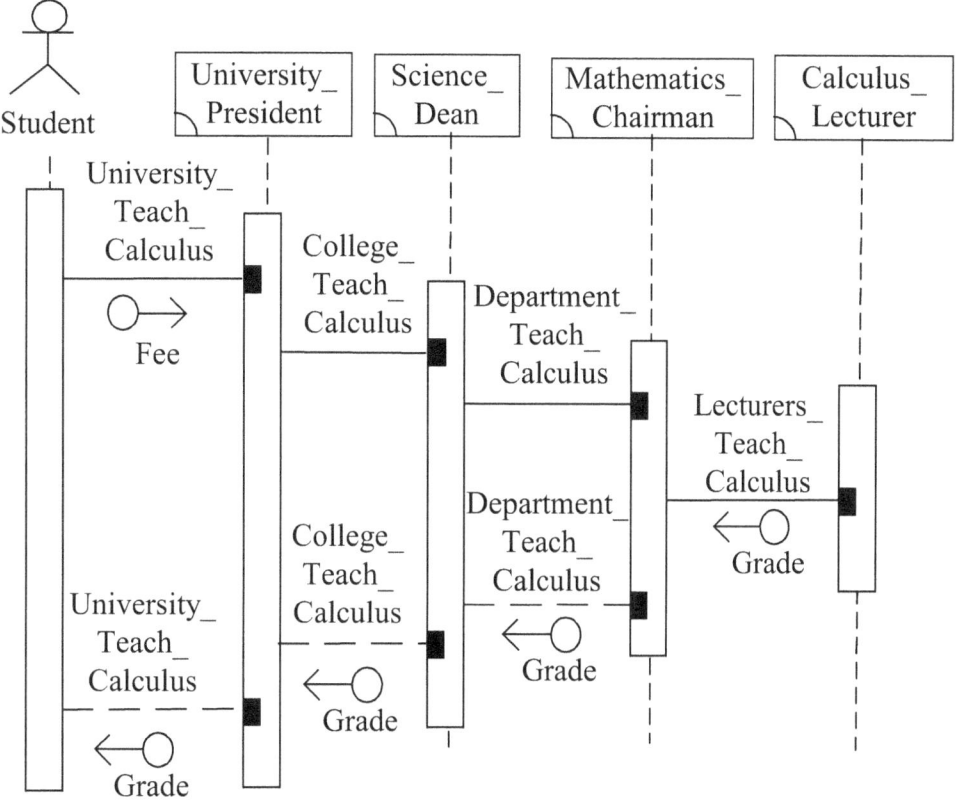

Figure 20-55   Implementation's IFD of the *Study_Calculus_Course* Behavior

Figure 20-56 shows the implementation's IFD of the S*tudy_Algebra_Course* behavior. First, actor *Student* interacts with the *University_President* component through the *University_Teach_Algebra* operation call interaction, carrying the *Fee* input parameter. Next, component *University_President* interacts with the *Science_Dean* component through the *College_Teach_Algebra* operation call interaction. Continuingly, component *Science_Dean* interacts with the *Mathematics_Chairman* component through the *Department_Teach_Algebra* operation call interaction. Again, component *Mathematics_Chairman* interacts with the *Algebra_Lecturer* component through the *Lecturers_Teach_Algebra* operation call interaction, carrying the *Grade* output parameter. Continuingly, component *Science_Dean* interacts with the *Mathematics_Chairman* component through the *Department_Teach_Algebra* operation return interaction, carrying the *Grade* output parameter. Continuingly, component *University_President* interacts with the *Science_Dean* component through the *College_Teach_Algebra* operation return interaction, carrying the *Grade* output parameter. Finally, actor *Student* interacts with the *University_President* component through the *University_Teach_Algebra* operation return interaction, carrying the *Grade* output parameter.

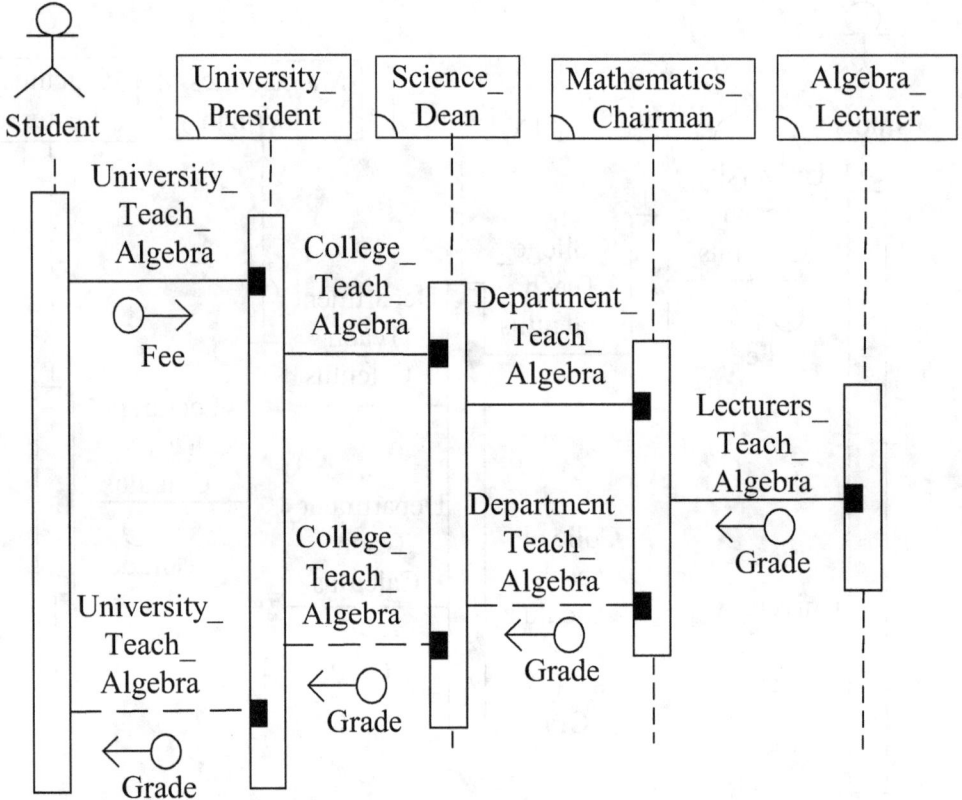

Figure 20-56   Implementation's IFD of the *Study_Algebra_Course* Behavior

Figure 20-57 shows the implementation's IFD of the *Study_Mechanics_Course* behavior. First, actor *Student* interacts with the *University_President* component through the *University_Teach_Mechanics* operation call interaction, carrying the *Fee* input parameter. Next, component *University_President* interacts with the *Science_Dean* component through the *College_Teach_Mechanics* operation call interaction. Continuingly, component *Science_Dean* interacts with the *Physics_Chairman* component through the *Department_Teach_Mechanics* operation call interaction. Again, component *Physics_Chairman* interacts with the *Mechanics_Lecturer* component through the *Lecturers_Teach_Mechanics* operation call interaction, carrying the *Grade* output parameter. Continuingly, component *Science_Dean* interacts with the *Physics_Chairman* component through the *Department_Teach_Mechanics* operation return interaction, carrying the *Grade* output parameter. Continuingly, component *University_President* interacts with the *Science_Dean* component through the *College_Teach_Mechanics* operation return interaction, carrying the *Grade* output parameter. Finally, actor *Student* interacts with the *University_President* component through the *University_Teach_Mechanics* operation return interaction, carrying the *Grade* output parameter.

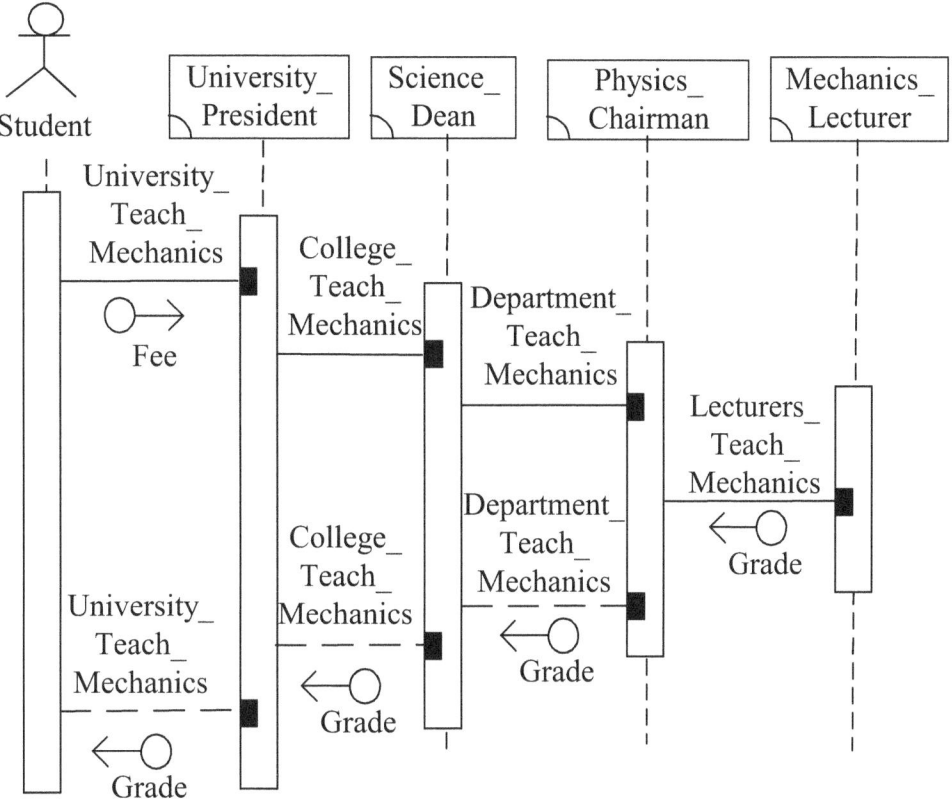

Figure 20-57   Implementation's IFD of the *Study_Mechanics_Course* Behavior

Figure 20-58 shows the implementation's IFD of the *Study_Quantum_Course* behavior. First, actor *Student* interacts with the *University_President* component through the *University_Teach_Quantum* operation call interaction, carrying the *Fee* input parameter. Next, component *University_President* interacts with the *Science_Dean* component through the *College_Teach_Quantum* operation call interaction. Continuingly, component *Science_Dean* interacts with the *Physics_Chairman* component through the *Department_Teach_Quantum* operation call interaction. Again, component *Physics_Chairman* interacts with the *Quantum_Lecturer* component through the *Lecturers_Teach_Quantum* operation call interaction, carrying the *Grade* output parameter. Continuingly, component *Science_Dean* interacts with the *Physics_Chairman* component through the *Department_Teach_Quantum* operation return interaction, carrying the *Grade* output parameter. Continuingly, component *University_President* interacts with the *Science_Dean* component through the *College_Teach_Quantum* operation return interaction, carrying the *Grade* output parameter. Finally, actor *Student* interacts with the *University_President* component through the *University_Teach_Quantum* operation return interaction, carrying the *Grade* output parameter.

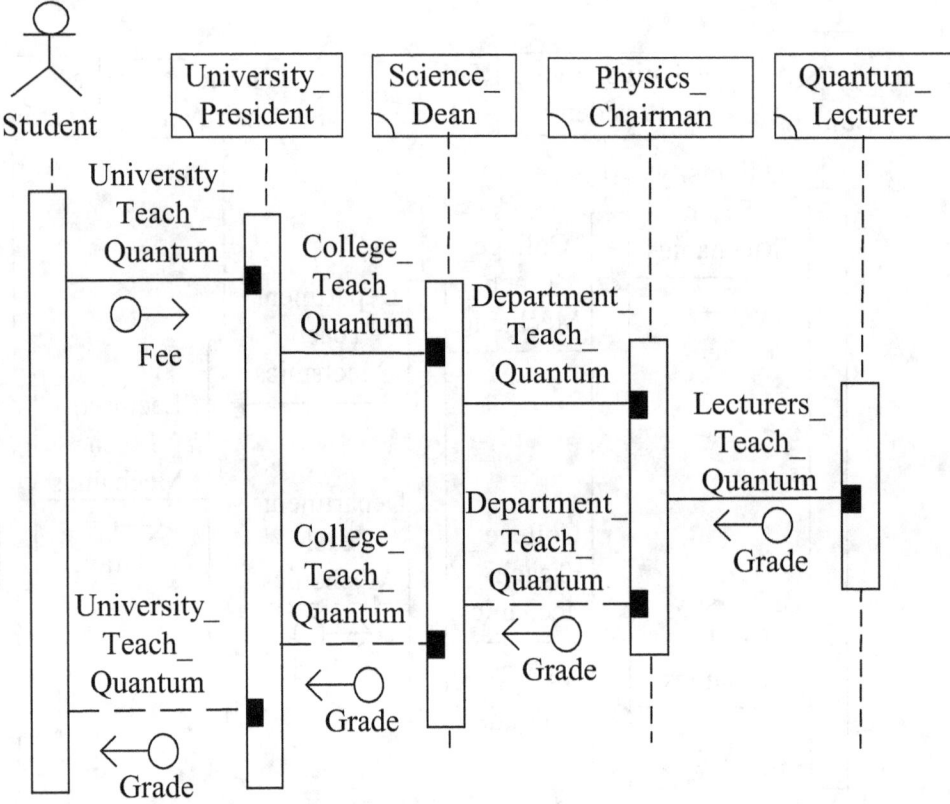

Figure 20-58   Implementation's IFD of the *Study_Quantum_Course* Behavior

Figure 20-59 shows the implementation's IFD of the S*tudy_Database_Course* behavior. First, actor *Student* interacts with the *University_President* component through the *University_Teach_Database* operation call interaction, carrying the *Fee* input parameter. Next, component *University_President* interacts with the *Management_Dean* component through the *College_Teach_Database* operation call interaction. Continuingly, component *Management_Dean* interacts with the *MIS_Chairman* component through the *Department_Teach_Database* operation call interaction. Again, component *MIS_Chairman* interacts with the *Database_Lecturer* component through the *Lecturers_Teach_Database* operation call interaction, carrying the *Grade* output parameter. Continuingly, component *Management_Dean* interacts with the *MIS_Chairman* component through the *Department_Teach_Database* operation return interaction, carrying the *Grade* output parameter. Continuingly, component *University_President* interacts with the *Management_Dean* component through the *College_Teach_Database* operation return interaction, carrying the *Grade* output parameter. Finally, actor *Student* interacts with the *University_President* component through the *University_Teach_Database* operation return interaction, carrying the *Grade* output parameter.

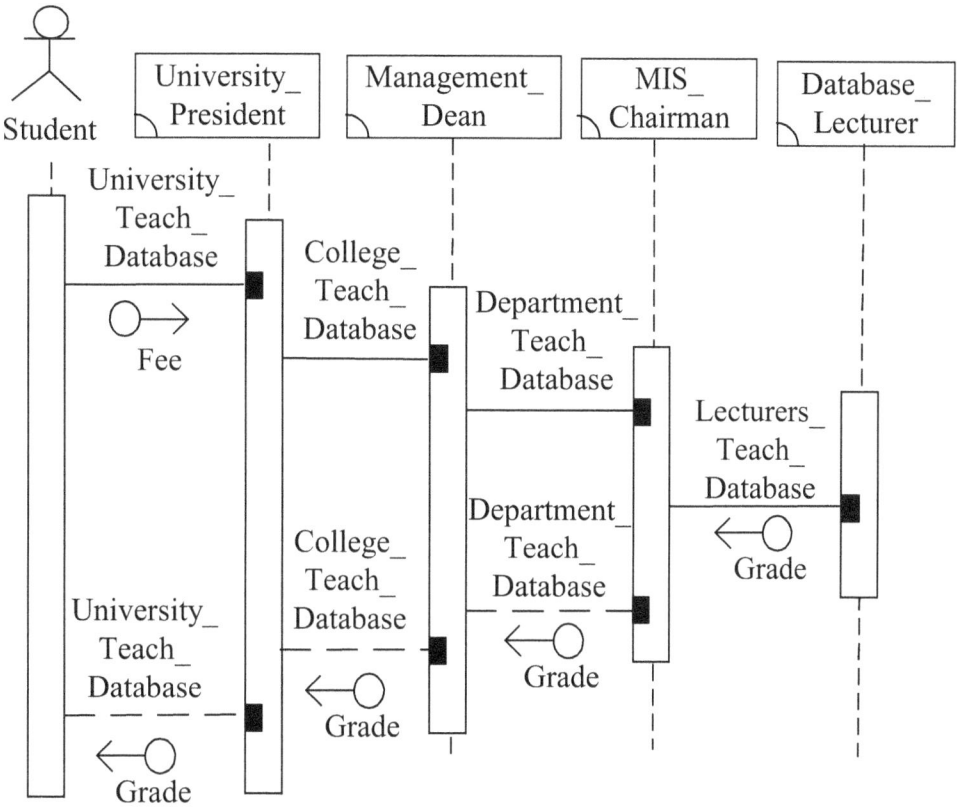

Figure 20-59   Implementation's IFD of the *Study_Database_Course* Behavior

Figure 20-60 shows the implementation's IFD of the *Study_Networking_Course* behavior. First, actor *Student* interacts with the *University_President* component through the *University_Teach_Networking* operation call interaction, carrying the *Fee* input parameter. Next, component *University_President* interacts with the *Management_Dean* component through the *College_Teach_Networking* operation call interaction. Continuingly, component *Management_Dean* interacts with the *MIS_Chairman* component through the *Department_Teach_Networking* operation call interaction. Again, component *MIS_Chairman* interacts with the *Networking_Lecturer* component through the *Lecturers_Teach_Networking* operation call interaction, carrying the *Grade* output parameter. Continuingly, component *Management_Dean* interacts with the *MIS_Chairman* component through the *Department_Teach_Networking* operation return interaction, carrying the *Grade* output parameter. Continuingly, component *University_President* interacts with the *Management_Dean* component through the *College_Teach_Networking* operation return interaction, carrying the *Grade* output parameter. Finally, actor *Student* interacts with the *University_President* component through the *University_Teach_Networking* operation return interaction, carrying the *Grade* output parameter.

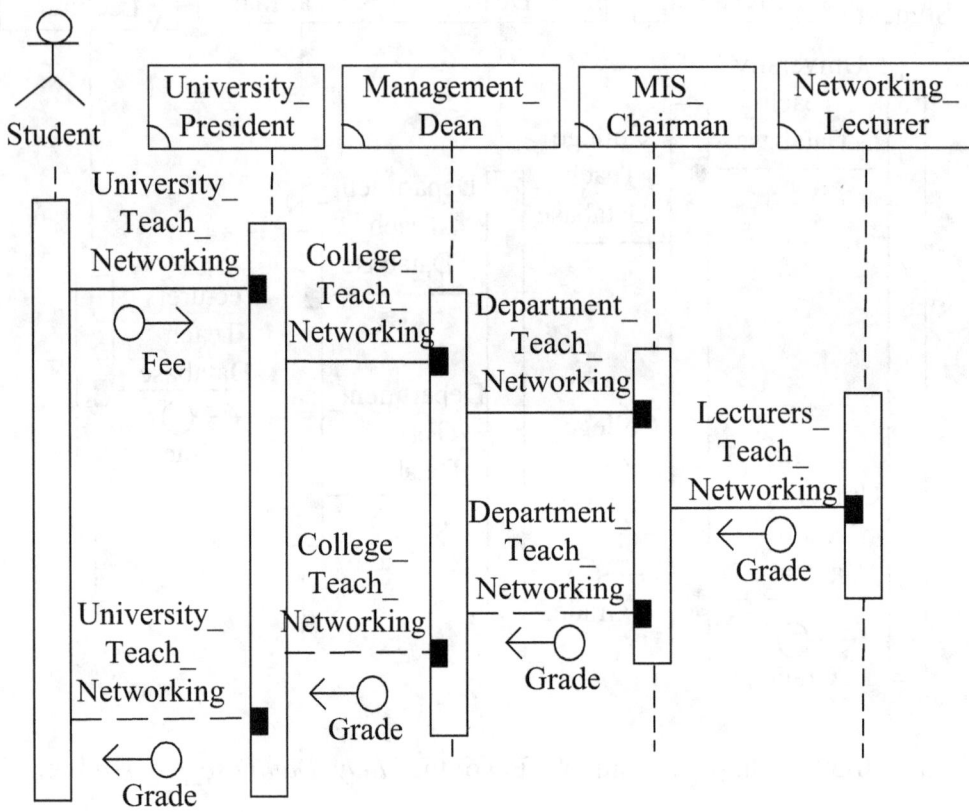

Figure 20-60   Implementation's IFD of the *Study_Networking_Course* Behavior

Figure 20-61 shows the implementation's IFD of the *Study_Economics_Course* behavior. First, actor *Student* interacts with the *University_President* component through the *University_Teach_Economics* operation call interaction, carrying the *Fee* input parameter. Next, component *University_President* interacts with the *Management_Dean* component through the *College_Teach_Economics* operation call interaction. Continuingly, component *Management_Dean* interacts with the *Finance_Chairman* component through the *Department_Teach_Economics* operation call interaction. Again, component *Finance_Chairman* interacts with the *Economics_Lecturer* component through the *Lecturers_Teach_Economics* operation call interaction, carrying the *Grade* output parameter. Continuingly, component *Management_Dean* interacts with the *Finance_Chairman* component through the *Department_Teach_Economics* operation return interaction, carrying the *Grade* output parameter. Continuingly, component

University_President interacts with the *Management_Dean* component through the *College_Teach_Economics* operation return interaction, carrying the *Grade* output parameter. Finally, actor *Student* interacts with the *University_President* component through the *University_Teach_Economics* operation return interaction, carrying the *Grade* output parameter.

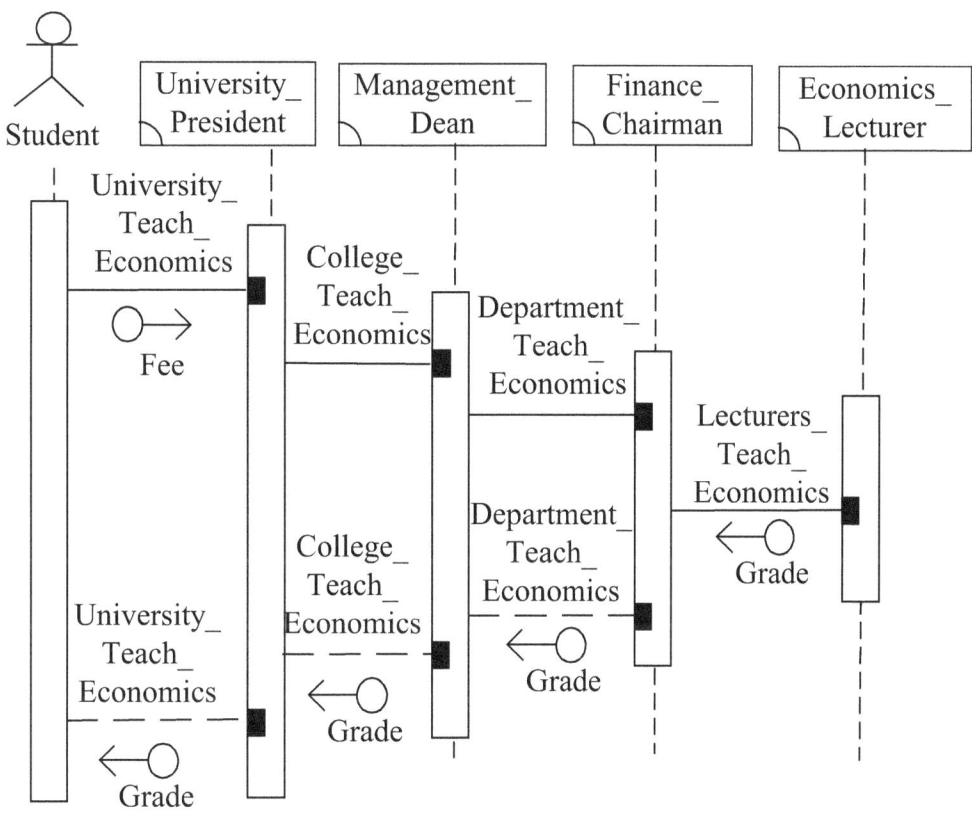

Figure 20-61   Implementation's IFD of the *Study_Economics_Course* Behavior

Figure 20-62 shows the implementation's IFD of the *Study_Accounting_Course* behavior. First, actor *Student* interacts with the *University_President* component through the *University_Teach_Accounting* operation call interaction, carrying the *Fee* input parameter. Next, component *University_President* interacts with the *Management_Dean* component through the *College_Teach_Accounting* operation call interaction. Continuingly, component *Management_Dean* interacts with the *Finance_Chairman* component through the *Department_Teach_Accounting* operation call interaction. Again, component *Finance_Chairman* interacts with the *Accounting_Lecturer* component through the *Lecturers_Teach_Accounting* operation call interaction, carrying the *Grade* output parameter. Continuingly, component *Management_Dean* interacts with the

Finance_Chairman component through the *Department_Teach_Accounting* operation return interaction, carrying the *Grade* output parameter. Continuingly, component *University_President* interacts with the *Management_Dean* component through the *College_Teach_Accounting* operation return interaction, carrying the *Grade* output parameter. Finally, actor *Student* interacts with the *University_President* component through the *University_Teach_Accounting* operation return interaction, carrying the *Grade* output parameter.

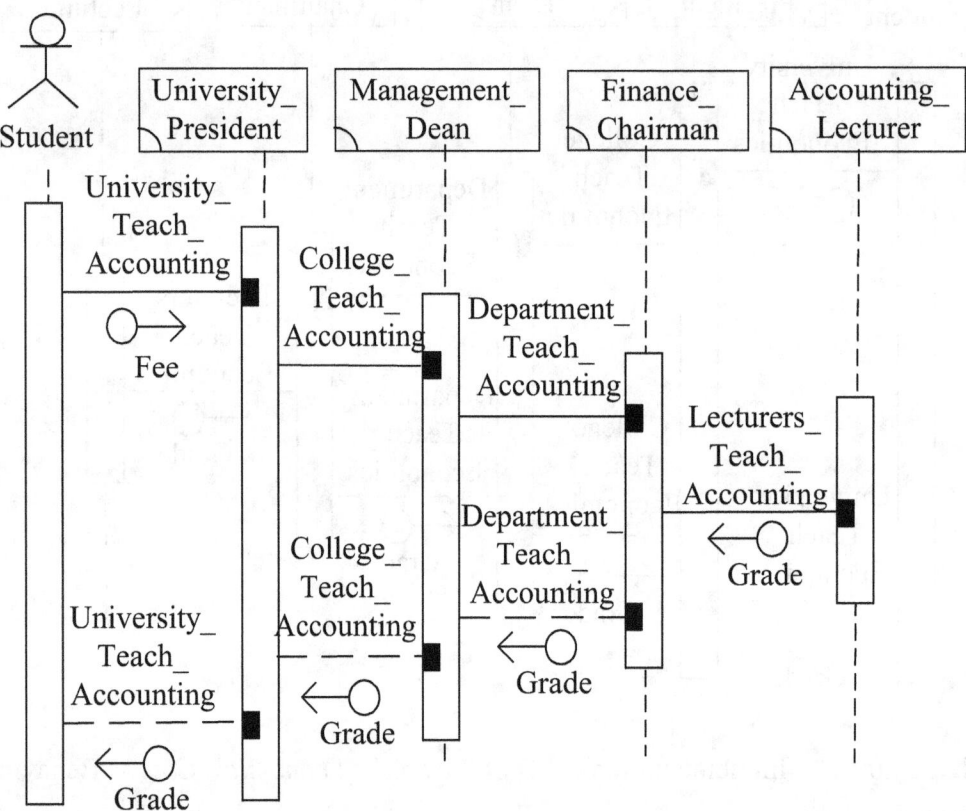

Figure 20-62   Implementation's IFD of the *Study_Accounting_Course* Behavior

Chapter 21: Robot -- Knowledge Architecture

This chapter examines the *Robot* which represents a case study of knowledge architecture [Kend06, Sowa99] for knowledge systems modeling and architecting. Timothy is a six years old boy. His parents have just bought him a robot as a Christmas gift. Timothy is eager to describe the knowledge architecture, or knowledge representation, to his best friends about what a robot is. In general, a person uses a remote controller to make the robot write and walk. The overall behavior of the *Robot* is prominently represented by two behaviors: *Writing* and *Walking* as shown in Figure 21-1.

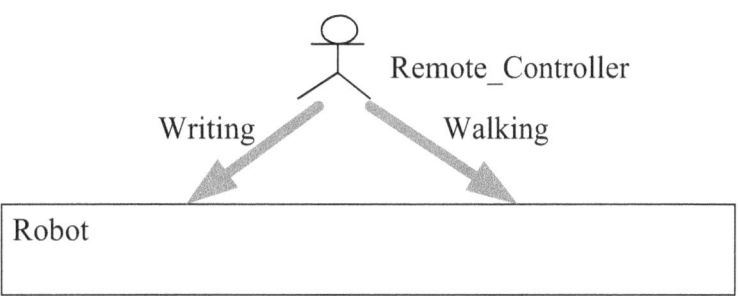

Figure 21-1   Two Behaviors of the *Robot*

Using the SBC knowledge multi-level (hierarchical) view, an architect goes through: a) analysis view and b) design view for the *Robot* knowledge systems architecture as shown in Figure 21-2.

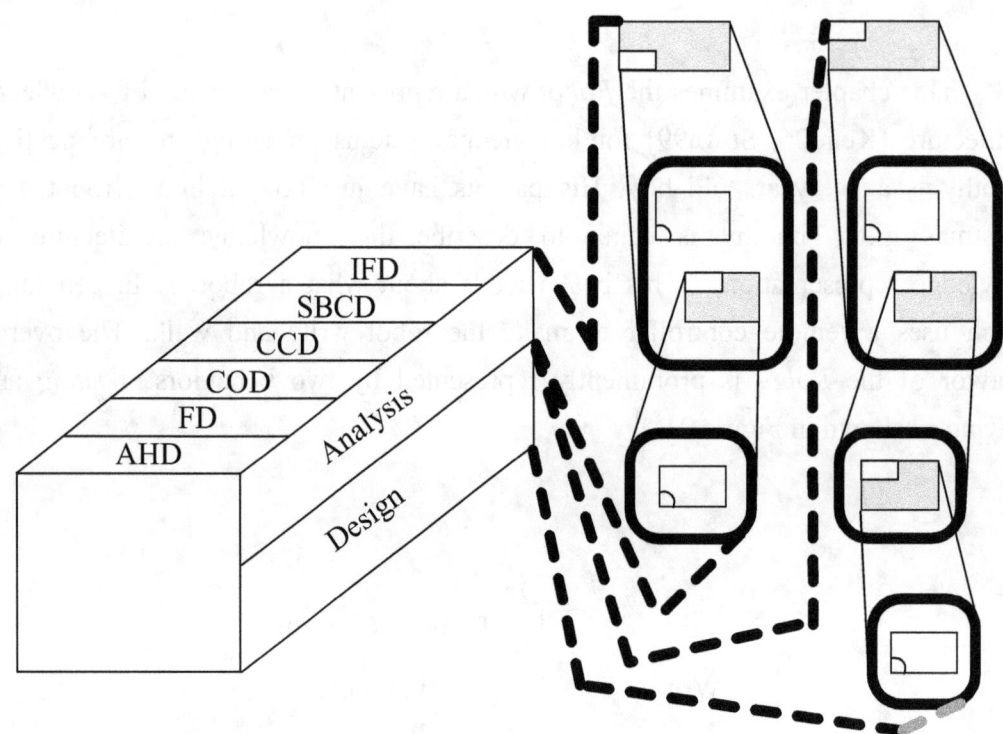

Figure 21-2 SBC Knowledge Multi-Level (Hierarchical) View

21-1 Analysis View of the Robot

In the SBC knowledge multi-level (hierarchical) view, an architect constructs the analysis' systems architecture for the analyzer to view. This analysis' systems architecture is called the analysis view as shown in Figure 21-3.

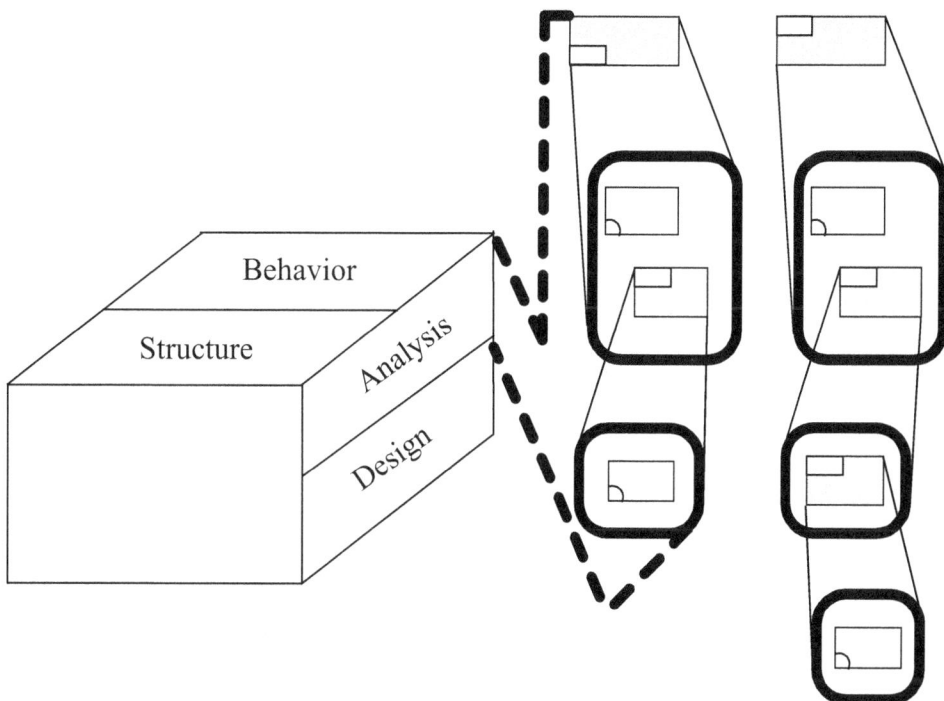

Figure 21-3 Analysis View

The analysis view consists of: a) analysis' systems structure and b) analysis' systems behavior.

21-1-1 Analysis' Systems Structure

The entire SBC analysis' systems structure includes: a) *Analysis' AHD*, b) *Analysis' FD*, c) *Analysis' COD* and d) *Analysis' CCD*.

We first draw the analysis' AHD of the *Robot*. As shown in Figure 21-4, *Robot* is composed of *Head* and *Limb*.

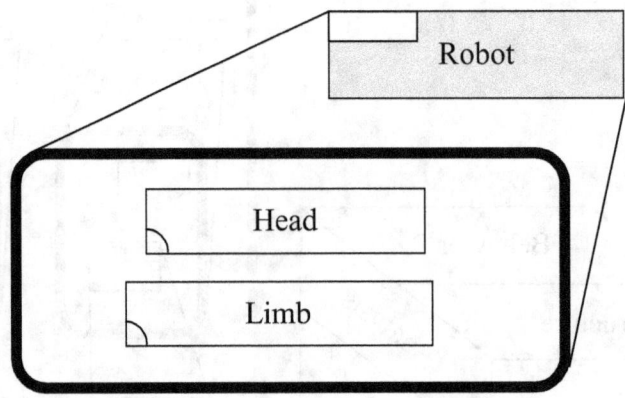

Figure 21-4   Analysis' AHD of the *Robot*

In Figure 21-4, *Robot* is an aggregated system while *Head* and *Limb* are non-aggregated systems.

Analysis' FD is the framework diagram we obtain after the analysis phase is finished. Figure 21-5 shows the analysis' FD of the *Robot*. In the figure, *Layer_2* contains the *Head* and *Limb* components.

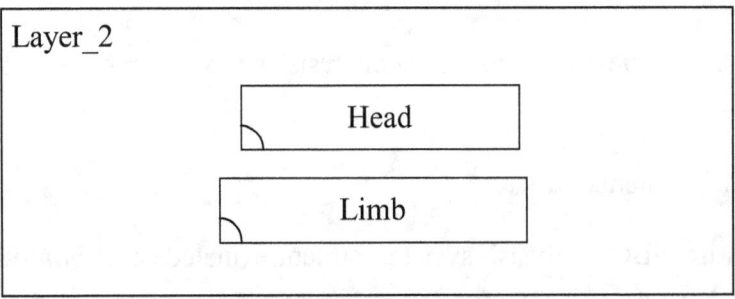

Figure 21-5   Analysis' FD of the *Robot*

Analysis' COD is the component operation diagram we obtain after the analysis phase is finished. Figure 21-6 shows the analysis' COD of the *Robot*. In the figure, component *Head* has two operations: *Receive_Write_Signal* and *Receive_Walk_Signal*; component *Limb* has two operation: *Move_Hand* and *Move_Foot*.

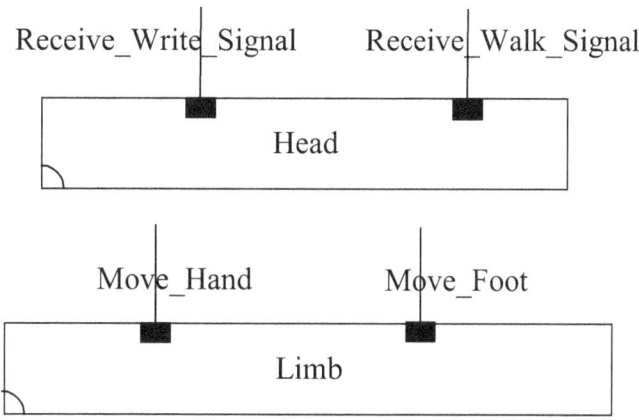

Figure 21-6 Analysis' COD of the *Robot*

Analysis' CCD is the component connection diagram we obtain after the analysis phase is finished. Figure 21-7 shows the analysis' CCD of the *Robot*.

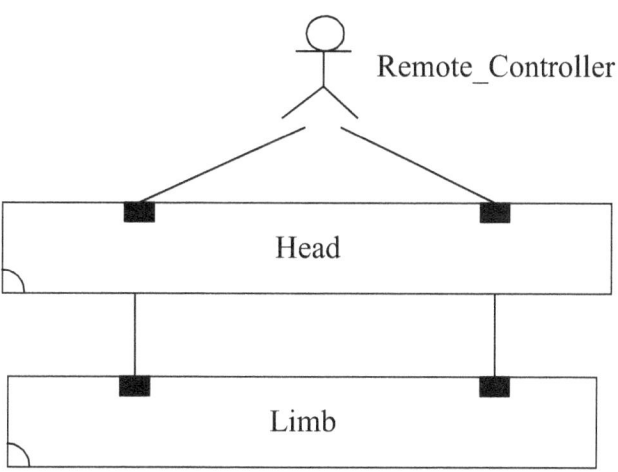

Figure 21-7 Analysis' CCD of the *Robot*

In Figure 21-7, actor *Remote_Controller* has two connections with the *Head* component; component *Head* also has two connections with the *Limb* component.

21-1-2 Analysis' Systems Behavior

The entire SBC analysis' systems behavior includes: a) *Analysis' SBCD* and b) *Analysis' IFD*.

Analysis' SBCD is the structure-behavior coalescence diagram we obtain after the analysis phase is finished. Figure 21-8 shows the analysis' SBCD of the *Robot* in which interactions among the *Remote_Controller* actor and the *Head, Limb* components shall draw forth the *Writing* and *Walking* behaviors.

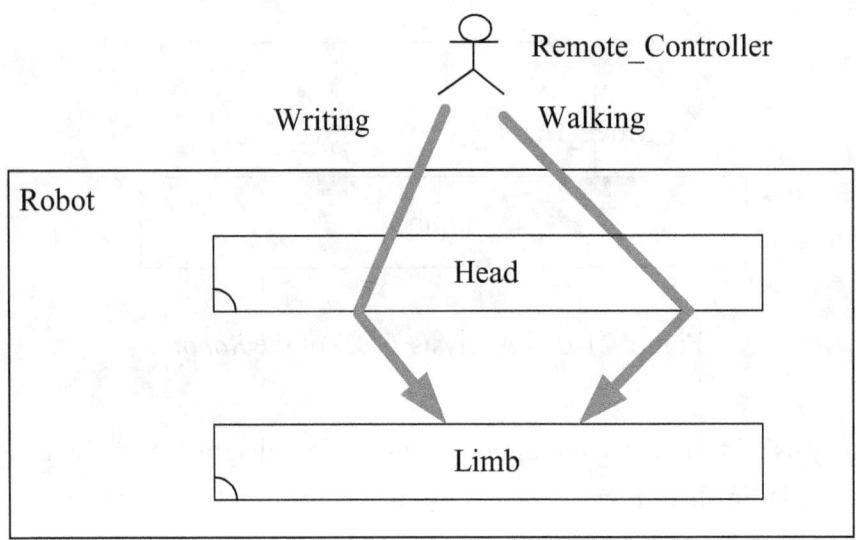

Figure 21-8   Analysis' SBCD of the *Robot*

The overall behavior of a knowledge system is contained in its individual behaviors. After the analysis phase is finished, the overall analysis' behavior of the *Robot* includes the *Writing* and *Walking* behaviors. In other words, the *Writing* and *Walking* behaviors together provide the overall analysis' behavior of the *Robot* after the analysis phase is finished.

Be noticed that the *Writing* and *Walking* behaviors are mutually independent of each other. They tend to be executed concurrently [Hoar85, Miln89, Miln99].

The major purpose of using the architectural approach, instead of separating the structure model from the behavior model, is to achieve a coalesced model. In Figure 21-8, we not only see its knowledge structure, also see at the same time its knowledge behavior in the analysis' SBCD of the *Robot*.

After the analysis phase is finished, the overall analysis' behavior of the *Robot* includes two behaviors: *Writing* and *Walking*. Each of them is described by an individual IFD. Figure 21-9 shows the analysis' IFD of the *Writing* behavior. First, actor *Remote_Controller* interacts with the *Head* component through the *Receive_Write_Signal* operation call interaction. Finally, component *Head* interacts with the *Limb* component through the *Move_Hand* operation call interaction.

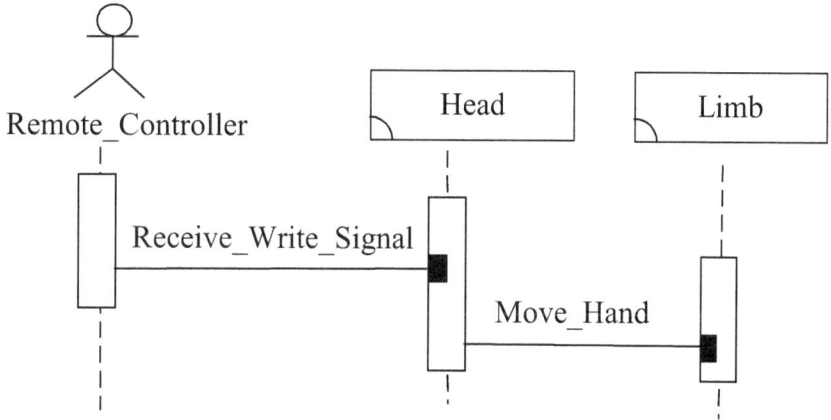

Figure 21-9   Analysis' IFD of the *Writing* Behavior

Figure 21-10 shows the analysis' IFD of the *Walking* behavior. First, actor *Remote_Controller* interacts with the *Head* component through the *Receive_Walk_Signal* operation call interaction. Finally, component *Head* interacts with the *Limb* component through the *Move_Foot* operation call interaction.

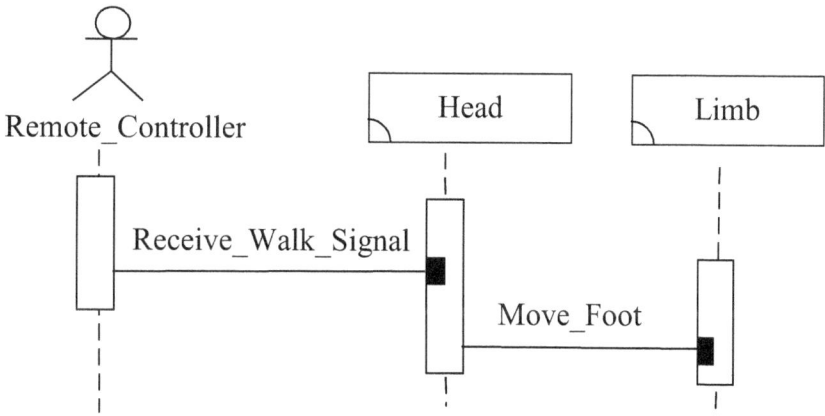

Figure 21-10   Analysis' IFD of the *Walking* Behavior

21-2 Design View of the Robot

In the SBC knowledge multi-level (hierarchical) view, an architect constructs the design's systems architecture for the designer to view. This design's systems architecture is called the design view as shown in Figure 21-11. Design view is one level down structural decomposition (with observation congruence verification) of the analysis view [Chao15a, Chao15b, Chao15c, Chao15d, Chao15e].

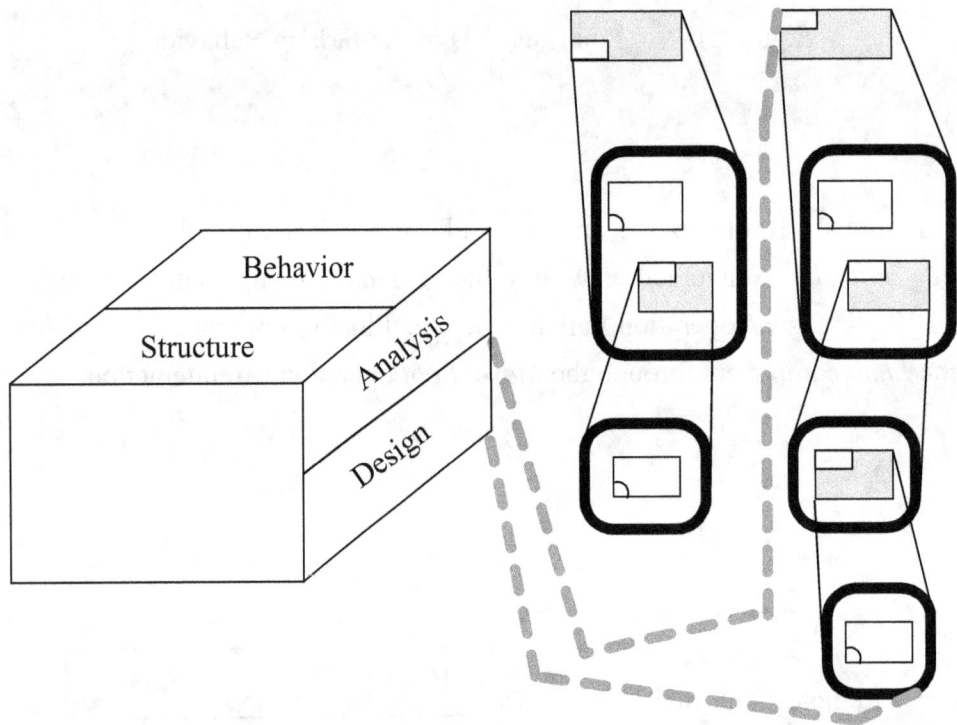

Figure 21-11   Design View

The design view consists of: a) design's systems structure and b) design's systems behavior.

21-2-1 Design's Systems Structure

The entire SBC design's systems structure includes: a) *Design's AHD*, b) *Design's FD*, c) *Design's COD* and d) *Design's CCD*.

We first draw the design's AHD of the *Robot*. As shown in Figure 21-12, *Robot* is composed of *Head* and *Limb*; *Limb* is composed of *Hands* and *Feet*.

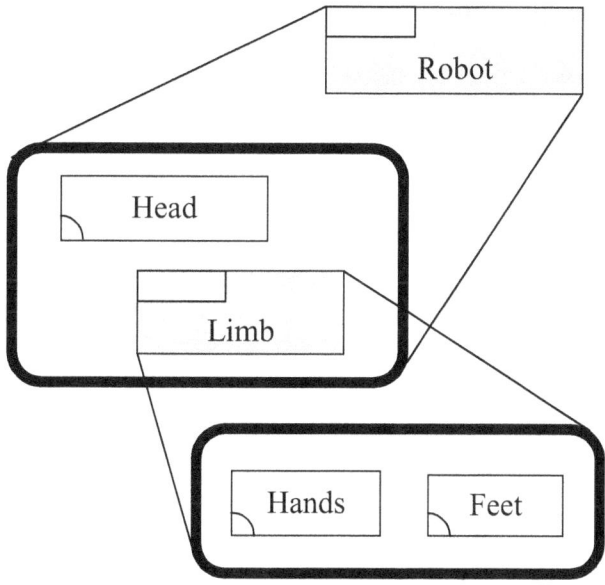

Figure 21-12   Design's AHD of the *Robot*

In Figure 21-12, *Robot* and *Limb* are aggregated systems while *Head*, *Hands* and *Feet* are non-aggregated systems.

Design's FD is the framework diagram we obtain after the design phase is finished. Figure 21-13 shows the design's FD of the *Robot*. In the figure, *Layer_2* contains the *Head* component; *Layer_1* contains the *Hands* and *Feet* components.

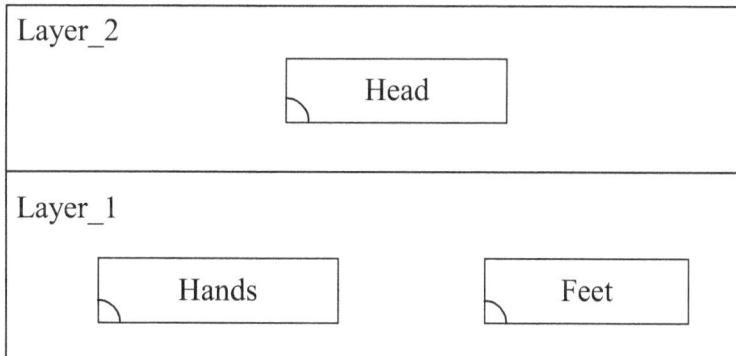

Figure 21-13   Design's FD of the *Robot*

Design's COD is the component operation diagram we obtain after the design phase is finished. Figure 21-14 shows the design's COD of the *Robot*. In the figure, component *Head* has two operations: *Receive_Write_Signal* and

Receive_Walk_Signal; component *Hands* has one operation: *Move_Hand*; component *Feet* has one operation: *Move_Foot*.

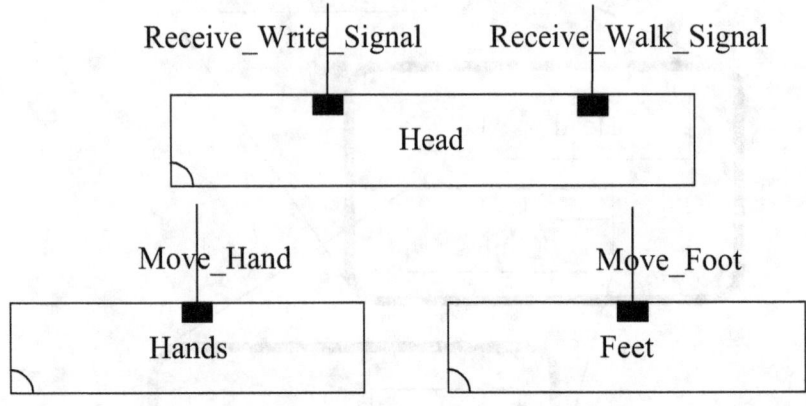

Figure 21-14   Design's COD of the *Robot*

Design's CCD is the component connection diagram we obtain after the design phase is finished. Figure 21-15 shows the design's CCD of the *Robot*.

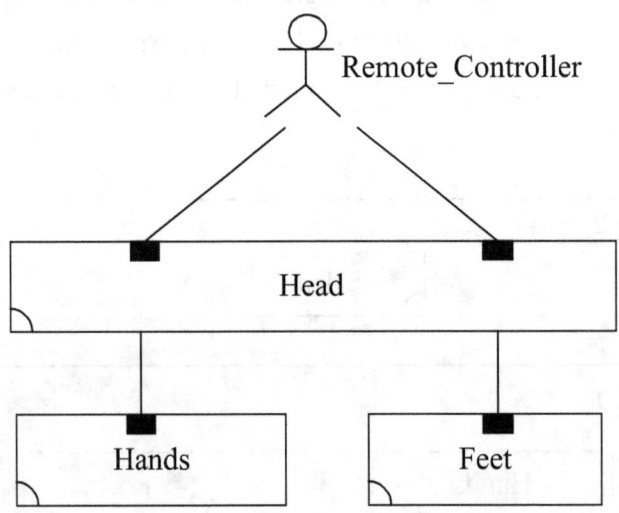

Figure 21-15   Design's CCD of the *Robot*

In Figure 21-15, actor *Remote_Controller* has a connection with the *Head* component; component *Head* has a connection with each one of the *Hands* and *Feet* components.

21-2-2 Design's Systems Behavior

The entire SBC design's systems behavior includes: a) *Design's SBCD* and b) *Design's IFD*.

Design's SBCD is the structure-behavior coalescence diagram we obtain after the design phase is finished. Figure 21-16 shows the design's SBCD of the *Robot* in which interactions among the *Remote_Controller* actor and the *Head*, Hand, *Feet* components shall draw forth the *Writing* and *Walking* behaviors.

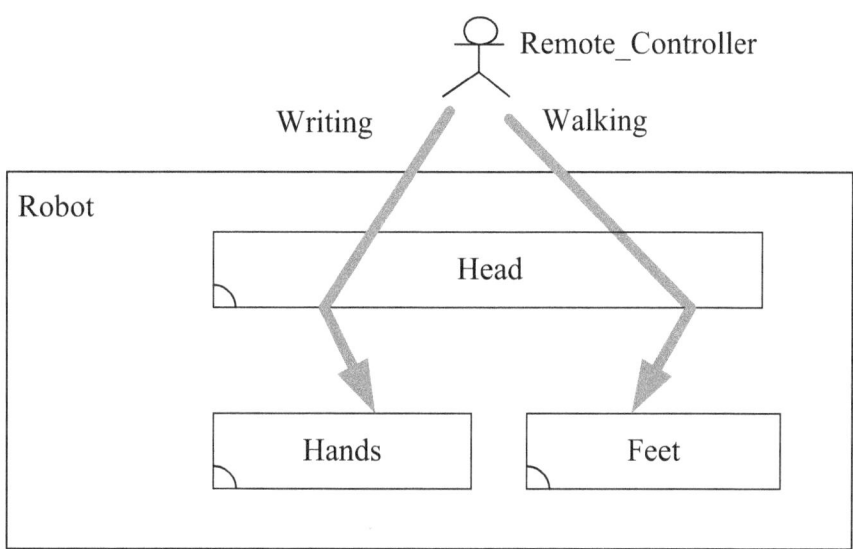

Figure 21-16   Designer's SBCD of the *Robot*

The overall behavior of a knowledge system is contained in its individual behaviors. After the design phase is finished, the overall design's behavior of the *Robot* includes the *Writing* and *Walking* behaviors. In other words, the *Writing* and *Walking* behaviors together provide the overall design's behavior of the *Robot* after the design phase is finished.

Be noticed that the *Writing* and *Walking* behaviors are mutually independent of each other. They tend to be executed concurrently [Hoar85, Miln89, Miln99].

The major purpose of using the architectural approach, instead of separating the structure model from the behavior model, is to achieve a coalesced model. In Figure 21-16, we not only see its knowledge structure, also see at the same time its knowledge behavior in the design's SBCD of the *Robot*.

After the design phase is finished, the overall design's behavior of the *Robot* includes two behaviors: *Writing* and *Walking*. Each of them is described by an individual IFD. Figure 21-17 shows the design's IFD of the *Writing* behavior. First, actor *Remote_Controller* interacts with the *Head* component through the *Receive_Write_Signal* operation call interaction. Finally, component *Head* interacts with the *Hands* component through the *Move_Hand* operation call interaction.

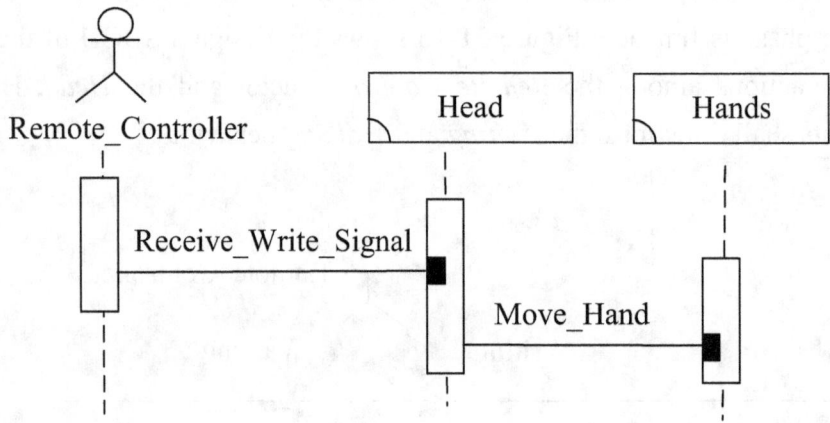

Figure 21-17   Design's IFD of the *Writing* Behavior

Figure 21-18 shows the design's IFD of the *Walking* behavior. First, actor *Remote_Controller* interacts with the *Head* component through the *Receive_Walk_Signal* operation call interaction. Finally, component *Head* interacts with the *Feet* component through the *Move_Foot* operation call interaction.

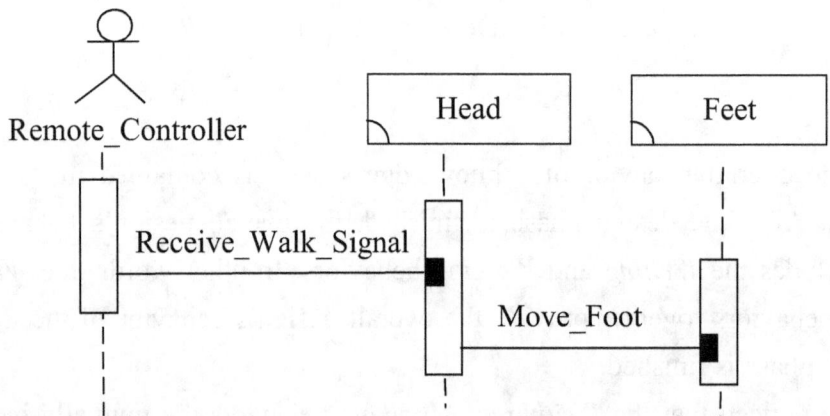

Figure 21-18   Design's IFD of the *Walking* Behavior

Chapter 22: Strategic Thinking of an Airline -- Thinking Architecture

This chapter examines the *Strategic Thinking of an Airline* which represents a case study of thinking architecture [Kapl04, Kapl08] for thinking systems modeling and architecting. An airline company chooses a "quick ground turnaround" strategy so that it would enable the company to reduce costs by operating with fewer planes and increase revenue growth by attracting more customers with on-time departures and arrivals. Both reducing costs and increasing revenue growth will help the airline company to expand shareholder value. The overall behavior of the *Strategic Thinking of an Airline* is significantly represented by two behaviors: *Productivity_Improvement* and *Revenue_Growth* as shown in Figure 22-1.

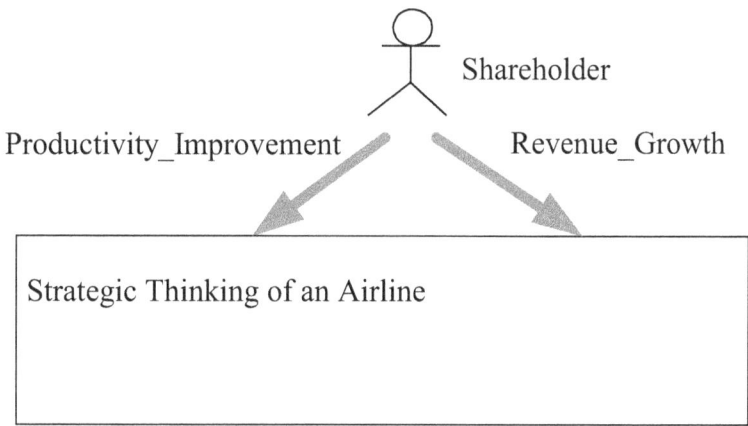

Figure 22-1 Two Behaviors of the *Strategic Thinking of an Airline*

Using the SBC thinking multi-level (hierarchical) view, an architect goes through: a) concept view, b) analysis view, c) design view and d) implementation view for the *Strategic Thinking of an Airline* thinking systems architecture as shown in Figure 22-2.

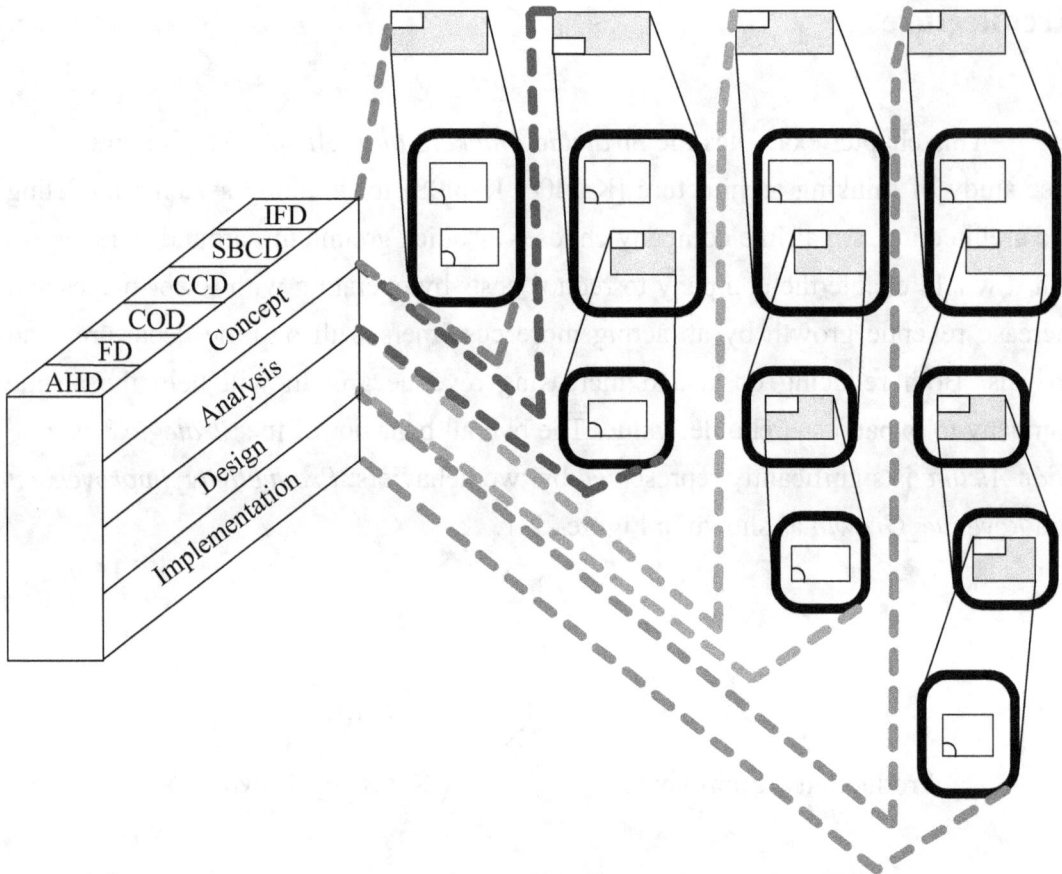

Figure 22-2   SBC Thinking Multi-Level (Hierarchical) View

22-1 Concept View of the Strategic Thinking of an Airline

In the SBC thinking multi-level (hierarchical) view, an architect constructs the concept's systems architecture for the concept to view. This concept's systems architecture is called the concept view as shown in Figure 22-3.

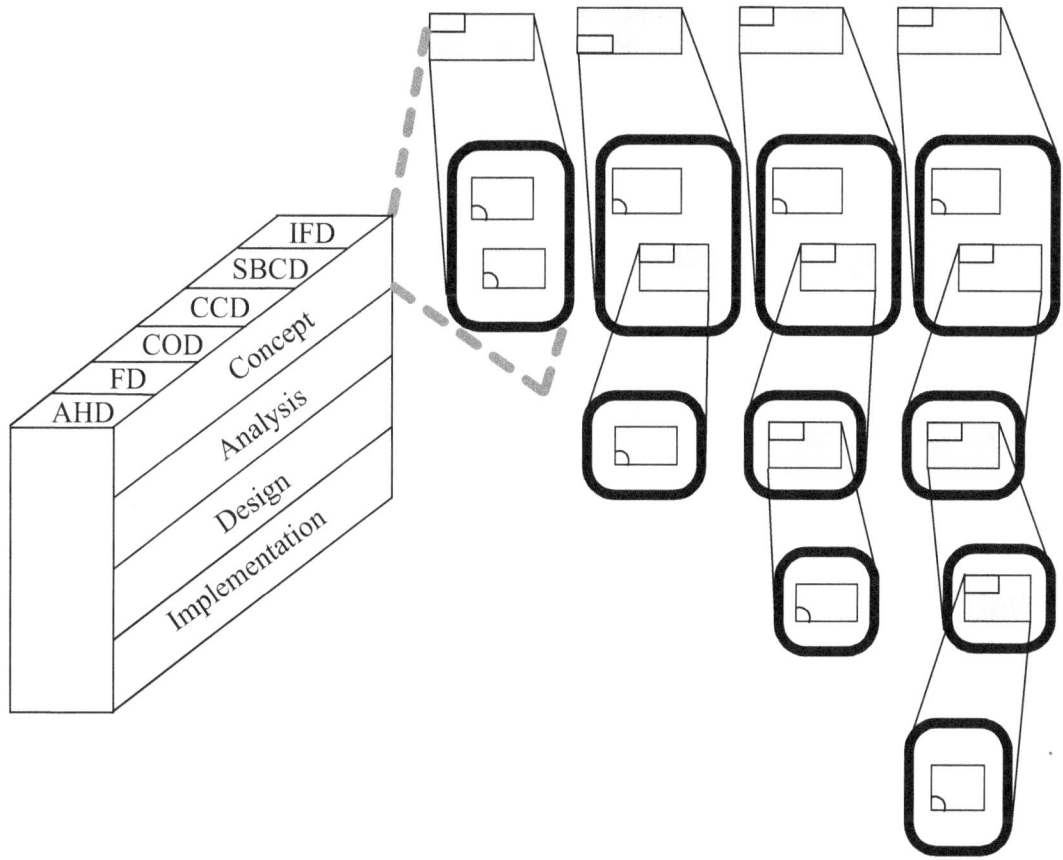

Figure 22-3   Concept View

The concept view consists of: a) concept's systems structure and b) concept's systems behavior.

22-1-1 Concept's Systems Structure

The entire SBC concept's systems structure includes: a) *Concept's AHD*, b) *Concept's FD*, c) *Concept's COD* and d) *Concept's CCD*.

We first draw the concept's AHD of the *Strategic Thinking of an Airline*. As shown in Figure 22-4, *Strategic Thinking of an Airline* is composed of *Financial_Productivity*, *Financial_Growth* and *S_Subsystem_3*.

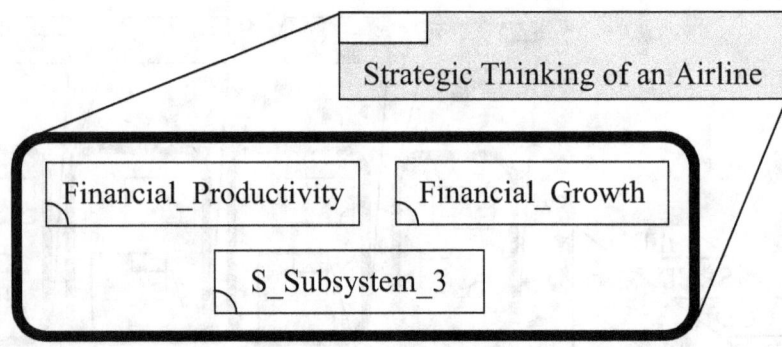

Figure 22-4   Concept's AHD of the *Strategic Thinking of an Airline*

In Figure 22-4, *Strategic Thinking of an Airline* is an aggregated system while *Financial_Productivity*, *Financial_Growth* and *S_Subsystem_3* are non-aggregated systems.

Concept's FD is the framework diagram we obtain after the concept phase is finished. Figure 22-5 shows the concept's FD of the *Strategic Thinking of an Airline*. In the figure, *Financial_Layer* contains the *Financial_Productivity*, *Financial_Growth* and *S_Subsystem_3* components.

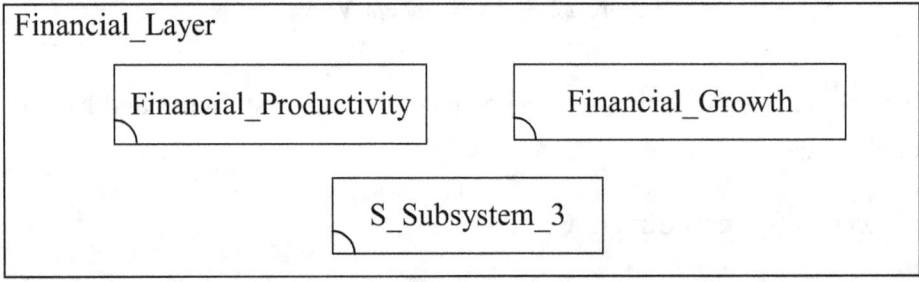

Figure 22-5   Concept's FD of the *Strategic Thinking of an Airline*

Concept's COD is the component operation diagram we obtain after the concept phase is finished. Figure 22-6 shows the concept's COD of the *Strategic Thinking of an Airline*. In the figure, component *Financial_Productivity* has one operation: *Achieve_Fewer_Planes*; component *Financial_Growth* has one operation: *Expand_Revenue_Opportunities*; component *S_Subsystem_3* has two operations: *Fast_Ground_Turnaround* and *Lowest_Prices*.

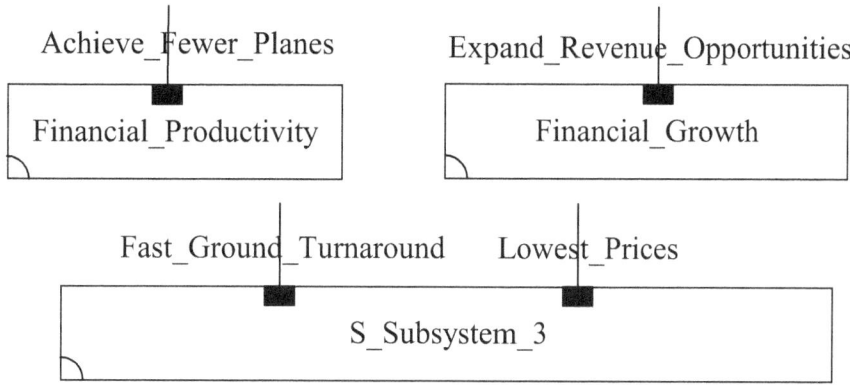

Figure 22-6 Concept's COD of the *Strategic Thinking of an Airline.*

Concept's CCD is the component connection diagram we obtain after the concept phase is finished. Figure 22-7 shows the concept's CCD of the *Strategic Thinking of an Airline*.

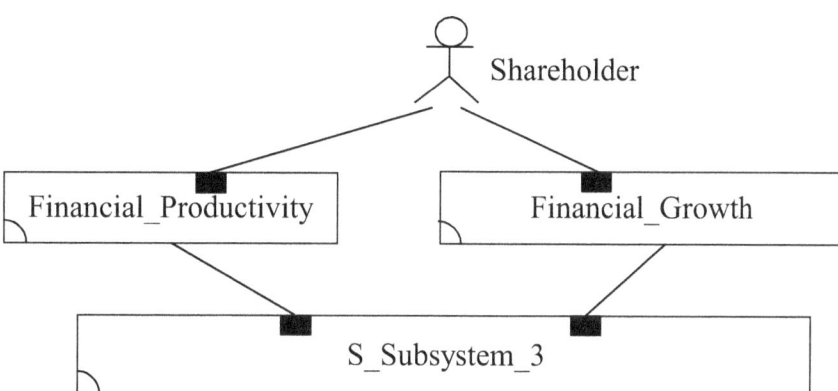

Figure 22-7 Concept's CCD of the *Strategic Thinking of an Airline*

In Figure 22-7, actor *Shareholder* has a connection with each one of the *Financial_Productivity* and *Financial_Growth* components. Both the *Financial_Productivity* and *Financial_Growth* components have a connection with the *S_Subsystem_3* component.

22-1-2 Concept's Systems Behavior

The entire SBC concept's systems behavior includes: a) *Concept's SBCD* and b) *Concept's IFD*.

Concept's SBCD is the structure-behavior coalescence diagram we obtain after the concept phase is finished. Figure 22-8 shows the concept's SBCD of the *Strategic Thinking of an Airline* in which interactions among the *Shareholder* actor and the *Financial_Productivity, Financial_Growth, S_Subsystem_3* components shall draw forth the *Productivity_Improvement* and *Revenue_Growth* behaviors.

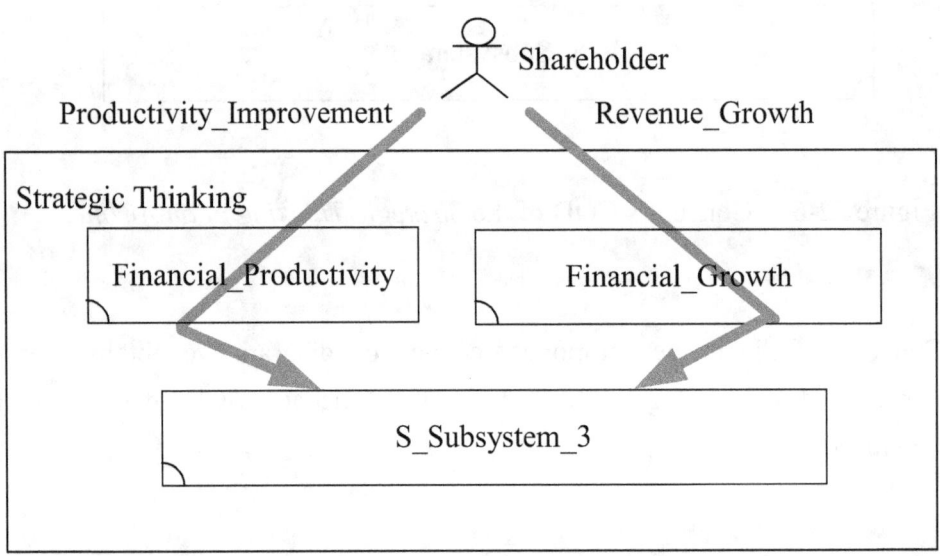

Figure 22-8   Concept's SBCD of the *Strategic Thinking of an Airline*

The overall behavior of a thinking system is contained in its individual behaviors. After the concept phase is finished, the overall concept's behavior of the *Strategic Thinking of an Airline* includes the *Productivity_Improvement* and *Revenue_Growth* behaviors. In other words, the *Productivity_Improvement* and *Revenue_Growth* behaviors together provide the overall concept's behavior of the *Strategic Thinking of an Airline* after the concept phase is finished.

Be noticed that the *Productivity_Improvement* behavior and the *Revenue_Growth* behavior are mutually independent of each other. They shall be executed concurrently [Hoar85, Miln89, Miln99].

The major purpose of using the architectural approach, instead of separating the structure model from the behavior model, is to achieve a coalesced model. In Figure 22-8, we not only see its structure, also see at the same time its behavior in the concept's SBCD of the *Strategic Thinking of an Airline*.

After the concept phase is finished, the overall concept's behavior of the *Strategic Thinking of an Airline* includes two behaviors: *Productivity_Improvement* and *Revenue_Growth*. Each of them is described by an individual IFD. Figure 22-9 shows the concept's IFD of the *Productivity_Improvement* behavior. First, actor *Shareholder* interacts with the *Financial_Productivity* component through the *Achieve_Fewer_Planes* operation call interaction. Finally, component *Financial_Productivity* interacts with the *S_Subsystem_3* component through the *Fast_Ground_Turnaround* operation call interaction.

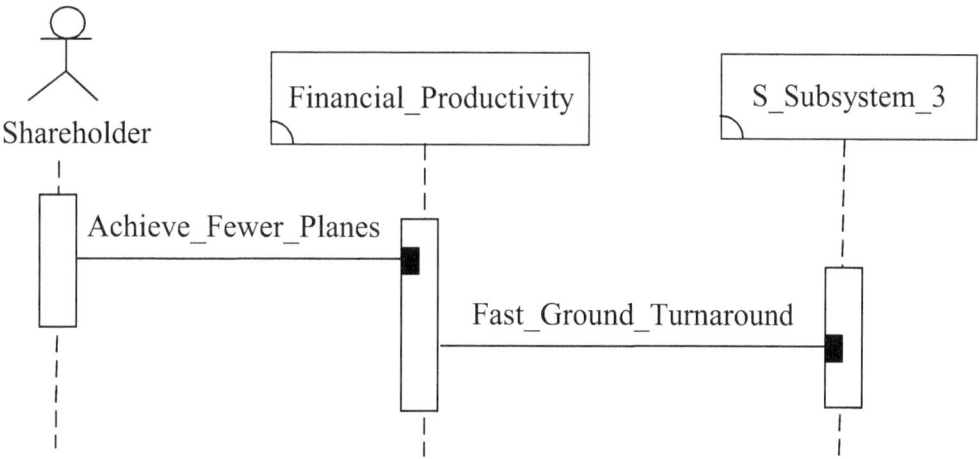

Figure 22-9   Concept's IFD of the *Productivity_Improvement* Behavior

Figure 22-10 shows the concept's IFD of the *Revenue_Growth* behavior. First, actor *Shareholder* interacts with the *Financial_Growth* component through the *Expand_Revenue_Opportunities* operation call interaction. Finally, component *Financial_Growth* interacts with the *S_Subsystem_3* component through the *Lowest_Prices* operation call interaction.

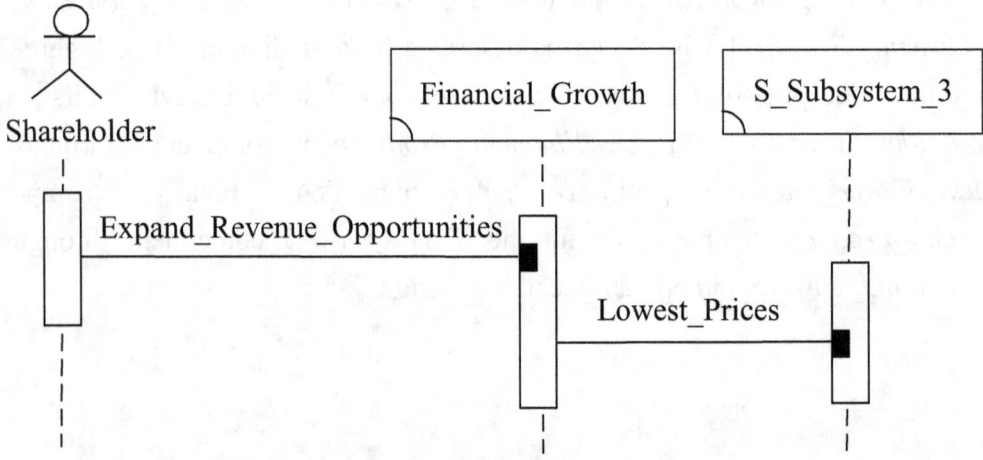

Figure 22-10   Concept's IFD of the *Revenue_Growth* Behavior

22-2 Analysis View of the Strategic Thinking of an Airline

In the SBC thinking multi-level (hierarchical) view, an architect constructs the analysis' systems architecture for the analyzer to view. This analysis' systems architecture is called the analysis view as shown in Figure 22-11. Analysis view is one level down structural decomposition (with observation congruence verification) of the concept view [Chao15a, Chao15b, Chao15c, Chao15d, Chao15e].

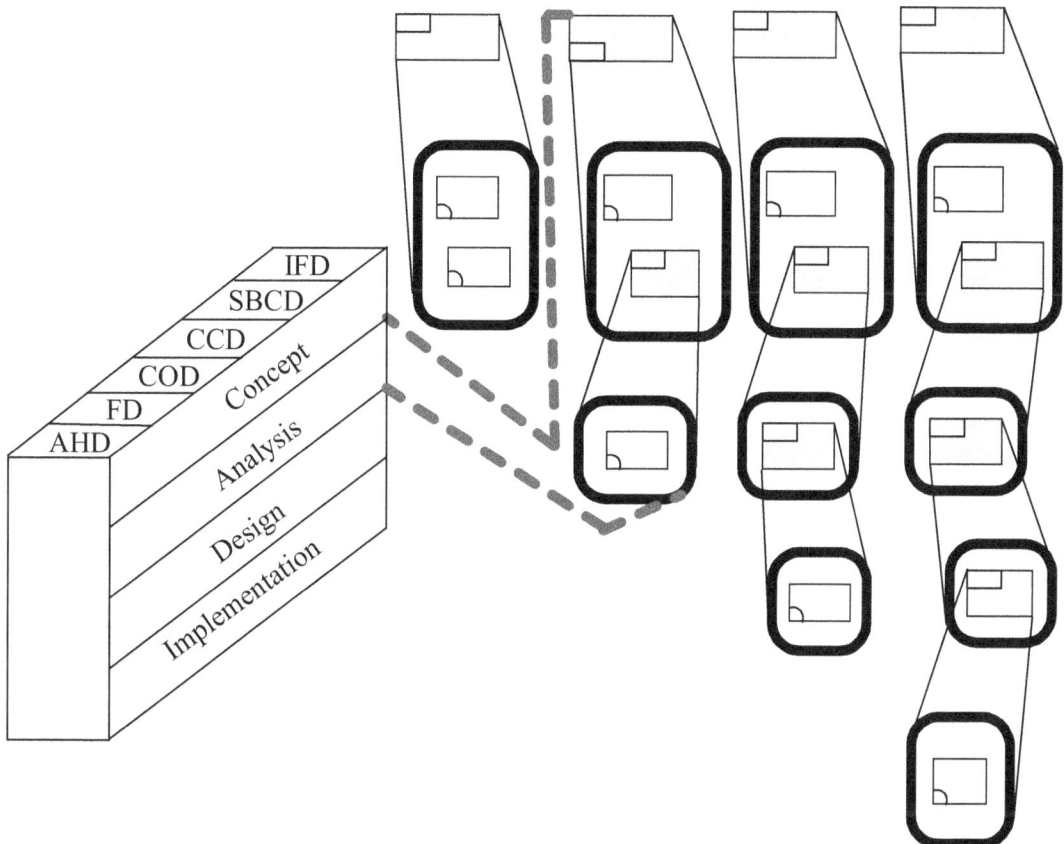

Figure 22-11   Analysis View

The analysis view consists of: a) analysis' systems structure and b) analysis' systems behavior.

22-2-1 Analysis' Systems Structure

The entire SBC analysis' systems structure includes: a) *Analysis' AHD*, b) *Analysis' FD*, c) *Analysis' COD* and d) *Analysis' CCD*.

We first draw the analysis' AHD of the *Strategic Thinking of an Airline*. As shown in Figure 22-12, *Strategic Thinking of an Airline* is composed of *Financial_Productivity*, *Financial_Growth* and *S_Subsystem_3*; *S_Subsystem_3* is composed of *Product_Attribute* and *S_Subsystem_2*.

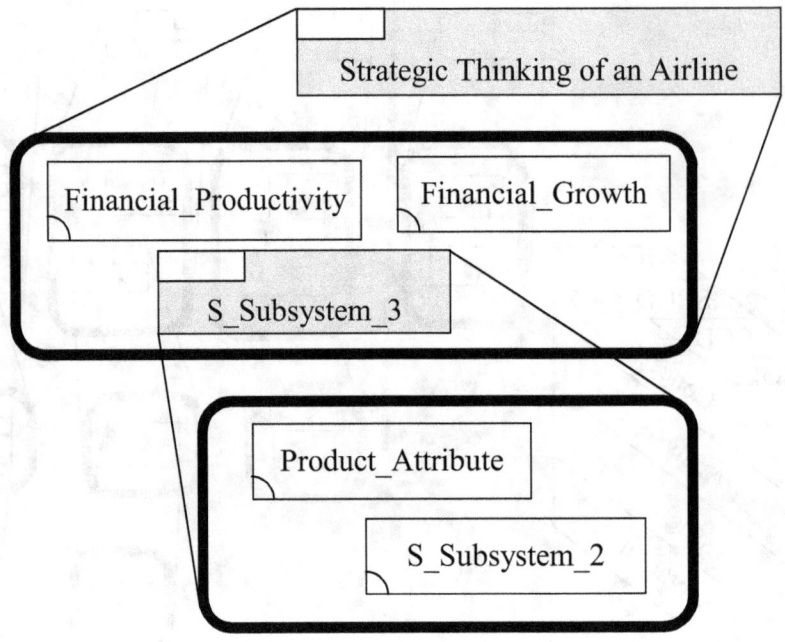

Figure 22-12 Analysis' AHD of the *Strategic Thinking of an Airline*

In Figure 22-12, *Strategic Thinking of an Airline* and *S_Subsystem_3* are aggregated systems while *Financial_Productivity*, *Financial_Growth*, *Product_Attribute* and *S_Subsystem_2* are non-aggregated systems.

Analysis' FD is the framework diagram we obtain after the analysis phase is finished. Figure 22-13 shows the analysis' FD of the *Strategic Thinking of an Airline*. In the figure, *Financial_Layer* contains the *Financial_Productivity* and *Financial_Growth* components; *Customer_Layer* contains the *Product_Attribut* and *S_Subsystem_2* components.

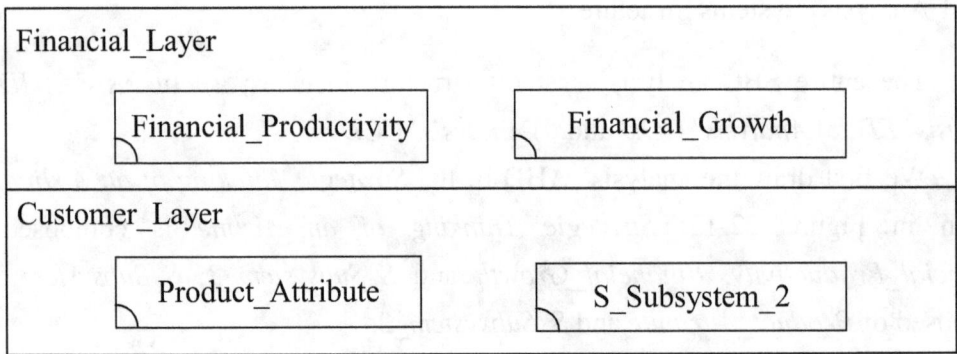

Figure 22-13 Analysis' FD of the *Strategic Thinking of an Airline*

Analysis' COD is the component operation diagram we obtain after the analysis phase is finished. Figure 22-14 shows the analysis' COD of the *Strategic Thinking of an Airline*. In the figure, component *Financial_Productivity* has one operation: *Achieve_Fewer_Planes*; component *Financial_Growth* has one operation: *Expand_Revenue_Opportunities*; component *Product_Attribute* has one operation: *Lowest_Prices*; component *S_Subsystem_2* has one operation: *Fast_Ground_Turnaround*.

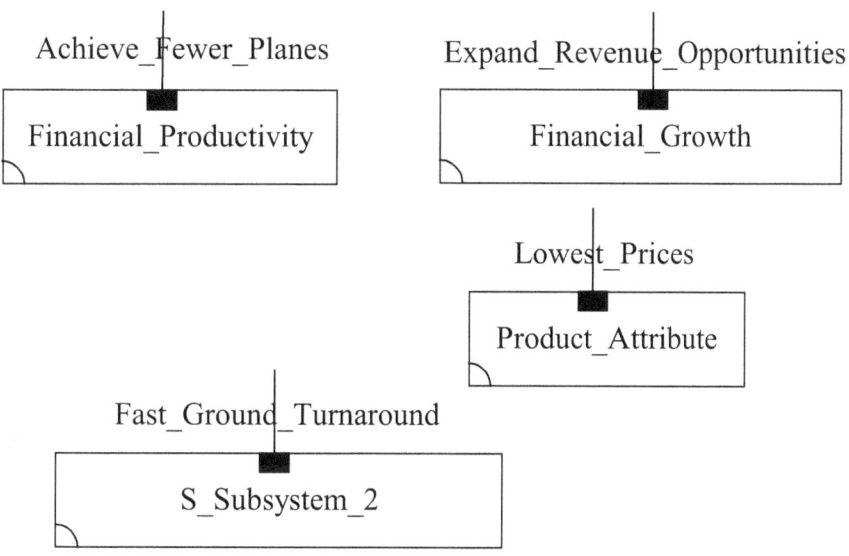

Figure 22-14   Analysis' COD of the *Strategic Thinking of an Airline*

Analysis' CCD is the component connection diagram we obtain after the analysis phase is finished. Figure 22-15 shows the analysis' CCD of the *Strategic Thinking of an Airline*.

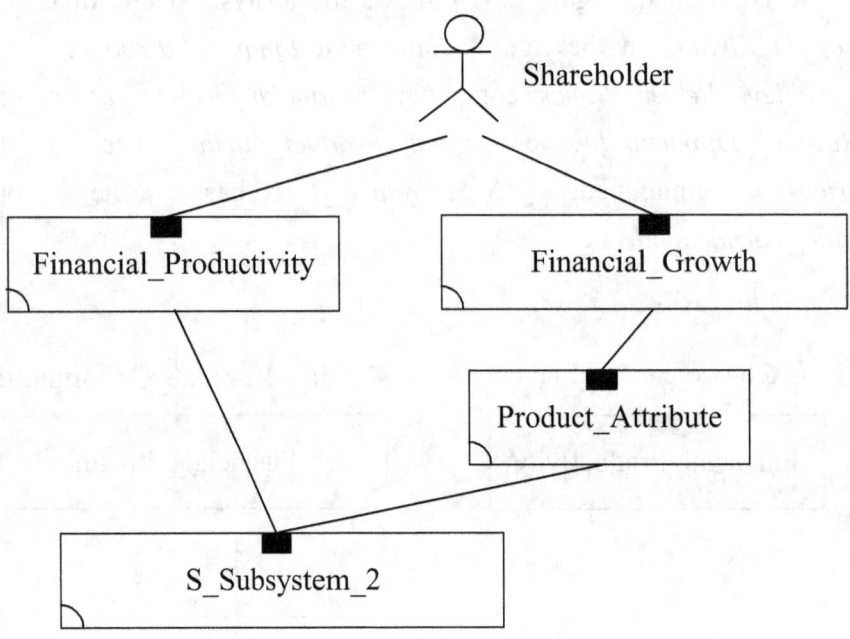

Figure 22-15　Analysis' CCD of the *Strategic Thinking of an Airline*

In Figure 22-15, actor *Shareholder* has a connection with each one of the *Financial_Productivity* and *Financial_Growth* components; component *Financial_Growth* has a connection with the *Product_Attribute* component. Both the *Financial_Productivity* and *Product_Attribute* components have a connection with the *S_Subsystem_2* component.

22-2-2 Analysis' Systems Behavior

The entire SBC analysis' systems behavior includes: a) *Analysis' SBCD* and b) *Analysis' IFD*.

Analysis' SBCD is the structure-behavior coalescence diagram we obtain after the analysis phase is finished. Figure 22-16 shows the analysis' SBCD of the *Strategic Thinking of an Airline* in which interactions among the *Shareholder* actor and the *Financial_Productivity*, *Financial_Growth*, *Product_Attribute*, *S_Subsystem_2* components shall draw forth the *Productivity_Improvement* and *Revenue_Growth* behaviors.

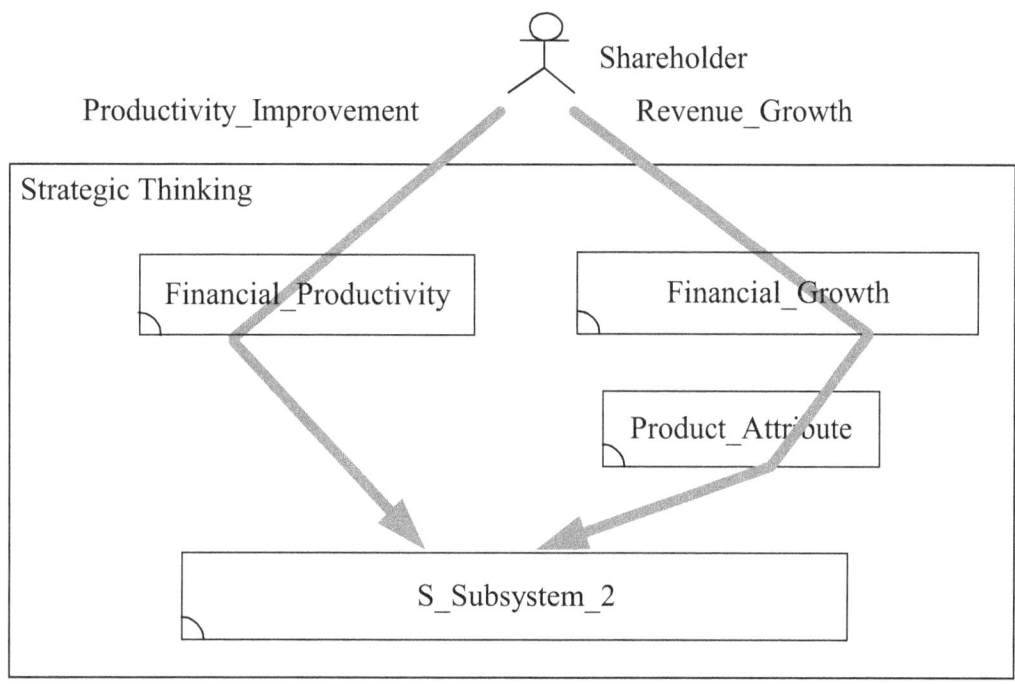

Figure 22-16 Analysis' SBCD of the *Strategic Thinking of an Airline*

The overall behavior of a thinking system is contained in its individual behaviors. After the analysis phase is finished, the overall analysis' behavior of the *Strategic Thinking of an Airline* includes the *Productivity_Improvement* and *Revenue_Growth* behaviors. In other words, the *Productivity_Improvement* and *Revenue_Growth* behaviors together provide the overall analysis' behavior of the *Strategic Thinking of an Airline* after the analysis phase is finished.

Be noticed that the *Productivity_Improvement* behavior and the *Revenue_Growth* behavior are mutually independent of each other. They shall be executed concurrently [Hoar85, Miln89, Miln99].

The major purpose of using the architectural approach, instead of separating the structure model from the behavior model, is to achieve a coalesced model. In Figure 22-16, we not only see its structure, also see at the same time its behavior in the analysis' SBCD of the *Strategic Thinking of an Airline*.

After the analysis phase is finished, the overall analysis' behavior of the *Strategic Thinking of an Airline* includes two behaviors: *Productivity_Improvement* and *Revenue_Growth*. Each of them is described by an individual IFD. Figure 22-17 shows the analysis' IFD of the *Productivity_Improvement* behavior. First, actor *Shareholder* interacts with the *Financial_Productivity* component through the

Achieve_Fewer_Planes operation call interaction. Finally, component *Financial_Productivity* interacts with the *S_Subsystem_2* component through the *Fast_Ground_Turnaround* operation call interaction.

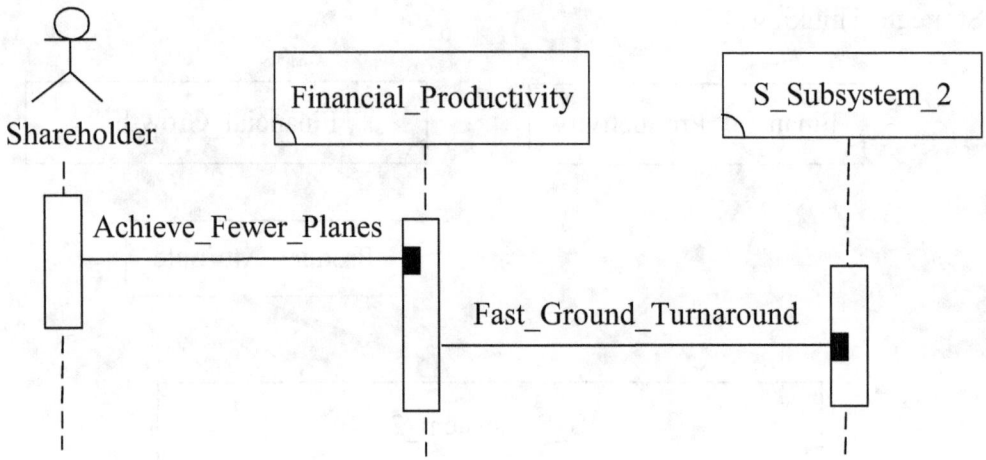

Figure 22-17   *Analysis' IFD of the Productivity_Improvement* Behavior

Figure 22-18 shows the analysis' IFD of the *Revenue_Growth* behavior. First, actor *Shareholder* interacts with the *Financial_Growth* component through the *Expand_Revenue_Opportunities* operation call interaction. Next, component *Financial_Growth* interacts with the *Product_Attribute* component through the *Lowest_Prices* operation call interaction. Finally, component *Product_Attribute* interacts with the *S_Subsystem_2* component through the *Fast_Ground_Turnaround* operation call interaction.

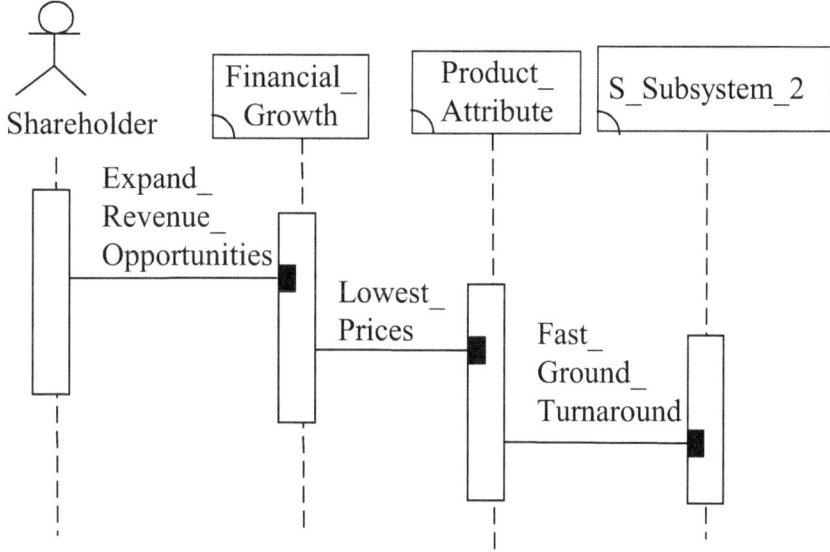

Figure 22-18 Analysis' IFD of the *Revenue_Growth* Behavior

22-3 Design View of the Strategic Thinking of an Airline

In the SBC thinking multi-level (hierarchical) view, an architect constructs the design's systems architecture for the designer to view. This design's systems architecture is called the design view as shown in Figure 22-19. Design view is one level down structural decomposition (with observation congruence verification) of the analysis view [Chao15a, Chao15b, Chao15c, Chao15d, Chao15e].

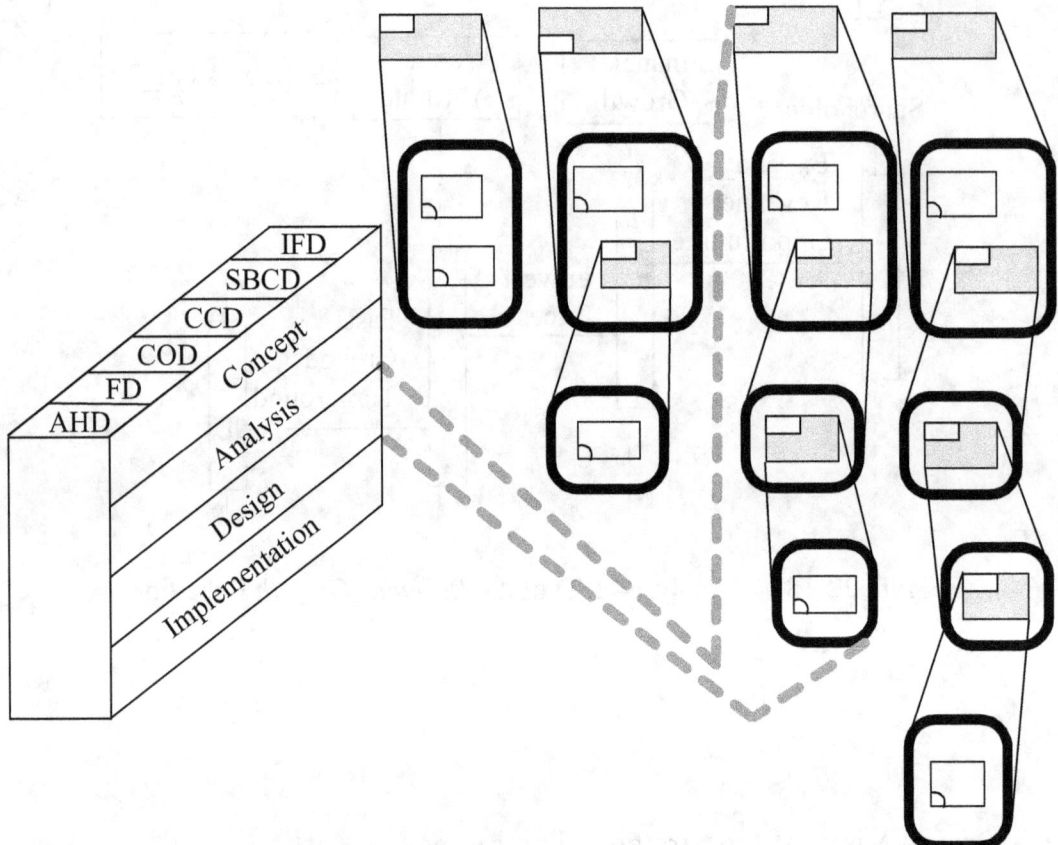

Figure 22-19   Design View

The design view consists of: a) design's systems structure and b) design's systems behavior.

22-3-1 Design's Systems Structure

The entire SBC design's systems structure includes: a) *Design's AHD*, b) *Design's FD*, c) *Design's COD* and d) *Design's CCD*.

We first draw the design's AHD of the *Strategic Thinking of an Airline*. As shown in Figure 22-20, *Strategic Thinking of an Airline* is composed of *Financial_Productivity*, *Financial_Growth* and *S_Subsystem_3*; *S_Subsystem_3* is composed of *Product_Attribute* and *S_Subsystem_2*; *S_Subsystem_2* is composed of *Innovation_Process* and *S_Subsystem_1*; *S_Subsystem_1* is composed of *Human_Capital*, *Information_Capital* and *Organization_Capital*.

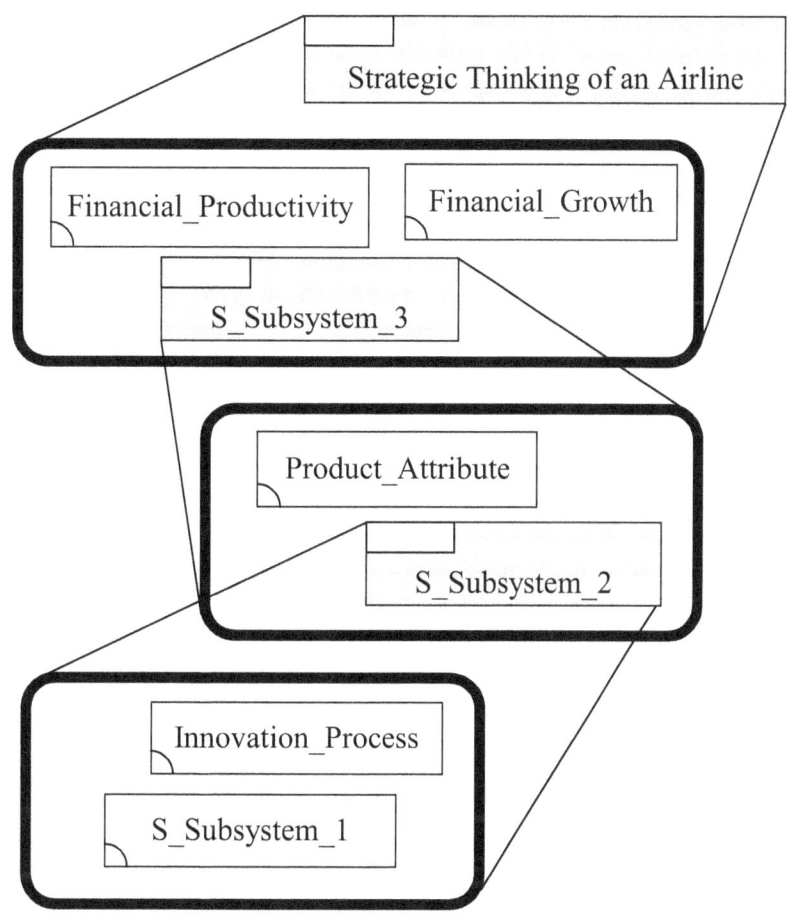

Figure 22-20 Design's AHD of the *Strategic Thinking of an Airline*

In Figure 22-20, *Strategic Thinking of an Airline*, *S_Subsystem_3* and *S_Subsystem_2* are aggregated systems while *Financial_Productivity*, *Financial_Growth*, *Product_Attribute*, *Innovation_Process* and *S_Subsystem_1* are non-aggregated systems.

Design's FD is the framework diagram we obtain after the design phase is finished. Figure 22-21 shows the design's FD of the *Strategic Thinking of an Airline*. In the figure, *Financial_Layer* contains the *Financial_Productivity* and *Financial_Growth* components; *Customer_Layer* contains the *Product_Attribute* component; *Internal_Layer* contains the *Innovation_Process* and *S_Subsystem_1* component.

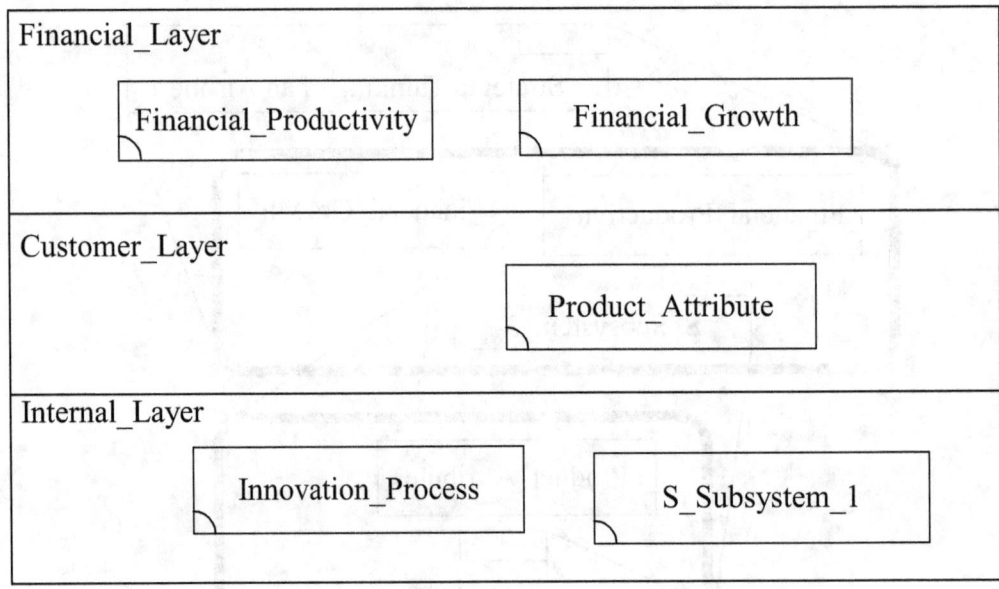

Figure 22-21   Design's FD of the *Strategic Thinking of an Airline*

Design's COD is the component operation diagram we obtain after the design phase is finished. Figure 22-22 shows the design's COD of the *Strategic Thinking of an Airline*. In the figure, component *Financial_Productivity* has one operation: *Achieve_Fewer_Planes*; component *Financial_Growth* has one operation: *Expand_Revenue_Opportunities*; component *Product_Attribute* has one operation: *Lowest_Prices*; component *Innovation_Process* has one operation: *Fast_Ground_Turnaround*; component *S_Subsystem_1* has three operations: *Strategic_Job*, *Strategic_Systems* and *Ground_Crew_Alignment*.

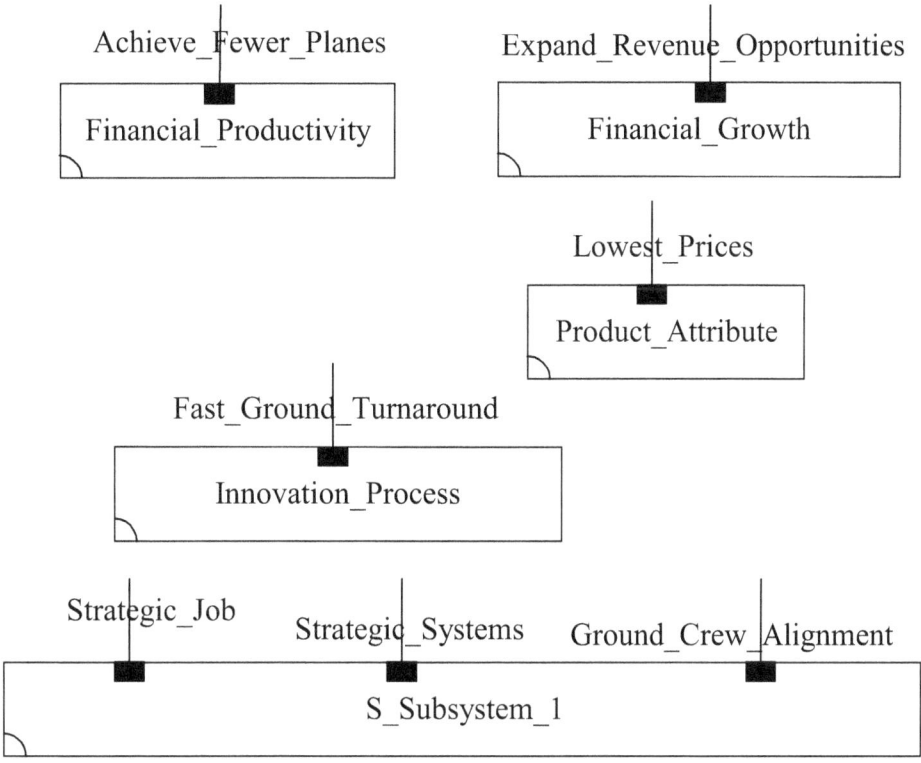

Figure 22-22   Design's COD of the *Strategic Thinking of an Airline*

Design's CCD is the component connection diagram we obtain after the design phase is finished. Figure 22-23 shows the design's CCD of the *Strategic Thinking of an Airline*.

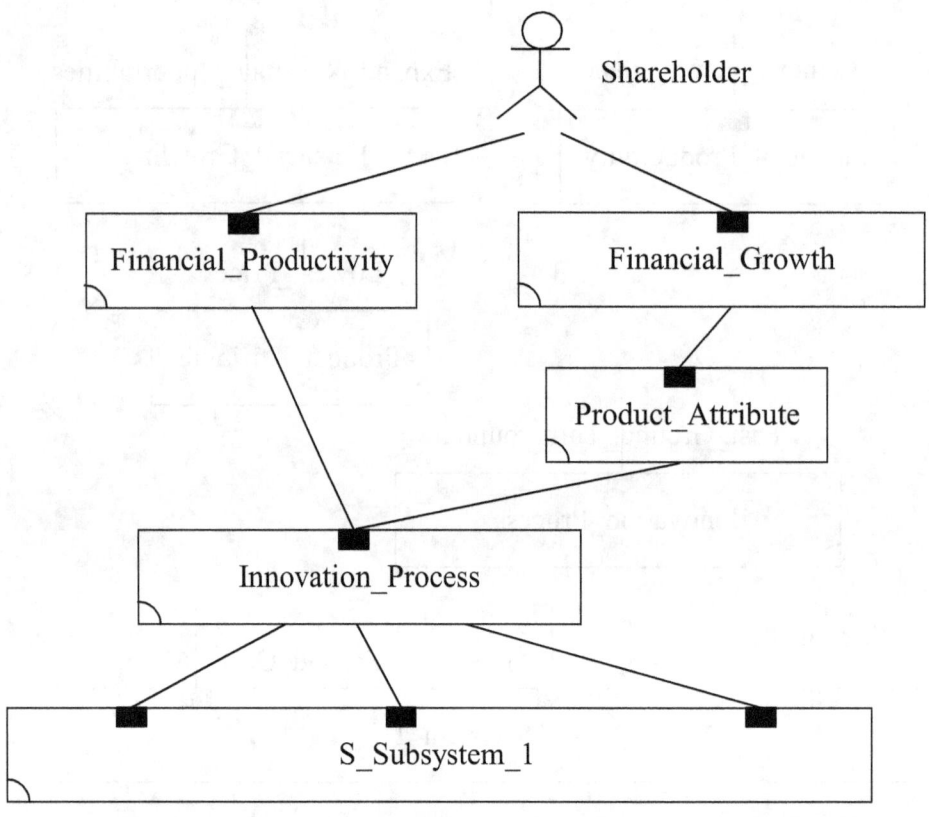

Figure 22-23   Design's CCD of the *Strategic Thinking of an Airline*

In Figure 22-23, actor *Shareholder* has a connection with each one of the *Financial_Productivity* and *Financial_Growth* components; component *Financial_Growth* has a connection with the *Product_Attribute* component. Both the *Financial_Productivity* and *Product_Attribute* components have a connection with the *Innovation_Process* component; component *Innovation_Process* has three connections with the *S_Subsystem_1* component.

22-3-2 Design's Systems Behavior

The entire SBC design's systems behavior includes: a) *Design's SBCD* and b) *Design's IFD*.

Design's SBCD is the structure-behavior coalescence diagram we obtain after the design phase is finished. Figure 22-24 shows the design's SBCD of the *Strategic Thinking of an Airline* in which interactions among the *Shareholder* actor and the *Financial_Productivity*, *Financial_Growth*, *Product_Attribute*, *Innovation_Process*, *S_Subsystem_1* components shall draw forth the *Productivity_Improvement* and *Revenue_Growth* behaviors.

311

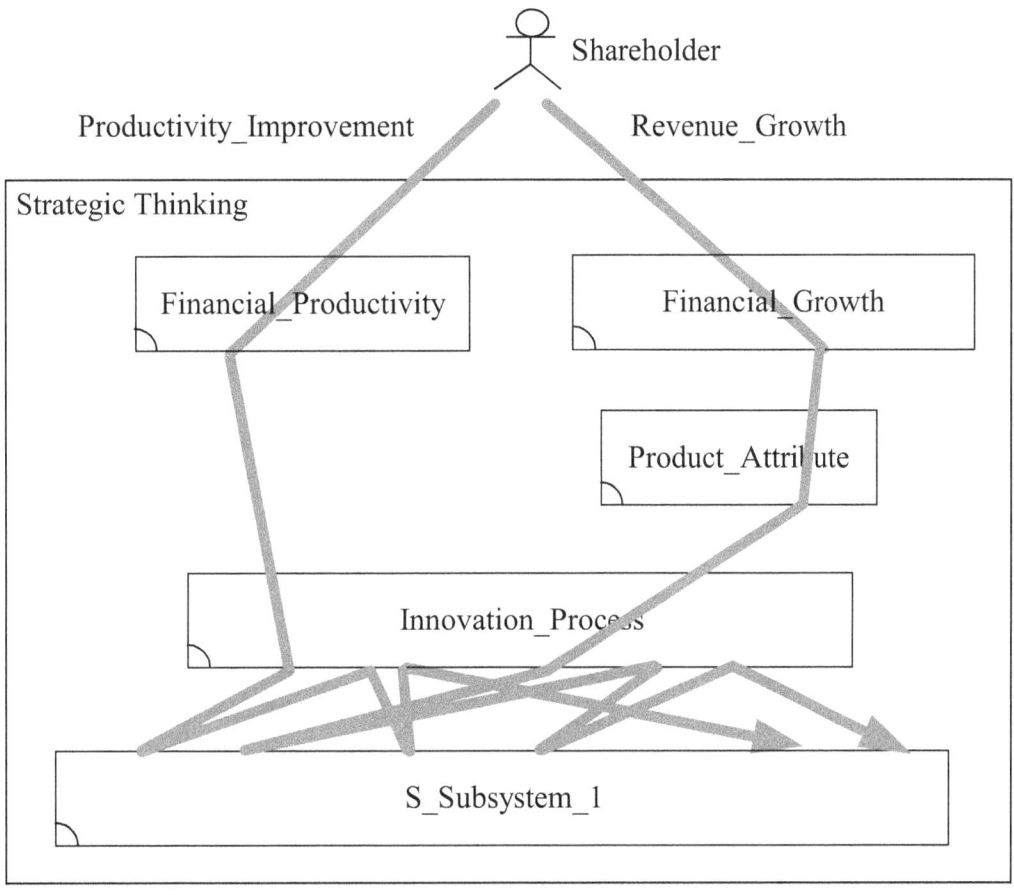

Figure 22-24 Design's SBCD of the *Strategic Thinking of an Airline*

The overall behavior of a thinking system is contained in its individual behaviors. After the design phase is finished, the overall design's behavior of the *Strategic Thinking of an Airline* includes the *Productivity_Improvement* and *Revenue_Growth* behaviors. In other words, the *Productivity_Improvement* and *Revenue_Growth* behaviors together provide the overall design's behavior of the *Strategic Thinking of an Airline* after the design phase is finished.

Be noticed that the *Productivity_Improvement* behavior and the *Revenue_Growth* behavior are mutually independent of each other. They shall be executed concurrently [Hoar85, Miln89, Miln99].

The major purpose of using the architectural approach, instead of separating the structure model from the behavior model, is to achieve a coalesced model. In Figure 22-24, we not only see its structure, also see at the same time its behavior in the design's SBCD of the *Strategic Thinking of an Airline*.

After the design phase is finished, the overall design's behavior of the *Strategic Thinking of an Airline* includes two behaviors: *Productivity_Improvement* and *Revenue_Growth*. Each of them is described by an individual IFD. Figure 22-25 shows the design's IFD of the *Productivity_Improvement* behavior. First, actor *Shareholder* interacts with the *Financial_Productivity* component through the *Achieve_Fewer_Planes* operation call interaction. Next, component *Financial_Productivity* interacts with the *Innovation_Process* component through the *Fast_Ground_Turnaround* operation call interaction. Continuingly, component *Innovation_Process* interacts with the *S_Subsystem_1* component through the *Strategic_Job* operation call interaction. Again, component *Innovation_Process* interacts with the *S_Subsystem_1* component through the *Strategic_Systems* operation call interaction. Finally, component *Innovation_Process* interacts with the *S_Subsystem_1* component through the *Ground_Crew_Alignment* operation call interaction.

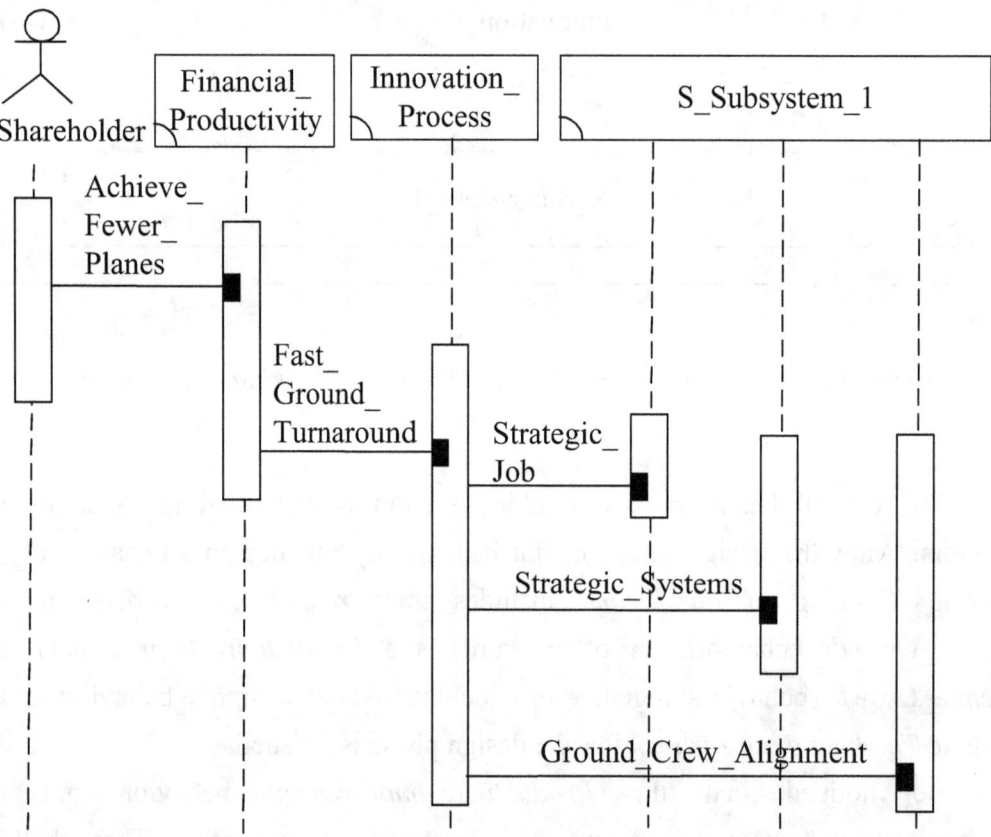

Figure 22-25   Design's IFD of the *Productivity_Improvement* Behavior

Figure 22-26 shows the design's IFD of the *Revenue_Growth* behavior. First, actor *Shareholder* interacts with the *Financial_Growth* component through the *Expand_Revenue_Opportunities* operation call interaction. Next, component *Financial_Growth* interacts with the *Product_Attribute* component through the *Lowest_Prices* operation call interaction. Continuingly, component *Product_Attribute* interacts with the *Innovation_Process* component through the *Fast_Ground_Turnaround* operation call interaction. Continuingly, component *Innovation_Process* interacts with the *S_Subsystem_1* component through the *Strategic_Job* operation call interaction. Again, component *Innovation_Process* interacts with the *S_Subsystem_1* component through the *Strategic_Systems* operation call interaction. Finally, component *Innovation_Process* interacts with the *S_Subsystem_1* component through the *Ground_Crew_Alignment* operation call interaction.

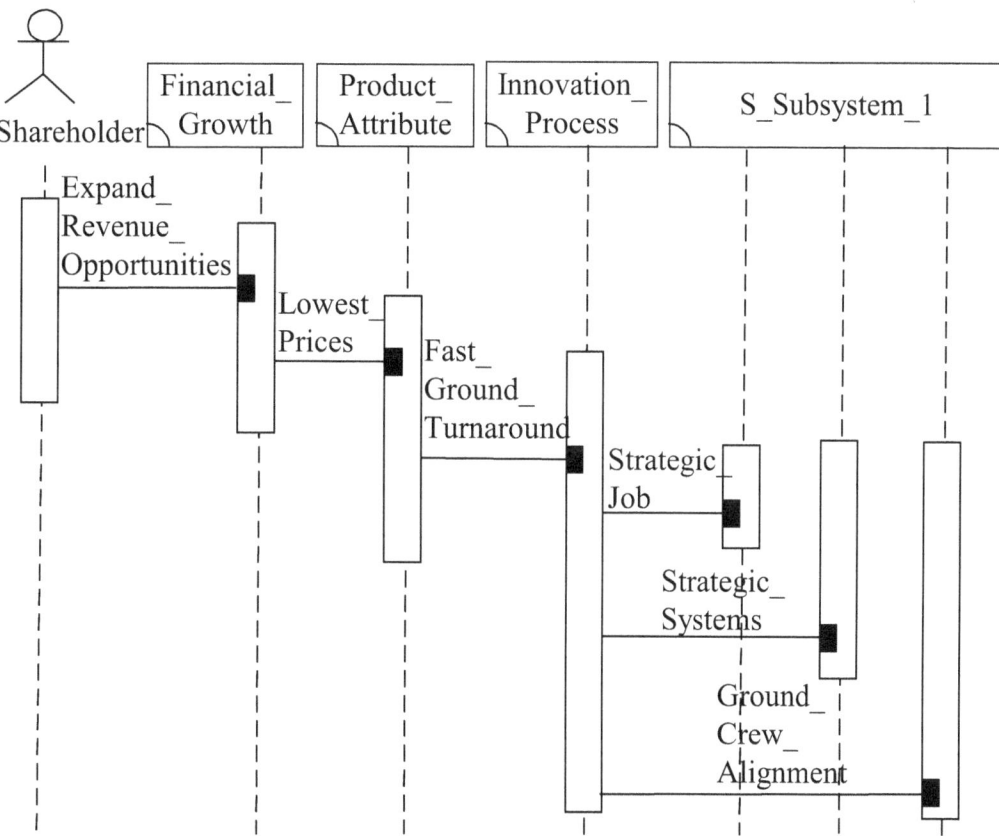

Figure 22-26   Design's IFD of the *Revenue_Growth* Behavior

22-4 Implementation View of the Strategic Thinking of an Airline

In the SBC thinking multi-level (hierarchical) view, an architect constructs the implementation's systems architecture for the implementer to view. This implementation's systems architecture is called the implementation view as shown in Figure 22-27. Implementation view is one level down structural decomposition (with observation congruence verification) of the design view [Chao15a, Chao15b, Chao15c, Chao15d, Chao15e].

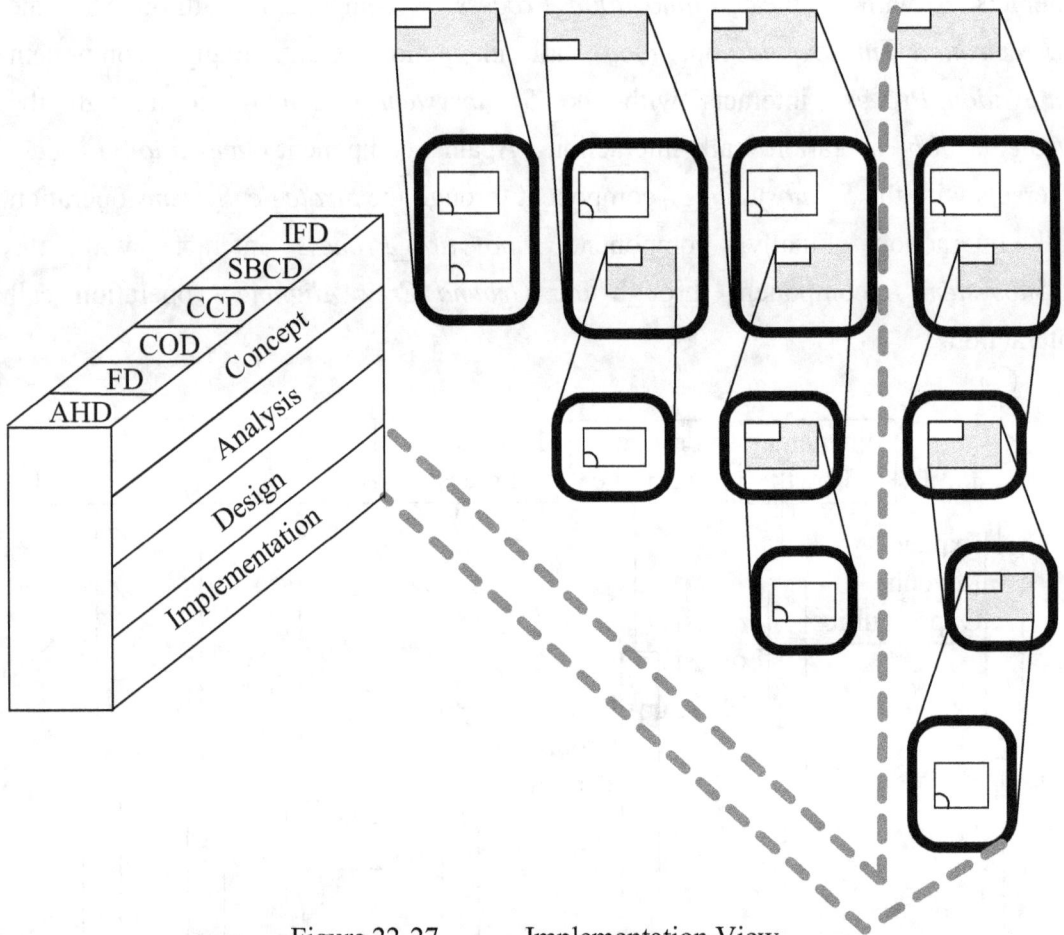

Figure 22-27   Implementation View

The implementation view consists of: a) implementation's systems structure and b) implementation's systems behavior.

22-4-1 Implementation's Systems Structure

The entire SBC implementation's systems structure includes: a) *Implementation's AHD*, b) *Implementation's FD*, c) *Implementation's COD* and d) *Implementation's CCD*.

We first draw the implementation's AHD of the *Strategic Thinking of an Airline*. As shown in Figure 22-28, *Strategic Thinking of an Airline* is composed of *Financial_Productivity*, *Financial_Growth* and *S_Subsystem_3*; *S_Subsystem_3* is composed of *Product_Attribute* and *S_Subsystem_2*; *S_Subsystem_2* is composed of

Innovation_Process and *S_Subsystem_1*; *S_Subsystem_1* is composed of *Human_Capital*, *Information_Capital* and *Organization_Capital*.

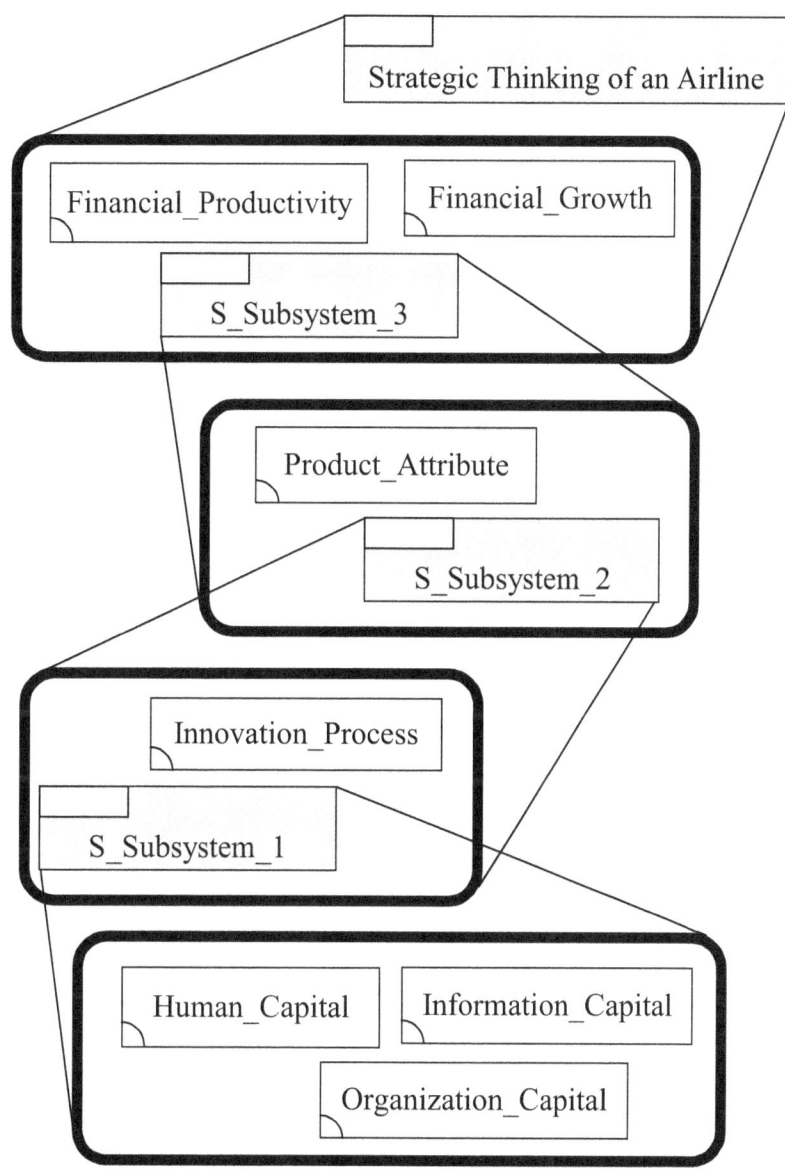

Figure 22-28   Implementation's AHD of the *Strategic Thinking of an Airline*

In Figure 22-28, *Strategic Thinking of an Airline*, *S_Subsystem_3*, *S_Subsystem_2* and *S_Subsystem_1* are aggregated systems while *Financial_Productivity*, *Financial_Growth*, *Product_Attribute*, *Innovation_Process*, *Human_Capital*, *Information_Capital* and *Organization_Capital* are non-aggregated systems.

Implementation's FD is the framework diagram we obtain after the implementation phase is finished. Figure 22-29 shows the implementation's FD of the *Strategic Thinking of an Airline*. In the figure, *Financial_Layer* contains the *Financial_Productivity* and *Financial_Growth* components; *Customer_Layer* contains the *Product_Attribute* component; *Internal_Layer* contains the *Innovation_Process* component; *Learning&Growth_Layer* contains the *Human_Capital, Information_Capital* and *Organization_Capital* components.

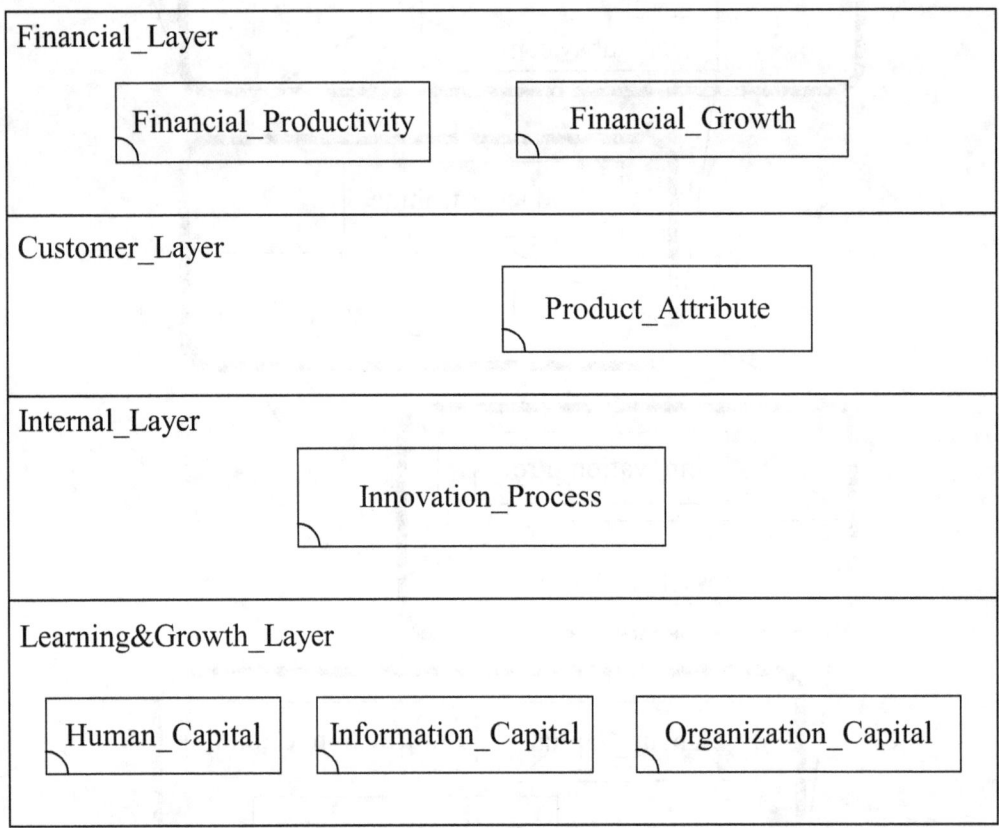

Figure 22-29   Implementation's FD of the *Strategic Thinking of an Airline*

Implementation's COD is the component operation diagram we obtain after the implementation phase is finished. Figure 22-30 shows the implementation's COD of the *Strategic Thinking of an Airline*. In the figure, component *Financial_Productivity* has one operation: *Achieve_Fewer_Planes*; component *Financial_Growth* has one operation: *Expand_Revenue_Opportunities*; component *Product_Attribute* has one operation: *Lowest_Prices*; component *Innovation_Process* has one operation: *Fast_Ground_Turnaround*; component *Human_Capital* has one operation: *Strategic_Job*; component *Information_Capital* has one operation: *Strategic_Systems*; component *Organization_Capital* has one operation: *Ground_Crew_Alignment*.

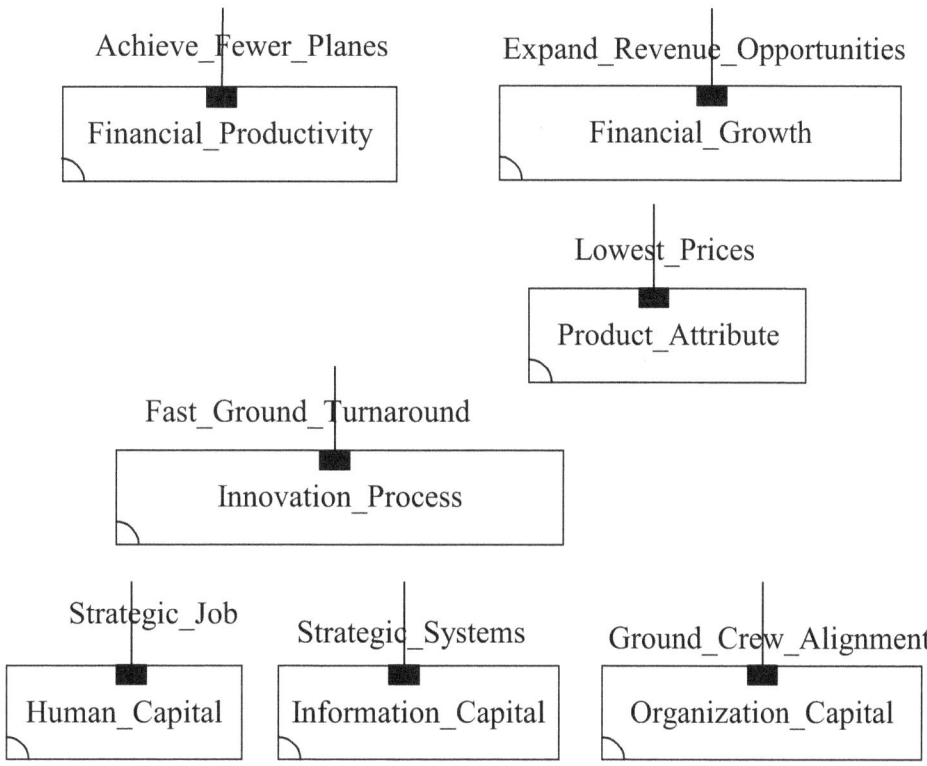

Figure 22-30 Implementation's COD of the *Strategic Thinking of an Airline*

Implementation's CCD is the component connection diagram we obtain after the implementation phase is finished. Figure 22-31 shows the implementation's CCD of the *Strategic Thinking of an Airline*.

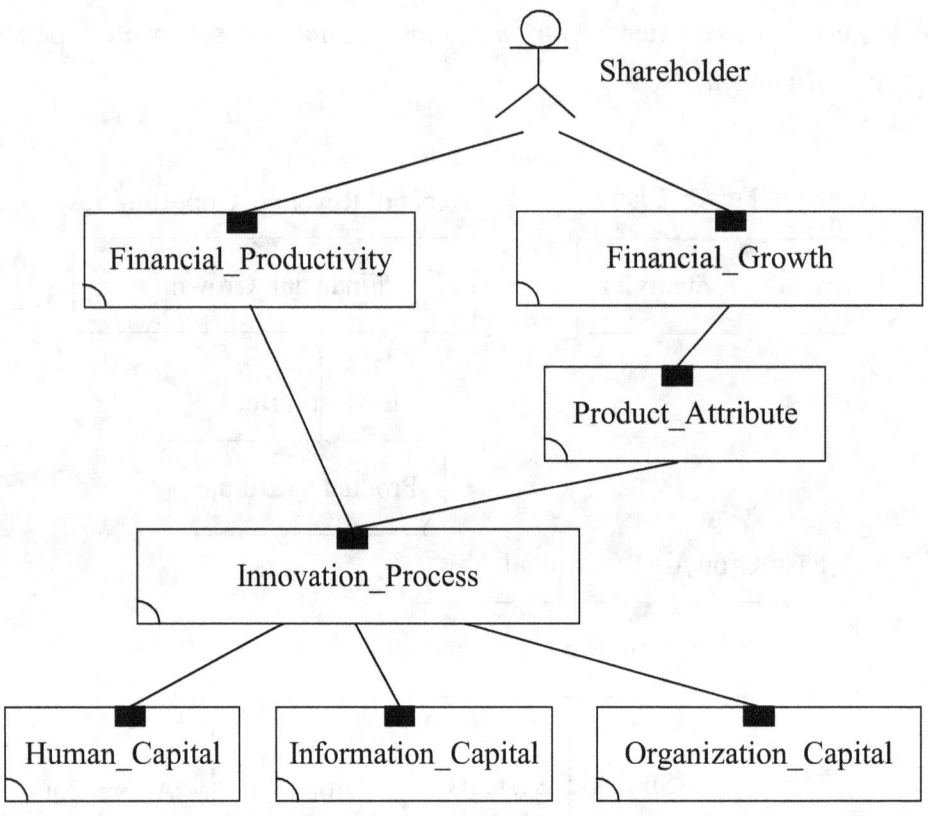

Figure 22-31   Implementation's CCD of the *Strategic Thinking of an Airline*

In Figure 22-31, actor *Shareholder* has a connection with each one of the *Financial_Productivity* and *Financial_Growth* components; component *Financial_Growth* has a connection with the *Product_Attribute* component. Both components *Financial_Productivity* and *Product_Attribute* have a connection with the *Innovation_Process* component; component *Innovation_Process* has a connection with each one of the *Human_Capital*, *Information_Capital* and *Organization_Capital* components.

22-4-2 Implementation's Systems Behavior

The entire SBC implementation's systems behavior includes: a) *Implementation's SBCD* and b) *Implementation's IFD*.

Implementation's SBCD is the structure-behavior coalescence diagram we obtain after the implementation phase is finished. Figure 22-32 shows the implementation's SBCD of the *Strategic Thinking of an Airline* in which interactions among the *Shareholder* actor and the *Financial_Productivity*, *Financial_Growth*, *Product_Attribute*, *Innovation_Process*, *Human_Capital*, *Information_Capital*,

Organization_Capital components shall draw forth the *Productivity_Improvement* and *Revenue_Growth* behaviors.

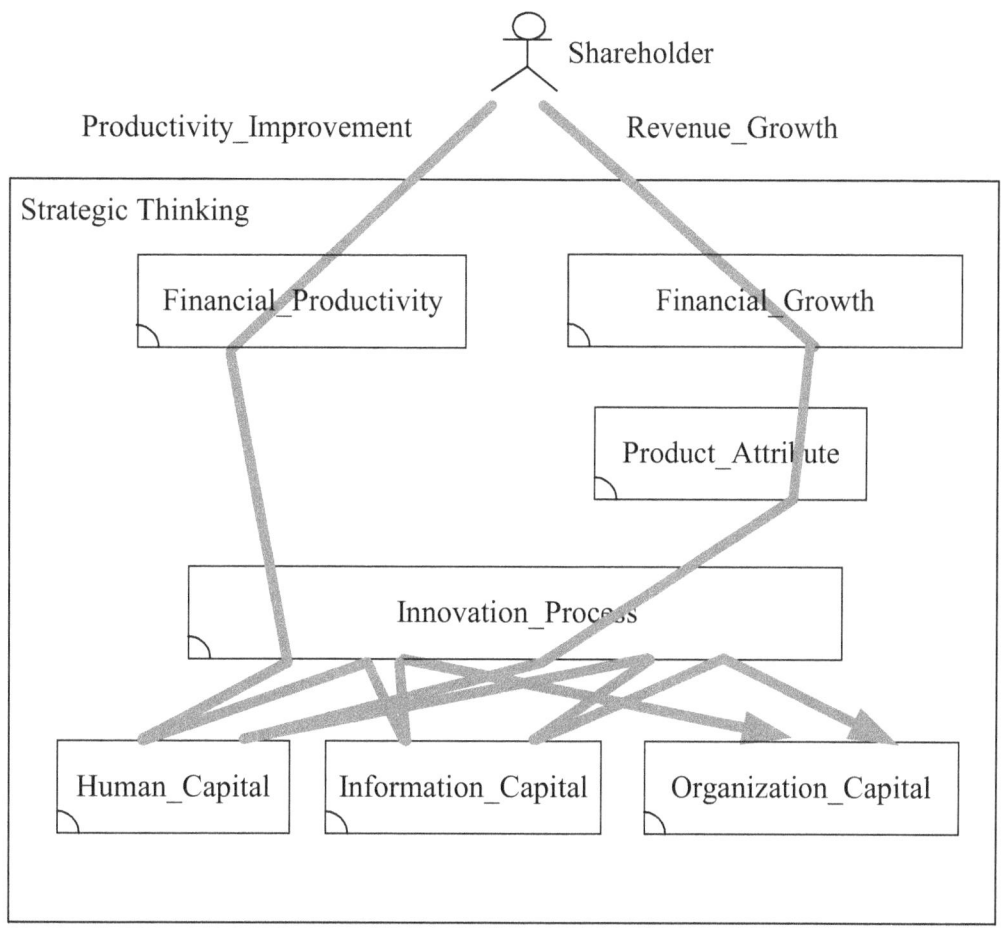

Figure 22-32　Implementation's SBCD of the *Strategic Thinking of an Airline*

The overall behavior of a thinking system is contained in its individual behaviors. After the implementation phase is finished, the overall implementation's behavior of the *Strategic Thinking of an Airline* includes the *Productivity_Improvement* and *Revenue_Growth* behaviors. In other words, the *Productivity_Improvement* and *Revenue_Growth* behaviors together provide the overall implementation's behavior of the *Strategic Thinking of an Airline* after the implementation phase is finished.

Be noticed that the *Productivity_Improvement* behavior and the *Revenue_Growth* behavior are mutually independent of each other. They shall be executed concurrently [Hoar85, Miln89, Miln99].

The major purpose of using the architectural approach, instead of separating the structure model from the behavior model, is to achieve a coalesced model. In Figure 22-32, we not only see its structure, also see at the same time its behavior in the implementation's SBCD of the *Strategic Thinking of an Airline*.

After the implementation phase is finished, the overall implementation's behavior of the *Strategic Thinking of an Airline* includes two behaviors: *Productivity_Improvement* and *Revenue_Growth*. Each of them is described by an individual IFD. Figure 22-33 shows the implementation's IFD of the *Productivity_Improvement* behavior. First, actor *Shareholder* interacts with the *Financial_Productivity* component through the *Achieve_Fewer_Planes* operation call interaction. Next, component *Financial_Productivity* interacts with the *Innovation_Process* component through the *Fast_Ground_Turnaround* operation call interaction. Continuingly, component *Innovation_Process* interacts with the *Human_Capital* component through the *Strategic_Job* operation call interaction. Again, component *Innovation_Process* interacts with the *Information_Capital* component through the *Strategic_Systems* operation call interaction. Finally, component *Innovation_Process* interacts with the *Organization_Capital* component through the *Ground_Crew_Alignment* operation call interaction.

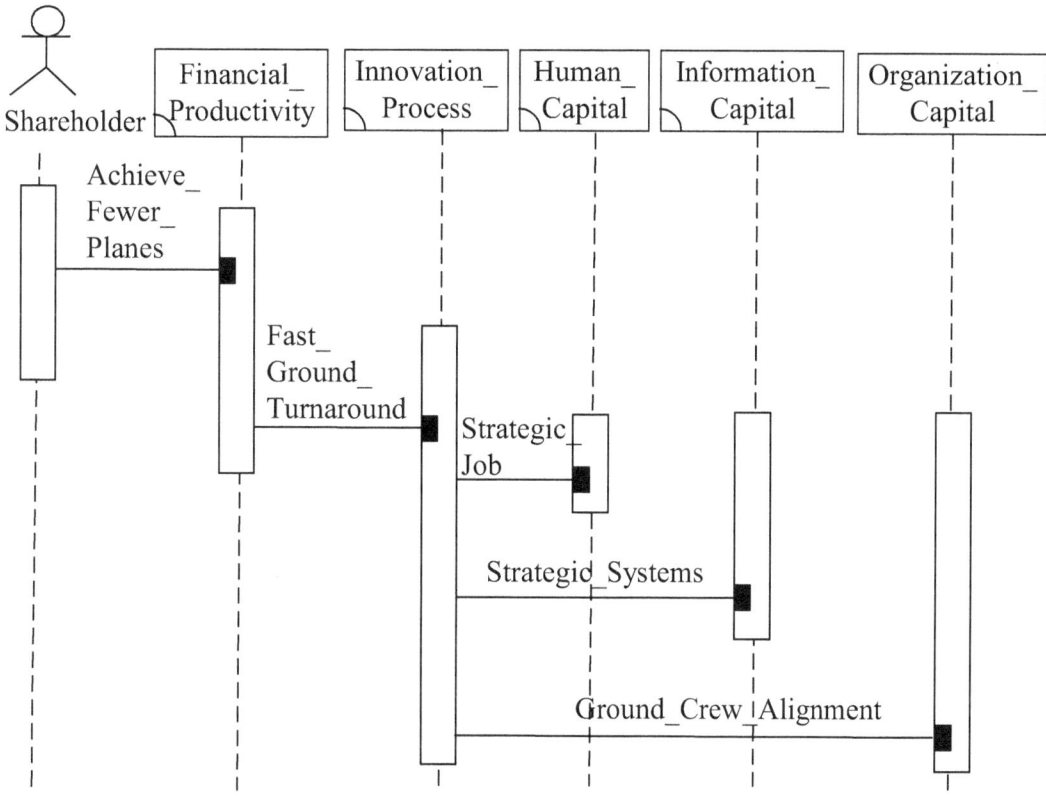

Figure 22-33 Implementation's IFD of the *Productivity_Improvement* Behavior

Figure 22-34 shows the implementation's IFD of the *Revenue_Growth* behavior. First, actor *Shareholder* interacts with the *Financial_Growth* component through the *Expand_Revenue_Opportunities* operation call interaction. Next, component *Financial_Growth* interacts with the *Product_Attribute* component through the *Lowest_Prices* operation call interaction. Continuingly, component *Product_Attribute* interacts with the *Innovation_Process* component through the *Fast_Ground_Turnaround* operation call interaction. Continuingly, component *Innovation_Process* interacts with the *Human_Capital* component through the *Strategic_Job* operation call interaction. Again, component *Innovation_Process* interacts with the *Information_Capital* component through the *Strategic_Systems* operation call interaction. Finally, component *Innovation_Process* interacts with the *Organization_Capital* component through the *Ground_Crew_Alignment* operation call interaction.

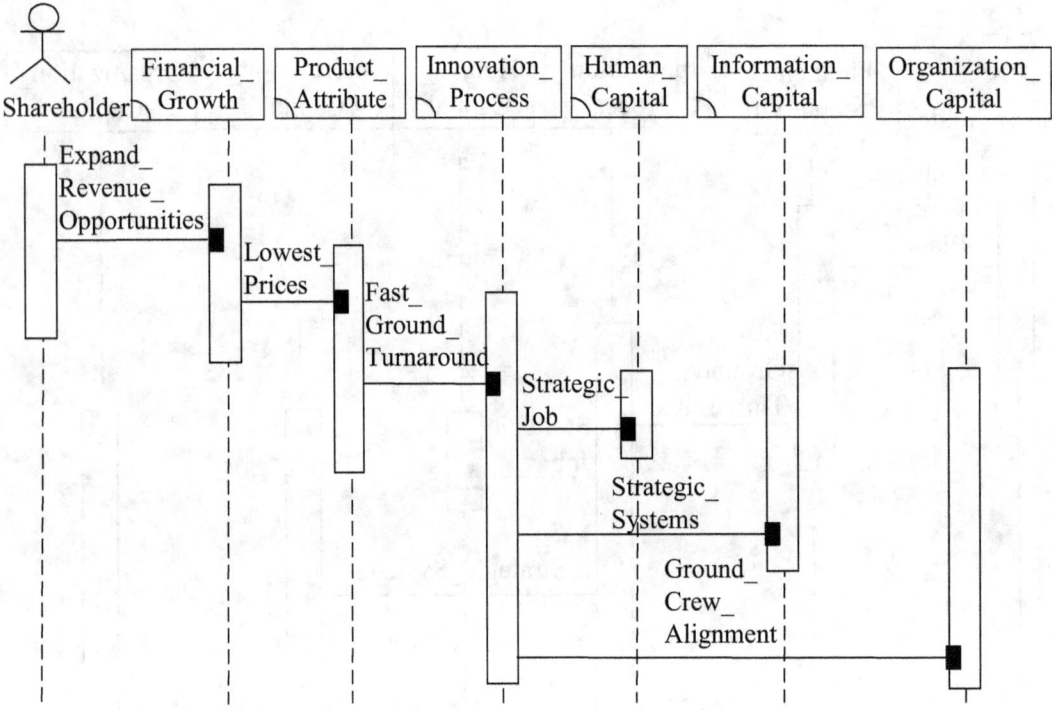

Figure 22-34   Implementation's IFD of the *Revenue_Growth* Behavior

APPENDIX A: SBC ARCHITECTURE DESCRIPTION LANGUAGE (SBC-ADL)

(1) Architecture Hierarchy Diagram

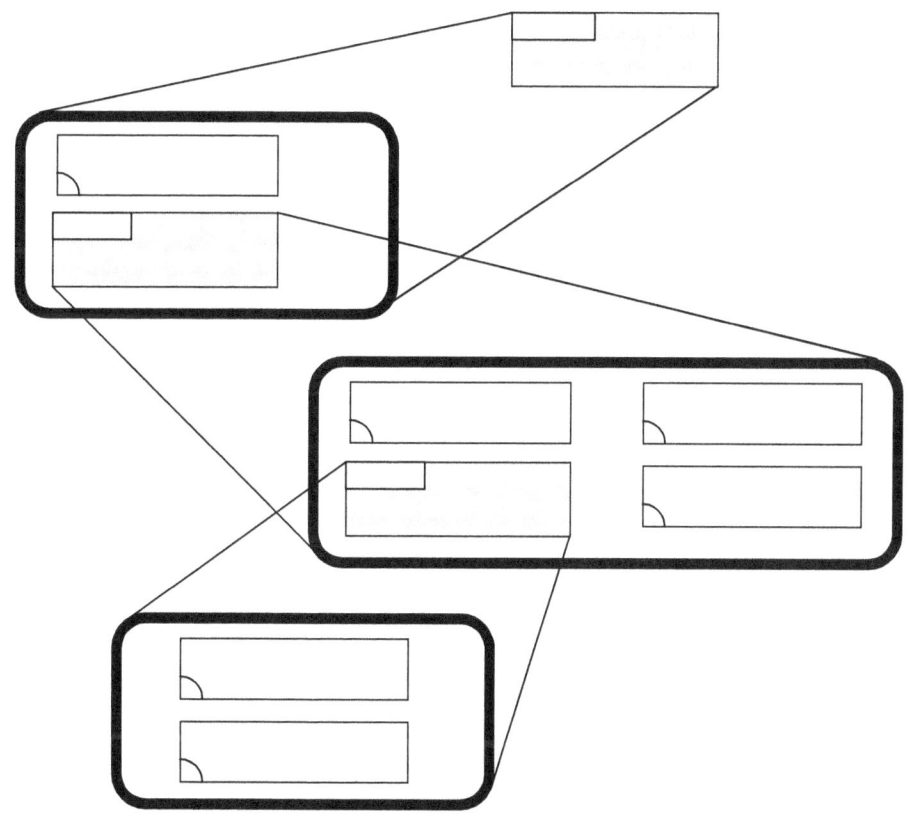

: Aggregated System

: Non-Aggregated System, Component

(2) Framework Diagram

```
┌─────────────────────────────────────────┐
│ Business Layer                          │
│      ┌──────────┐      ┌──────────┐     │
│      │          │      │          │     │
│      └──────────┘      └──────────┘     │
├─────────────────────────────────────────┤
│ Application Layer                       │
│            ┌──────────┐                 │
│            │          │                 │
│            └──────────┘                 │
├─────────────────────────────────────────┤
│ Data Layer                              │
│   ┌────────┐  ┌────────┐  ┌────────┐    │
│   │        │  │        │  │        │    │
│   └────────┘  └────────┘  └────────┘    │
├─────────────────────────────────────────┤
│ Technology Layer                        │
│      ┌──────────┐     ┌──────────┐      │
│      │          │     │          │      │
│      └──────────┘     └──────────┘      │
└─────────────────────────────────────────┘
```

┌──────────┐
│   │   : Component
└──────────┘

(3) Component Operation Diagram

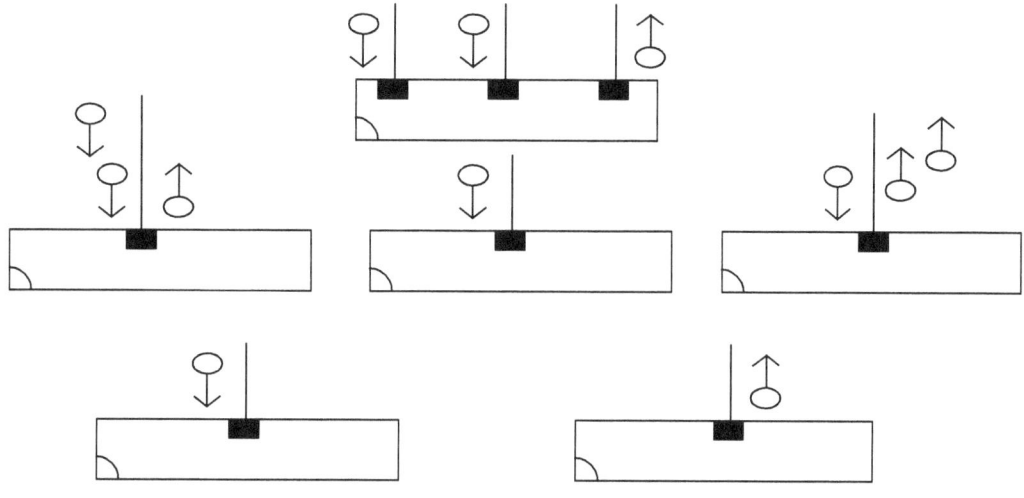

▪ with vertical line	: Operation
↓○	: Input Data
↑○	: Output Data
▭	: Component

(4) Component Connection Diagram

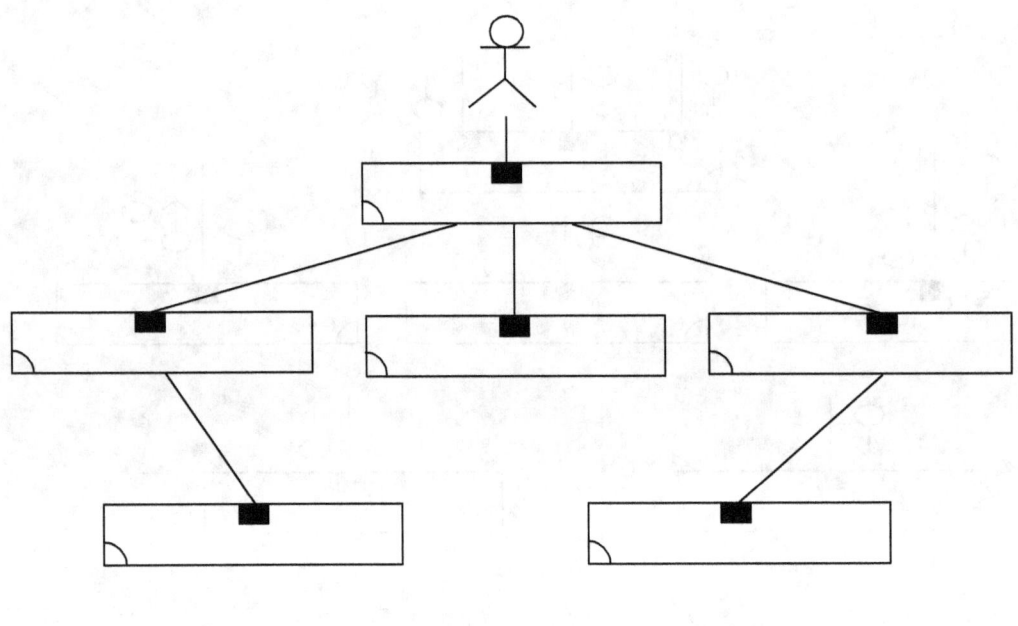

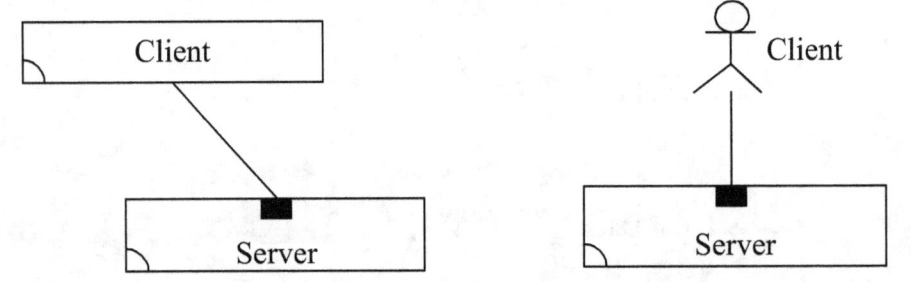

(5) Structure-Behavior Coalescence Diagram

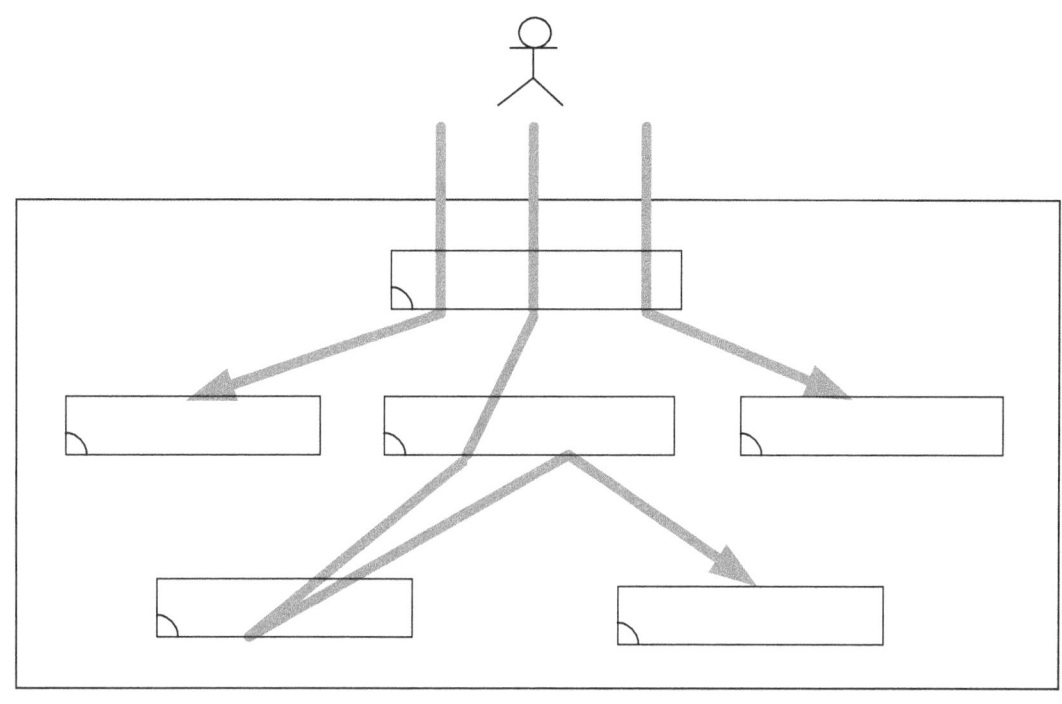

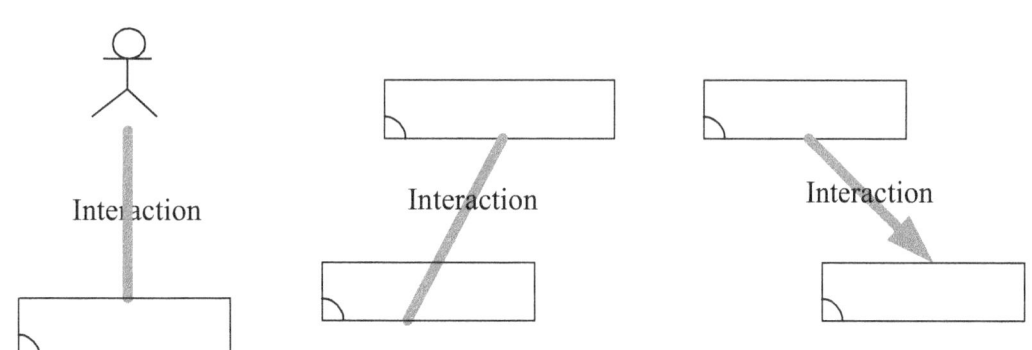

(6) Interaction Flow Diagram

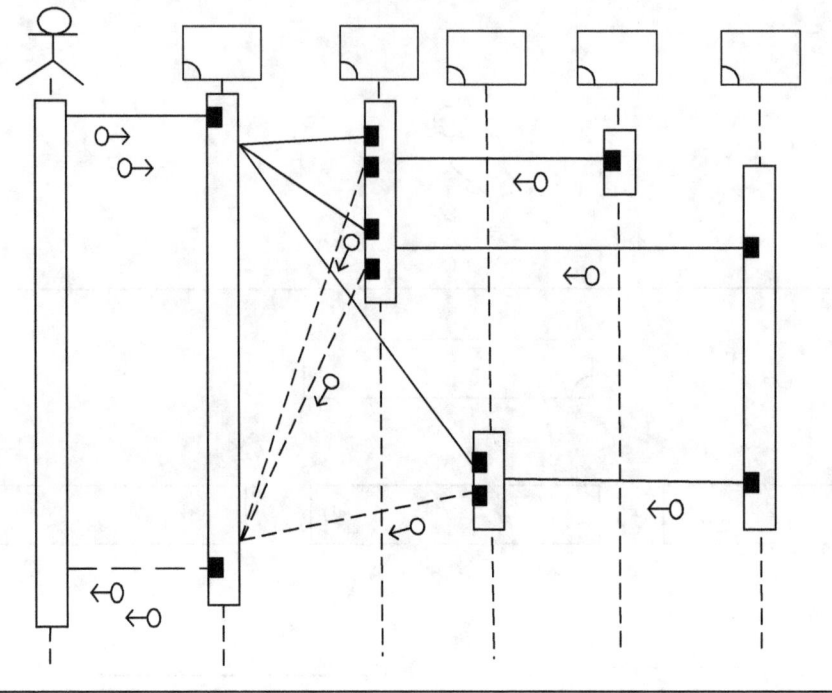

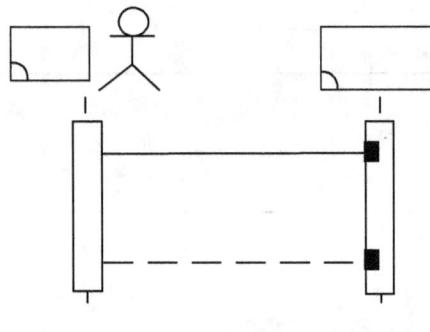

: Operation Call Interaction

: Operation Return Interaction

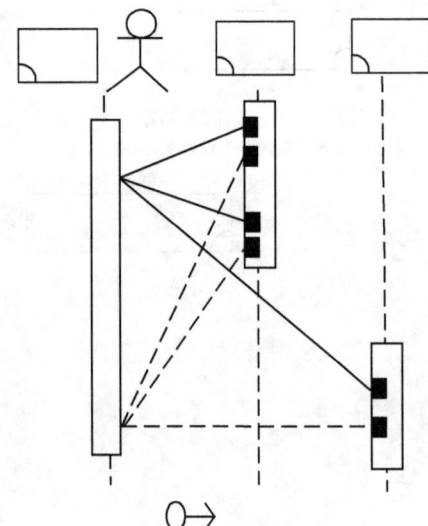

: Conditional Operation Call Interaction

: Conditional Operation Return Interaction

: Input Data

: Output Data

APPENDIX B: SBC ARCHITECTURE DEVELOPMENT METHOD (SBC-ADM)

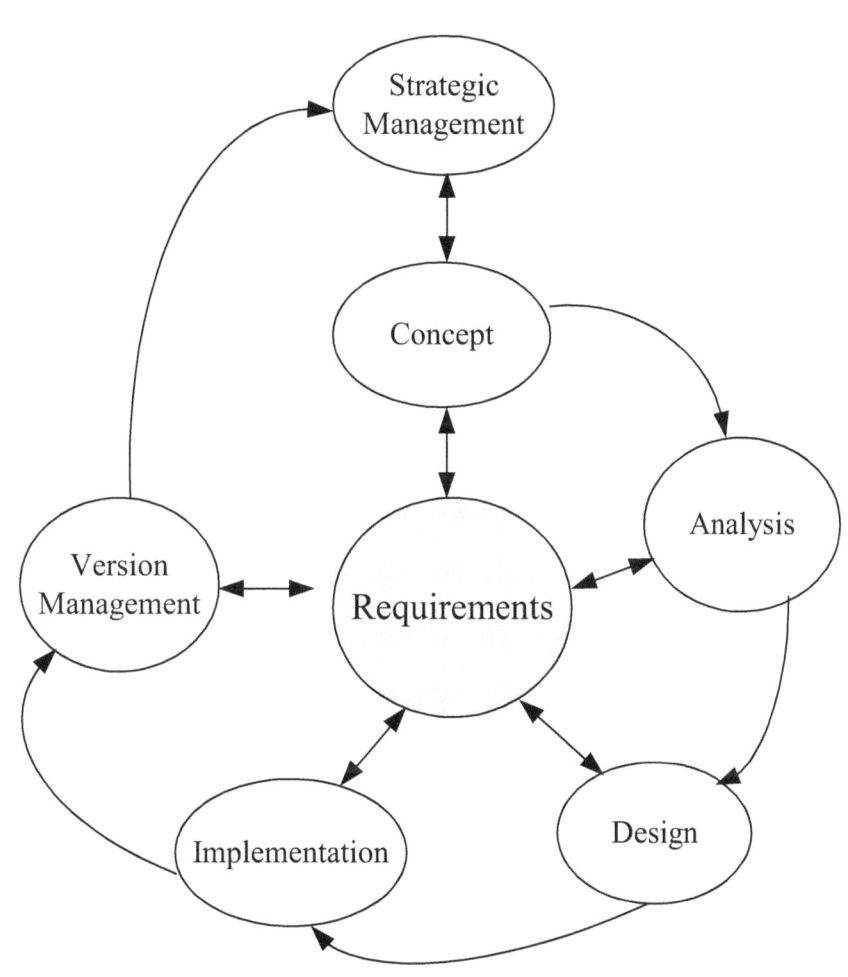

APPENDIX C: SBC VIEW MODEL (SBC-VM)

(1) SBC Hardware View Model

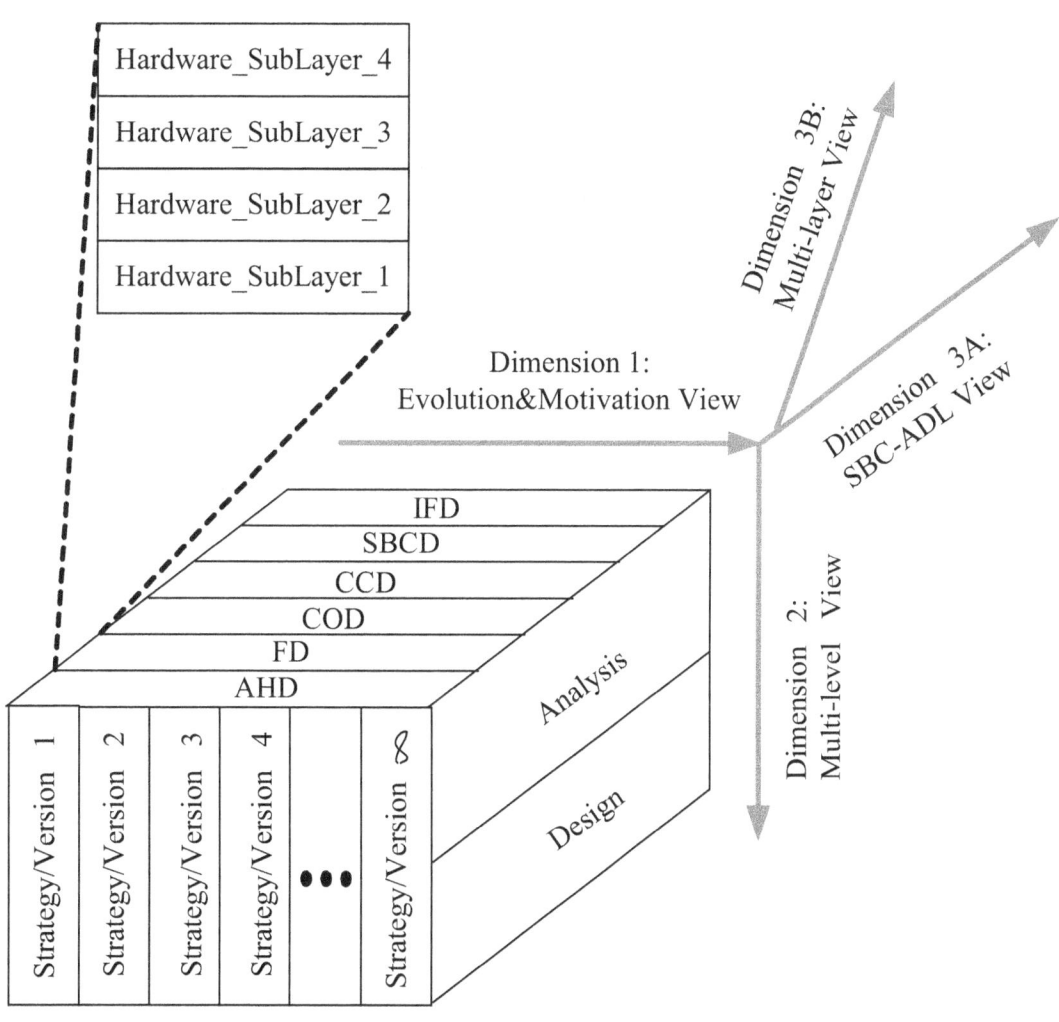

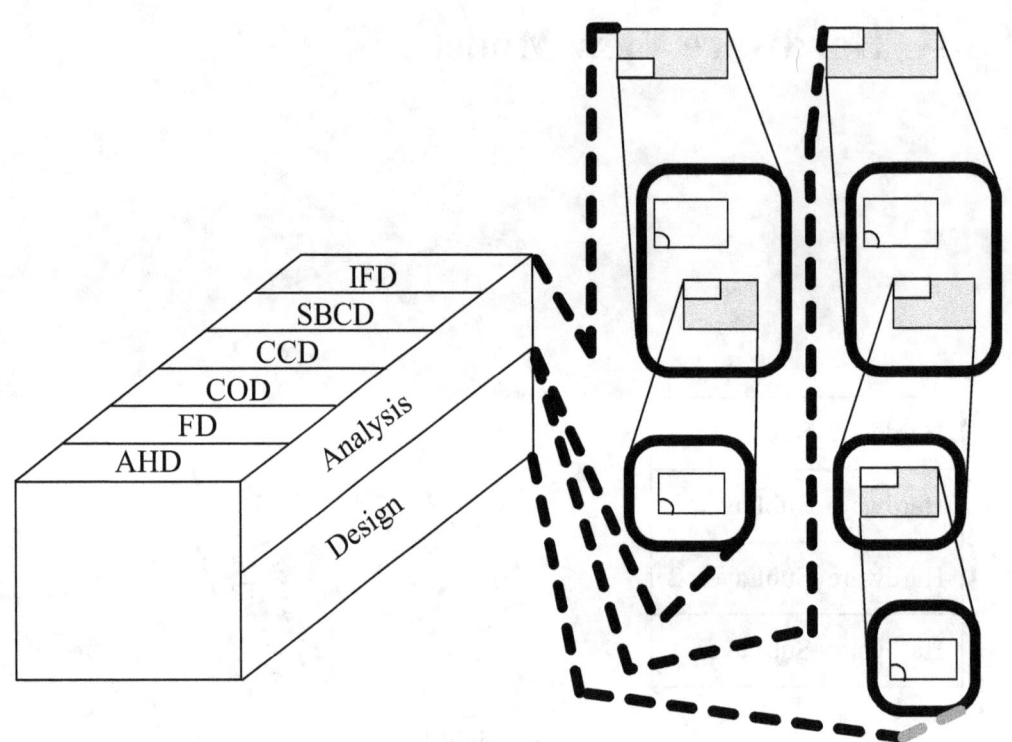

(2) SBC Software View Model

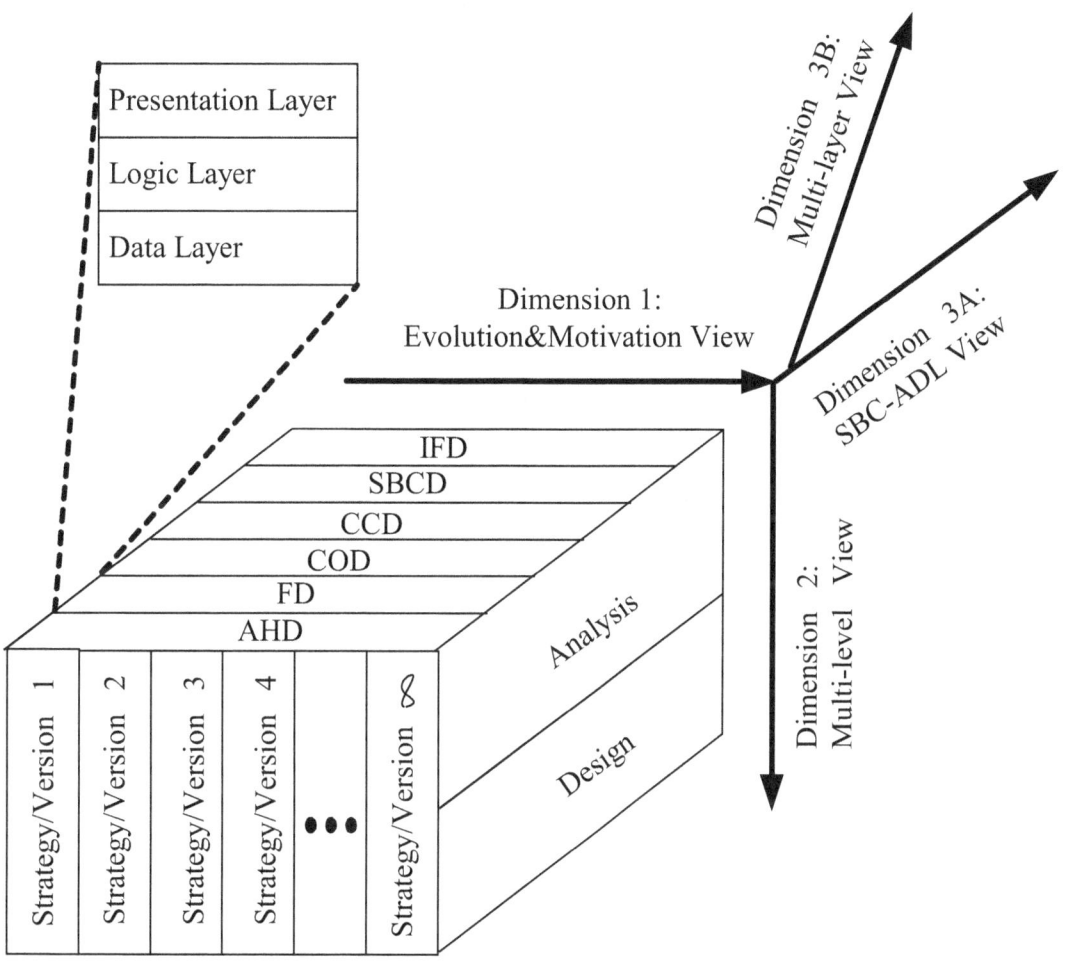

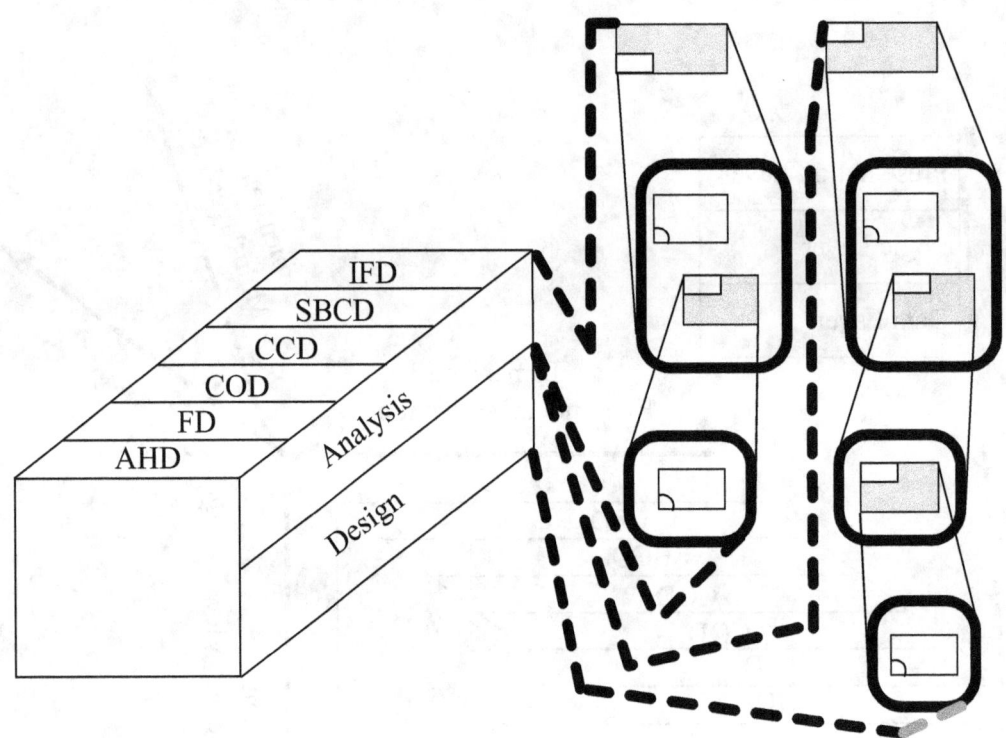

(3) SBC Enterprise View Model

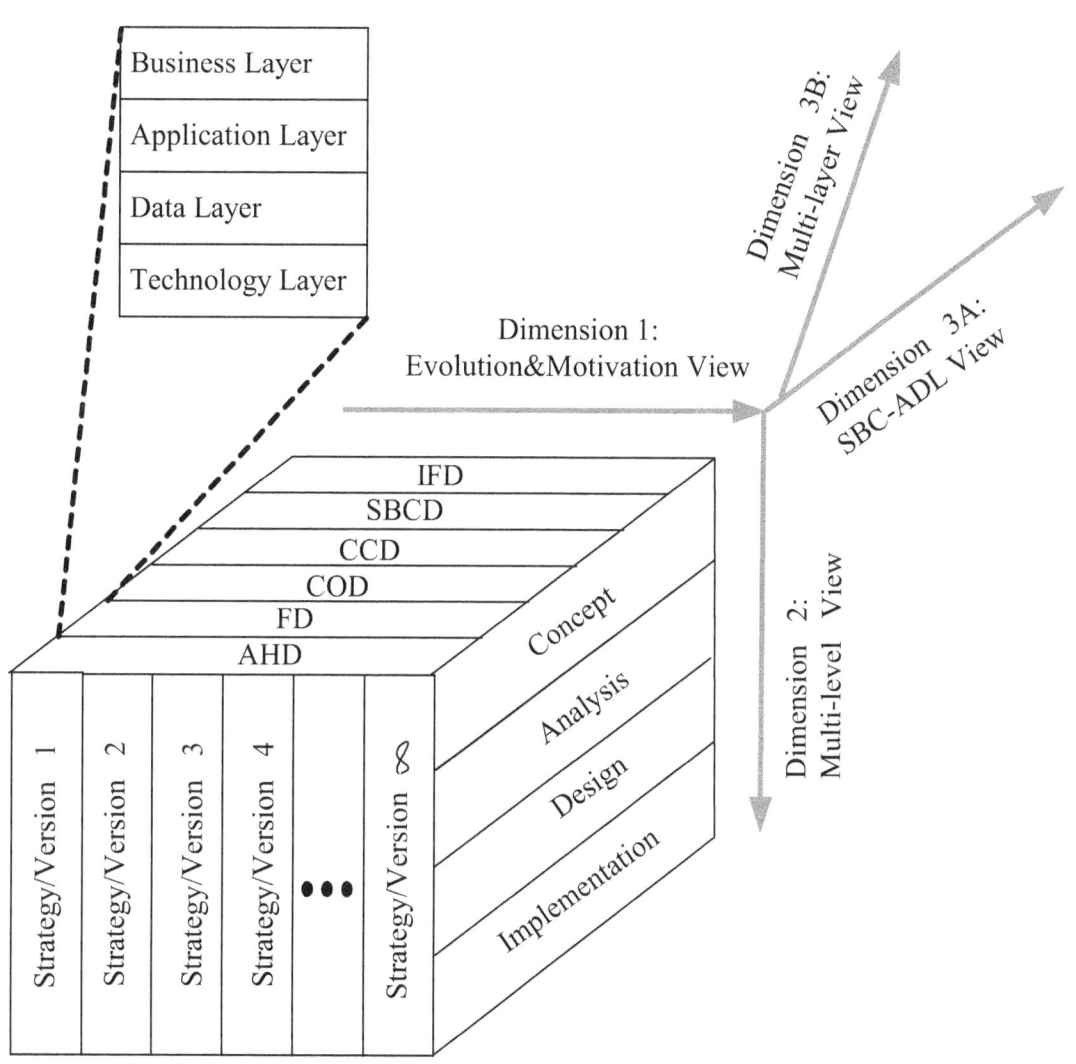

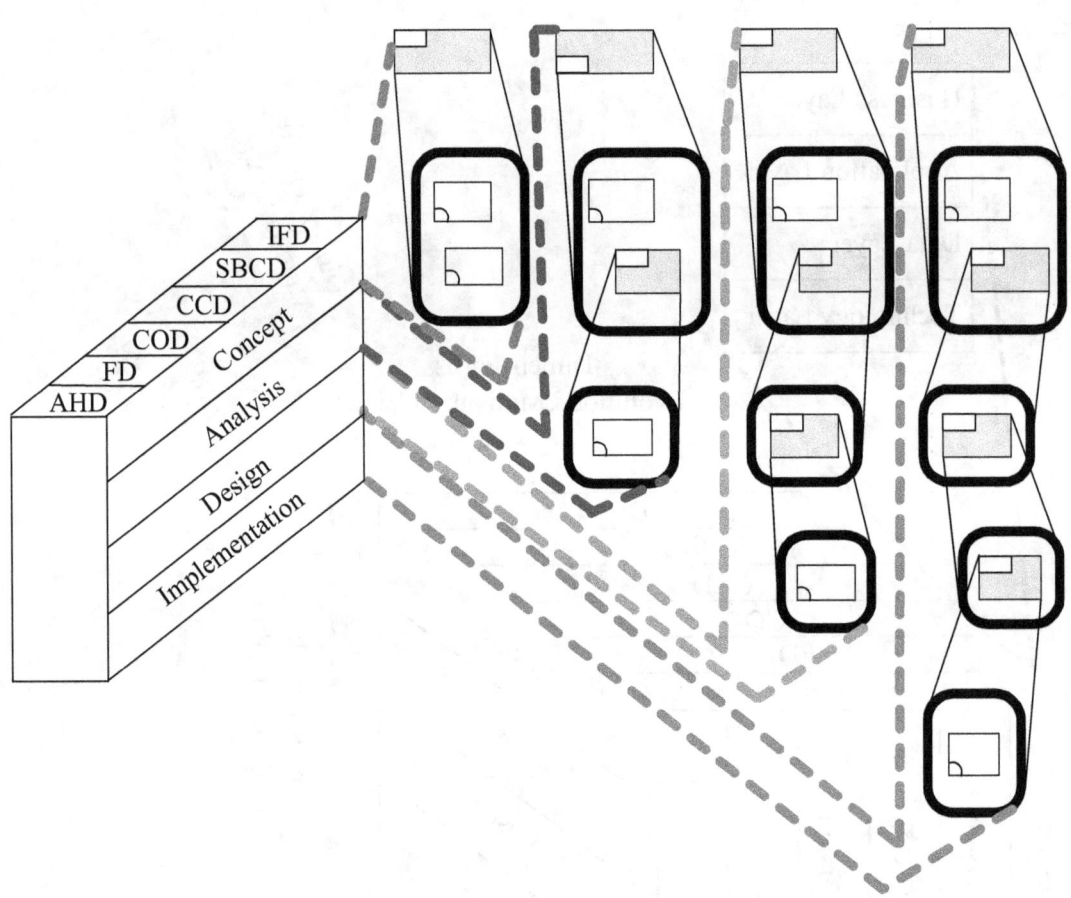

(4) SBC Knowledge View Model

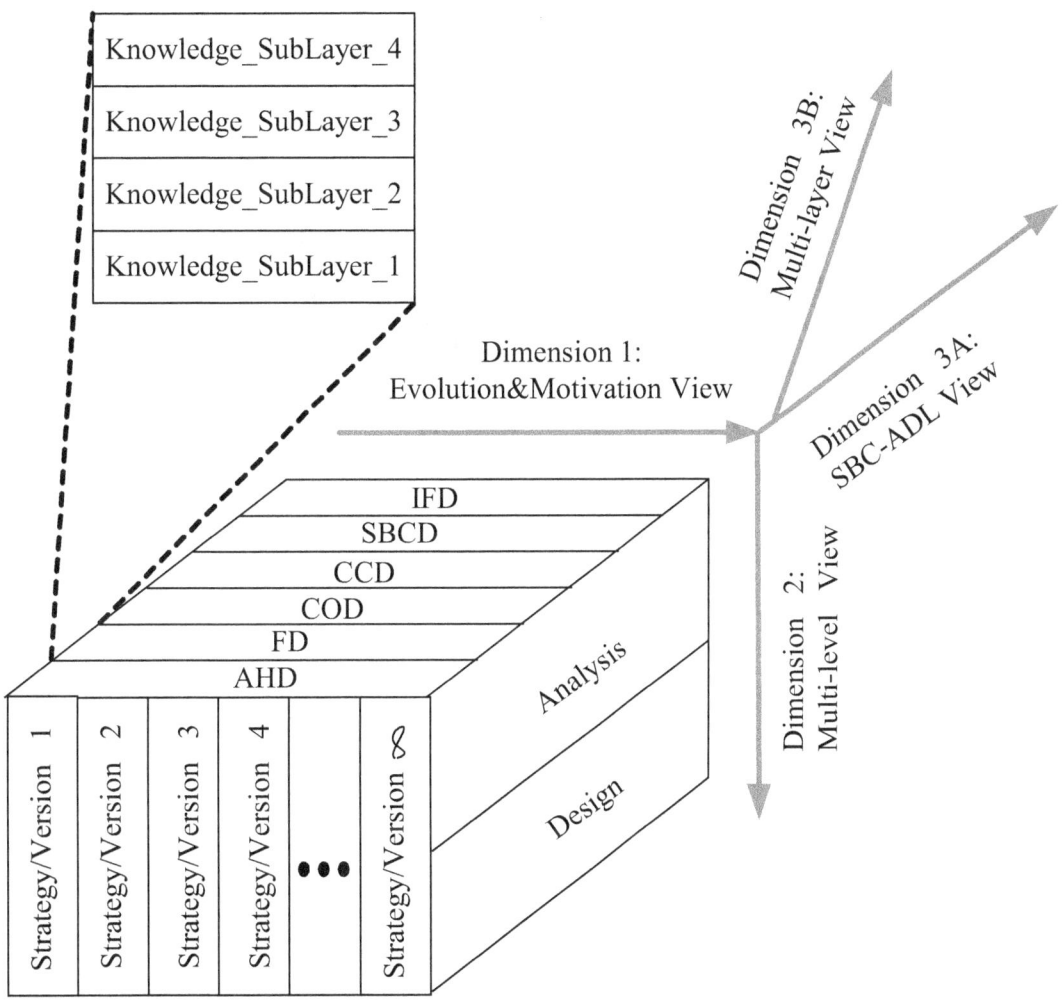

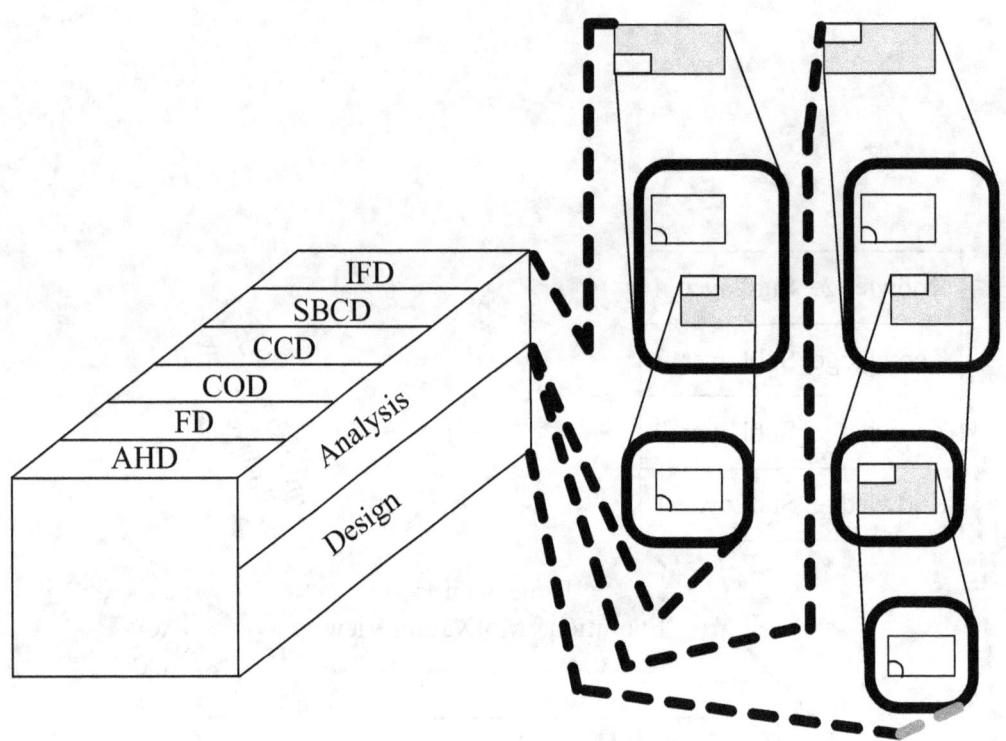

(5) SBC Thinking View Model

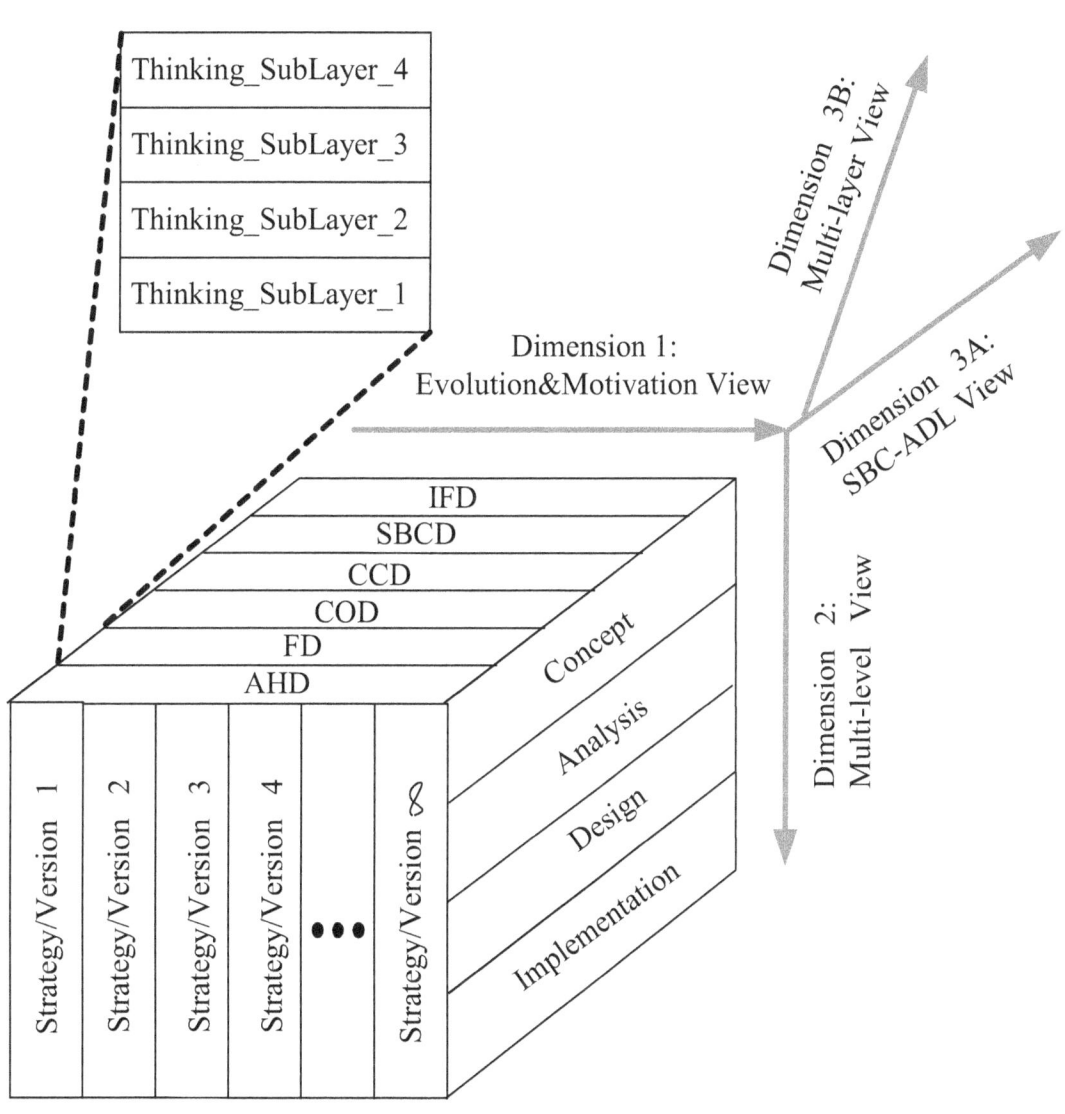

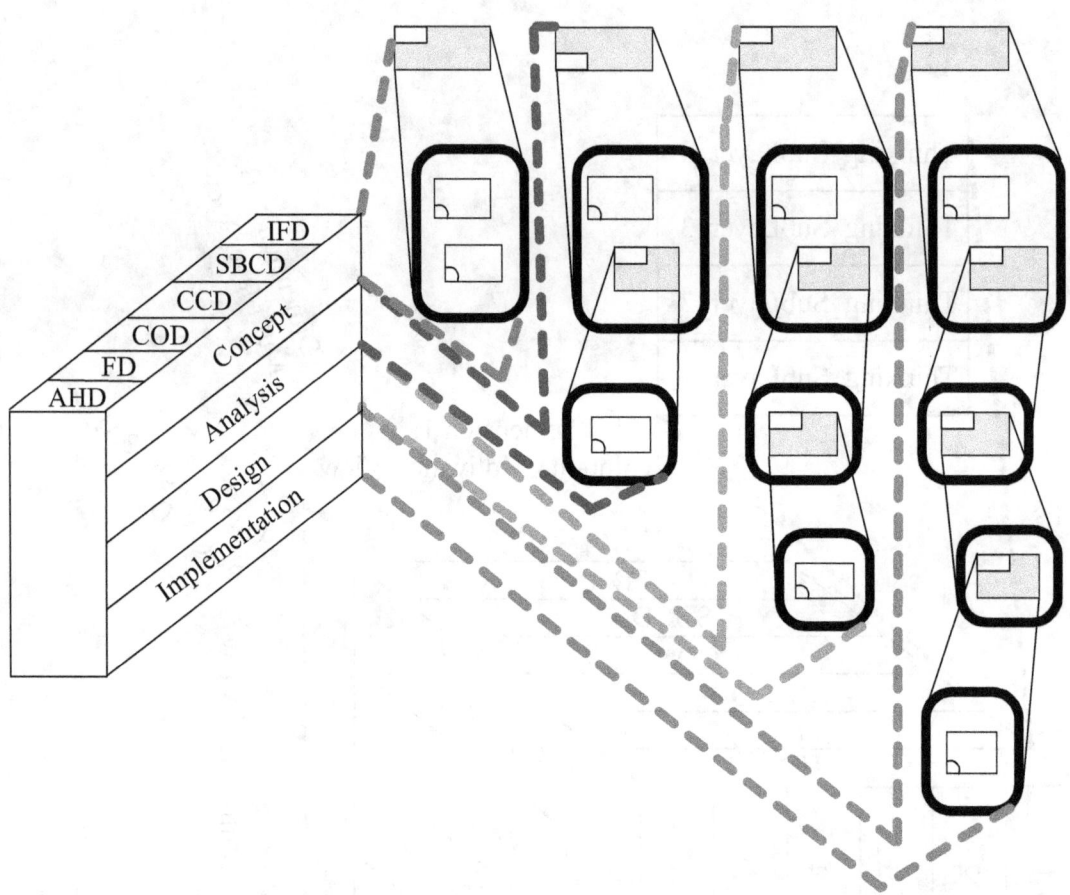

BIBLIOGRAPHY

[Bass03] Bass, L. et al., *Software Architecture in Practice*, 2nd Edition, Addison-Wesley, 2003.

[Bern05] Bernard, S., *An Introduction To Enterprise Architecture*, 2nd Edition, AuthorHouse, 2005.

[Burd10] Burd, S. D., *Systems Architecture*, 6th Edition, Cengage Learning, 2010.

[Chao14a] Chao, W. S., *Systems Thingking 2.0: Architectural Thinking Using the SBC Architecture Description Language*, CreateSpace Independent Publishing Platform, 2014.

[Chao14b] Chao, W. S., *General Systems Theory 2.0: General Architectural Theory Using the SBC Architecture*, CreateSpace Independent Publishing Platform, 2014.

[Chao15a] Chao, W. S., *A Process Algebra For Systems Architecture: The Structure-Behavior Coalescence Approach*, CreateSpace Independent Publishing Platform, 2015.

[Chao15b] Chao, W. S., *An Observation Congruence Model For Systems Architecture: The Structure-Behavior Coalescence Approach*, CreateSpace Independent Publishing Platform, 2015.

[Chao15c] Chao, W. S., *Variants of SBC Process Algebra: The Structure-Behavior Coalescence Approach*, CreateSpace Independent Publishing Platform, 2015.

[Chao15d] Chao, W. S., *Single-Queue SBC Observation Congruence Model For Systems Architecture: The Structure-Behavior Coalescence Approach*, CreateSpace Independent Publishing Platform, 2015.

[Chao15e] Chao, W. S., *Multi-Queue SBC Observation Congruence Model For*

Systems Architecture: The Structure-Behavior Coalescence Approach, CreateSpace Independent Publishing Platform, 2015.

[Chao16] Chao, W. S., *System: Contemporary Concept, Definition, and Language*, CreateSpace Independent Publishing Platform, 2016.

[Chao17a] Chao, W. S., *Channel-Based Single-Queue SBC Process Algebra For Systems Definition: General Architectural Theory at Work*, CreateSpace Independent Publishing Platform, 2017.

[Chao17b] Chao, W. S., *Channel-Based Multi-Queue SBC Process Algebra For Systems Definition: General Architectural Theory at Work*, CreateSpace Independent Publishing Platform, 2017.

[Chao17c] Chao, W. S., *Channel-Based Infinite-Queue SBC Process Algebra For Systems Definition: General Architectural Theory at Work*, CreateSpace Independent Publishing Platform, 2017.

[Chao17d] Chao, W. S., *Operation-Based Single-Queue SBC Process Algebra For Systems Definition: General Architectural Theory at Work*, CreateSpace Independent Publishing Platform, 2017.

[Chao17e] Chao, W. S., *Operation-Based Multi-Queue SBC Process Algebra For Systems Definition: Unification of Systems Structure and Systems Behavior*, CreateSpace Independent Publishing Platform, 2017.

[Chao17f] Chao, W. S., *Operation-Based Infinite-Queue SBC Process Algebra For Systems Definition: Unification of Systems Structure and Systems Behavior*, CreateSpace Independent Publishing Platform, 2017.

[Clem02] Clements, P. et al., *Documenting Software Architectures: Views and Beyond*, Addison Wesley, 2002.

[Clem10] Clements, P. et al., *Documenting Software Architectures: Views and Beyond*, 2nd Edition, Addison Wesley, 2010.

[Dam06] Dam, S., *DoD Architecture Framework: A Guide to Applying System Engineering to Develop Integrated Executable Architectures*, BookSurge Publishing, 2006.

[Date03] Date, C. J., *An Introduction to Database Systems*, 8th Edition, Addison Wesley, 2003.

[Denn08] Dennis, A. et al., *Systems Analysis and Design*, 4th Edition, Wiley, 2008.

[Dike01] Dikel, D. M. et al., *Software Architecture: Organizational Principles and Patterns*, Prentice-Hall, 2001.

[Dori95] Dori, D., "Object-Process Analysis: Maintaining the Balance between System Structure and Behavior," *Journal of Logic and Computation* 5(2), pp.227-249, 1995.

[Dori02] Dori, D., *Object-Process Methodology: A Holistic Systems Paradigm*, Springer Verlag, New York, 2002.

[Dori16] Dori, D., *Model-Based Systems Engineering with OPM and SysML*, Springer Verlag, New York, 2016.

[Eele09] Eeles, p. et al., *The Process of Software Architecting*, Addison-Wesley Professional, 2009.

[Engl09] Englander, I., *The Essential Guide to User Interface Design: An Introduction to GUI Design Principles and Techniques*, 4th Edition, Wiley, 2009.

[Elma10] Elmasri, R., *Fundamentals of Database Systems*, 6th Edition, Addison Wesley, 2010.

[Gali07] Galitz, W., *The Essential Guide to User Interface Design: An Introduction to GUI Design Principles and Techniques*, 3rd Edition Wiley, 2007.

[Hoar85] Hoare, C. A. R., *Communicating Sequential Processes*, Prentice-Hall, 1985.

[Kapl04] Kaplan, R. et al., *Strategy Maps: Converting Intangible Assets into Tangible*

Outcomes, 1st Edition, Harvard Business Press, 2004.

[Kapl08] Kaplan, R. et al., *The Execution Premium: Linking Strategy to Operations for Competitive Advantage*, 1st Edition, Harvard Business Press, 2008.

[Kend06] Kendal, S. et al., *An Introduction to Knowledge Engineering*, 1st Edition, Springer, 2006.

[Kend10] Kendall, K. et al., *Systems Analysis and Design*, 8th Edition, Prentice Hall, 2010.

[Maie09] Maier, M. W., *The Art of Systems Architecting*, 3rd Edition, CRC Press, 2009.

[Miln89] Milner, R., *Communication and Concurrency*, Prentice-Hall, 1989.

[Miln99] Milner, R., *Communicating and Mobile Systems: the π-Calculus*, 1st Edition, Cambridge University Press, 1999.

[Mino08] Minoli08, D., *Enterprise Architecture A to Z: Frameworks, Business Process Modeling, SOA, and Infrastructure Technology*, 1st Edition, Auerbach Publications, 2008.

[O'Rou03] O'Rourke, C. et al, *Enterprise Architecture Using the Zachman Framework*, 1st Edition, Course Technology, 2003.

[Paul01] Paulish, D. J., *Architecture-Centric Software Project Management: A Practical Guide*, Addison Wesley, 2001.

[Pele00] Peleg, M. et al., "The Model Multiplicity Problem: Experimenting with Real-Time Specification Methods". *IEEE Tran. on Software Engineering*. 26 (8), pp. 742–759, 2000.

[Prat00] Pratt, T. W. et al., *Programming Languages: Design and Implementation*, 4th Edition, Prentice Hall 2000.

[Pres09] Pressman, R. S., *Software Engineering: A Practitioner's Approach*, 7th Edition, McGraw-Hill, 2009.

[Putm00] Putman, J. R. et al., *Architecting with RM-ODP*, Prentice-Hall, 2000.

[Rayn09] Raynard, B., *TOGAF The Open Group Architecture Framework 100 Success Secrets*, Emereo Pty Ltd, 2009.

[Roza05] Rozansk, N. et al., *Software Systems Architecture: Working With Stakeholders Using Viewpoints and Perspectives*, Addison-Wesley, 2005.

[Rumb91] Rumbaugh, J. et al., *Object-Oriented Modeling and Design*, Prentice-Hall, 1991.

[Sche06] Schekkerman, J., *How to Survive in the Jungle of Enterprise Architecture Frameworks: Creating or Choosing an Enterprise Architecture Framework*, Trafford Publishing, 2006.

[Sche08] Schekkerman, J., *Enterprise Architecture Good Practices Guide: How to Manage the Enterprise Architecture Practice*, Trafford Publishing, 2008.

[Seth96] Sethi, R., *Programming Languages: Concepts and Constructs*, 2nd Edition, Addison-Wesley, 1996.

[Shaw96] Shaw, M. et al., *Software Architecture: Perspectives on an Emerging Discipline*, Prentice-Hall, 1996.

[Sode03] Soderborg, N.R. et al., "OPM-based Definitions and Operational Templates," *Communications of the ACM* 46(10), pp. 67-72, 2003.

[Somm06] Sommerville, I., *Software Engineering*, 8th Edition, Addison-Wesley, 2006.

[Sowa99] Sowa, J., *Knowledge Representation: Logical, Philosophical, and Computational Foundations*, 1st Edition, Brooks / Cole, 1999.

[Tayl09] Taylor, R. N. et al., *Software Architecture: Foundations, Theory, and Practice*, Wiley, 2009.

[Toga08] The Open Group, *TOGAF Version 9 - A Manual (TOGAF Series)*, Van Haren Publishing, 9th Edition, 2008.

[Wall04] Wall, D., *Multi-Tier Application Programming with PHP: Practical Guide for Architects and Programmers*, Morgan Kaufmann, 2004.

[Wang99] Wang, J. et al., "Introducing Software Architecture Specification and Analysis in SAM Through an Example," *Information and Software Technology* 41, 1999, pp.451-467.

[Your99] Yourdon, E., *Death March: The Complete Software Developer's Guide to Surviving Mission Impossible Projects*, Prentice-Hall, 1999.

INDEX

A

actor, 55, 56

ADL. *See* architecture description language

ADM. *See* architecture development method

AF. *See* architecture framework

aggregated system, 54, 66, 76, 78, 79

AHD. *See* architecture hierarchy diagram

architecting, 31, 37

    evolutionally, 31

    iteratively, 31

architecture description language, 29

architecture development method, 32

architecture framework. *See* view model

architecture hierarchy diagram, 53, 54, 323

B

baseline architecture, 32

building block. *See* component

C

CCD. *See* component connection diagram

COD. *See* component operation diagram

component, 20, 27, 50, 53, 54

component connection diagram, 56, 85, 87, 326

component operation diagram, 55, 79, 82, 325

connection

    client, 87

    operation provider, 86

    operation user, 86

    server, 87

D

data flow diagram, 44

data type

    composite, 81, 84, 85

    primitive, 81, 84

DFD. *See* data flow diagram

E

enterprise architecture, 211

entity. *See* component

evolution&motivation view, 20, 35, 49, 62, 64

    strategy/version n view, 20

    strategy/version n+1 view, 20

external environment. *See* actor

F

FD. *See* framework diagram

FD view. *See* multi-layer view

framework diagram, 54, 76, 78, 324

function. *See* operation

G

graphical user interface, 190

GUI. *See* graphical user interface

H

handshake. *See* interaction

hardware architecture, 175

hierarchical view. *See* multi-level view

I

IFD. *See* interaction flow diagram
interaction, 27, 50, 57
   operation call, 97
   operation return, 97
interaction flow diagram, 58, 59, 60, 92, 328
interrelationship, 21

K

knowledge architecture, 279
knowledge repository, 29, 30
knowledge representation, 279

M

method. *See* operation
model multiplicity, 25
model multiplicity problem, 25
model singularity, 26, 43
multi-layer, 54, 76
   composition, 54, 76, 77
   decomposition, 54, 76, 77
multi-layer view, 54, 64, 77
   application layer view, 54, 64
   business layer view, 54, 64
   college layer view, 226, 243, 262
   customer layer view, 316
   data layer view, 54, 64, 202
   department layer view, 243, 262
   financial layer view, 316
   internal layer view, 316
   learning&growth layer view, 316
   lecturer layer view, 262
   logic layer view, 202
   presentation layer view, 202
   technology layer view, 54, 64
   university layer view, 214, 226, 243, 262
multi-level, 53, 71, 74
   composition, 53, 71, 74
   decomposition, 53, 71, 74
multi-level view, 20, 35, 36, 65, 175, 176, 181, 192, 193, 200, 211, 280, 286
   analysis view, 20, 66, 121, 176, 193, 224, 298
   concept view, 20, 66, 103
   design view, 20, 67, 139, 181, 200, 241, 286, 305
   implementation view, 20, 67, 157, 259, 314
multiple models. *See* model multiplicity
multiple views, 20, 21, 25, 28, 34
   analysis view, 20, 66, 224, 298
   behavior view, 20
   concept view, 20, 66
   design view, 20, 67, 181, 200, 241, 286, 305
   evolution&motivation view, 20
   implementation view, 20, 67, 259, 314
   input/output data view, 20
   multi-layer view, 54
   multi-level view, 20
   strategy/version n view, 20
   strategy/version n+1 view, 20
   structure view, 20
   systemic view, 20
multiple views coalescence, 26, 41
multi-tier. *See* multi-layer
multi-tier view. *See* multi-layer view
MVC. *See* multiple views coalescence
MVC architecture, 41, 42

N

non-aggregated system. *See* component

O

object. *See* component

operation, 79, 80

operation formula, 80
    input parameter, 80
    operation name, 80
    output parameter, 80

P

part. *See* component

physical system, 22

procedure. *See* operation

S

SBC. *See* structure-behavior coalescence

SBC architecture, 43
    SBC architecture description language, 51, 69
    SBC architecture development method, 61, 329
    SBC view model, 63, 64, 331

SBC architecture description language, 49, 50, 69
    architecture hierarchy diagram, 49, 50, 51, 53, 71, 323
    component connection diagram, 49, 50, 51, 53, 85, 326
    component operation diagram, 49, 50, 51, 53, 79, 325
    framework diagram, 49, 50, 51, 53, 76, 324
    interaction flow diagram, 50, 51, 53, 92, 328
    structure-behavior coalescence diagram, 49, 50, 51, 53, 89, 327

SBC architecture development method, 61, 329

SBC process algebra, 50

SBC view model
    evolution&motivation view, 35, 49, 62, 64
    multi-level view, 35, 103, 121, 139, 157
    systemic view, 35

SBC-ADL. *See* SBC architecture description language

SBC-ADM. *See* SBC architecture development method

SBCD. *See* structure-behavior coalescence diagram

SBC-PA. *See* SBC process algebra

SBC-VM. *See* SBC view model

SC. *See* structure chart

single model. *See* model singularity

SM. *See* systems model

software architecture, 189

SSA&D. *See* structured systems analysis and design

stakeholder, 26, 29

structure chart, 44

structure-behavior coalescence, 42, 45

structure-behavior coalescence diagram, 57, 58, 89, 90, 327

structured systems analysis and design, 44
    data flow diagram, 44
    structure chart, 44

systemic view, 20, 35, 36, 67
    behavior view, 20
    input/output data view, 20
    multi-layer view, 64
    structure view, 20

systems architect, 31

systems architecture, 26, 27, 28, 30
    definition, 26, 27, 49
    hardware architecture, 175
    maintainability, 38

systems behavior, 21
    interaction flow diagram, 53
    structure-behavior coalescence diagram, 53

systems model, 22, 25
    MVC architecture, 41
    SBC architecture, 46
    structured systems analysis and design, 44

systems architecture, 26, 27, 49
systems modeling and architecting, 175, 189, 211, 279, 291
systems structure, 20
    architecture hierarchy diagram, 53
    component connection diagram, 53
    component operation diagram, 53
    framework diagram, 53

T

target architecture, 32
thinking architecture, 291

V

view
    analysis view, 20
    behavior view, 20
    concept view, 20
    design view, 20
    evolution&motivation view, 20
    implementation view, 20
    input/output data view, 20
    multi-level view, 20
    strategy/version n view, 20
    strategy/version n+1 view, 20
    structure view, 20
    systemic view, 20
view model, 34, 35
virtual system, 22
VM. *See* view model

www.ingramcontent.com/pod-product-compliance
Lightning Source LLC
Chambersburg PA
CBHW081105170526
45165CB00008B/2330